SELECTED TRANSLATIONS IN MATHEMATICAL STATISTICS AND PROBABILITY

Volume 6

AUTHORS

S. A. Aĭvazjan
V. I. Arkin
A. A. Borovkov
D. M. Čibisov
G. I. Kalmykov
F. M. Kirillova
V. A. Kolemaev
N. N. Krasovskiĭ
Ju. V. Linnik
V. V. Petrov
È. L. Presman
Ju. V. Prohorov
Ju. A. Rozanov
V. V. Sazonov
A. N. Širjaev
M. G. Šur
O. V. Viskov
V. M. Zolotarev

Printed for
INSTITUTE OF MATHEMATICAL STATISTICS
by
AMERICAN MATHEMATICAL SOCIETY
Providence, Rhode Island
1966

These translations are selected by a joint committee of the
Institute of Mathematical Statistics
and the
American Mathematical Society

Published under Grant NSF-GN410
from the
National Science Foundation

First printing 1966
Second printing 1970

Library of Congress Card Number 61-9803
Standard Book Number 8218-1456-7

TABLE OF CONTENTS

A GENERALIZATION OF CRAMÉR'S LIMIT THEOREM

V. V. PETROV

Let a sequence of independent random variables $Z_1, Z_2, \cdots$ with the distribution functions $V_1(x), V_2(x), \cdots$ be given. One can assume without loss of generality that their (mathematical) expectations are zero:

$$\mathrm{E}Z_j = 0, \qquad j = 1, 2, \ldots.$$

We introduce the notation:

$$\mathrm{E}Z_j^2 = \sigma_j^2, \qquad \sum_{j=1}^{n} \sigma_j^2 = B_n, \qquad \mathrm{P}\left\{\frac{Z_1 + \ldots + Z_n}{\sqrt{B_n}} < x\right\} = F_n(x).$$

As it is known Ljapunov's limit theorem asserts that under very general conditions

$$F_n(x) \xrightarrow[n\to\infty]{} \Phi(x) = \frac{1}{\sqrt{2\pi}} \int_{-\infty}^{x} e^{-\frac{t^2}{2}}\, dt \tag{1}$$

uniformly in every finite interval. But if $x \to \infty$ for $n \to \infty$, one can conclude from (1) only that $F_n(x) \to 1$. On the other hand, in many problems the relative magnitude of $F_n(x)$ and $\Phi(x)$, for x increasing indefinitely with n, are of interest. In the case of a Bernoulli scheme this problem was investigated by A. Ja. Hinčin, N. V. Smirnov, P. Levy and H. Fréchet.

In a more general case of identically distributed variables $(V_j(x) \equiv V(x),\ j = 1, 2, \cdots)$. H. Cramér has shown that if for $|h| < A$ the integral

$$R = \int_{-\infty}^{\infty} e^{hy}\, dV(y)$$

converges (condition A), then for $x > 1$ and $x = o(\sqrt{n}/\ln n)$ we have:

$$\frac{1 - F_n(x)}{1 - \Phi(x)} = e^{\frac{x^3}{\sqrt{n}}\lambda\left(\frac{x}{\sqrt{n}}\right)}\left[1 + O\left(\frac{x \ln n}{\sqrt{n}}\right)\right],$$

$$\frac{F_n(-x)}{\Phi(-x)} = e^{-\frac{x^3}{\sqrt{n}}\lambda\left(-\frac{x}{\sqrt{n}}\right)}\left[1 + O\left(\frac{x \ln n}{\sqrt{n}}\right)\right],$$

where $\lambda(t)$ is a power series which is convergent for $|t|$ sufficiently small.

This result was generalized by W. Feller [2] for the case of not necessarily identically distributed random variables. Feller succeeded in eliminating the logarithmic factor, but the conditions of Feller's theorem in the particular case of identical distributions are more restrictive than Cramér's conditions. Thus Cramér's theorem does not follow from Feller's.

The present note, which is a sharpening of the author's results [3], contains a generalization of Cramér's theorem to the case of not necessarily identically distributed random variables from which Cramér's theorem does follow.

We shall assume that the following conditions are satisfied:

Condition 1. There exist positive numbers A, K and k such that in the circle $|h| < A$ we have the inequalities

$$k \leqslant \left| \int_{-\infty}^{\infty} e^{hy}\, dV_j(y) \right| \leqslant K, \qquad j = 1, 2, \ldots .$$

Condition 2. For all n

$$\frac{1}{n} \sum_{j=1}^{n} \sigma_j^2 \geqslant \delta > 0.$$

Condition 3. For any $\epsilon > 0$, a $C > 1$ can be found such that for all n

$$\frac{1}{n} \sum_{j=1}^{n} \int_{|y|>C} y^2\, dV_j(y) < \varepsilon.$$

Under these assumptions we obtain the following result:

Theorem 1. *Let conditions* 1–3 *be satisfied. Furthermore, let x be a real number depending on n and such that $x > 1$ and $x = o(\sqrt{n})$ for $n \to \infty$. Then*

$$\frac{1 - F_n(x)}{1 - \Phi(x)} = e^{\frac{x^3}{\sqrt{n}} \lambda_n \left(\frac{x}{\sqrt{n}} \right)} \left[1 + O\left(\frac{x}{\sqrt{n}} \right) \right], \tag{2}$$

$$\frac{F_n(-x)}{\Phi(-x)} = e^{-\frac{x^3}{\sqrt{n}} \lambda_n \left(-\frac{x}{\sqrt{n}} \right)} \left[1 + O\left(\frac{x}{\sqrt{n}} \right) \right], \tag{3}$$

where $\lambda_n(t)$ is a power series convergent for sufficiently small values of $|t|$ uniformly for all n.

In the particular case of identically distributed variables conditions 2 and 3 are clearly satisfied, and condition 1 coincides with Cramér's condition A. As will be evident from the proof, the series $\lambda_n(t)$ in this case reduces to the series $\lambda(t)$ which appears in Cramér's theorem, with coefficients independent of n. Hence Cramér's theorem follows from ours; moreover, the condition $x = o(\sqrt{n}/\ln n)$ is replaced in our case by a less restrictive condition $x = o(\sqrt{n})$ and the remainder term is also improved.

Other classes of random variables, beside the identically distributed ones, also satisfy the conditions of Theorem 1. For example, a sequence of uniformly bounded random variables with variances satisfying condition 2 is such a class.

We proceed to the proof of Theorem 1.

1. We put

$$R_j = \int_{-\infty}^{\infty} e^{hy}\, dV_j(y), \qquad j = 1, 2, \ldots .$$

We have:

$$\frac{d}{dh} \ln R_j = \frac{1}{R_j} \int_{-\infty}^{\infty} y e^{hy}\, dV_j(y).$$

We divide the domain of integration in the last integral into 3 parts by means of the points $-N$ and N where N is such that for $|y| > N$ we have $|y| < e^{A|y|/2}$. In view of condition 1 we obtain

$$\left| \frac{d}{dh} \ln R_j \right| \leqslant B$$

for $|h|$ $A_1 = A/2$. Furthermore

$$\frac{d}{dh} \ln R_j = \sum_{k=2}^{\infty} \frac{\gamma_{kj}}{(k-1)!} h^{k-1},$$

where γ_{kj} is the kth cumulant of the variable Z_j.

For $-A_2 < h \leqslant A_2$, where $A_2 < A_1$, we have:

$$\sum_{k=2}^{\infty} \frac{|\gamma_{kj}|}{(k-1)!} |h|^{k-1} \leqslant M, \tag{4}$$

where M is some constant.

2. We introduce the distribution function

$$W_n(x) = \mathbf{P}\{Z_1 + \ldots + Z_n < x\}.$$

We choose a real number h from the interval $(-A_2, A_2)$. We introduce independent random variables $\overline{Z}_1, \overline{Z}_2, \cdots$ with the distribution functions

$$\overline{V}_j(x) = \frac{1}{R_j} \int_{-\infty}^{x} e^{hy}\, dV_j(y), \qquad j = 1, 2, \ldots .$$

We denote:

$$\mathbf{E}\overline{Z}_j = \overline{m}_j, \qquad \mathbf{E}(\overline{Z}_j - \overline{m}_j)^2 = \overline{\sigma}_j^2, \qquad \sum_{j=1}^{n} \overline{m}_j = \overline{A}_n, \qquad \sum_{j=1}^{n} \overline{\sigma}_j^2 = \overline{B}_n,$$

$$\mathbf{P}\{\overline{Z}_1 + \ldots + \overline{Z}_n < x\} = \overline{W}_n(x), \qquad \mathbf{P}\left\{\frac{\overline{Z}_1 + \ldots + \overline{Z}_n - \overline{A}_n}{\sqrt{\overline{B}_n}} < x\right\} = \overline{F}_n(x).$$

Then we have:

$$\overline{F}_n(x) = \overline{W}_n\left(\overline{A}_n + x\sqrt{\overline{B}_n}\right), \qquad F_n(x) = W_n\left(x\sqrt{B_n}\right).$$

Considering the characteristic functions $V_j(Z)$, $\overline{V}_j(Z)$, $W_n(Z)$, $\overline{W}_n(Z)$ which correspond to the distribution functions $V_j(x)$, $\overline{V}_j(x)$, $W_n(x)$, $\overline{W}_n(x)$ we obtain, in view of the independence of the random variables

$$w_n(z) = v_1(z) \; \ldots \; v_n(z), \quad \overline{w}_n(z) = \overline{v}_1(z) \; \ldots \; \overline{v}_n(z).$$

Since

$$\overline{v}_j(z) = \int_{-\infty}^{\infty} e^{izy} \, d\overline{V}_j(y) = \frac{1}{R_j} v_j(z - ih),$$

we thus obtain

$$w_n(z) = R_1 \; \ldots \; R_n \overline{w}_n(z + ih),$$

which is equivalent to the relation

$$W_n(x) = R_1 \; \ldots \; R_n \int_{-\infty}^{x} e^{-hy} \, d\overline{W}_n(y).$$

After some elementary manipulations we obtain:

$$1 - F_n\left(\frac{\overline{A}_n}{\sqrt{B_n}}\right) = R_1 \; \ldots \; R_n e^{-h\overline{A}_n} \int_0^{\infty} e^{-h\sqrt{\overline{B}_n}\, y} \, d\overline{F}_n(y). \tag{5}$$

3. Regarding h as a real variable we obtain:

$$\overline{m}_j = \frac{d}{dh} \ln R_j = \sum_{k=2}^{\infty} \frac{\gamma_{kj}}{(k-1)!} h^{k-1}, \tag{6}$$

$$\overline{\sigma}_j^2 = \frac{d^2}{dh^2} \ln R_j = \sum_{k=2}^{\infty} \frac{\gamma_{kj}}{(k-2)!} h^{k-2}. \tag{7}$$

Consider the equation

$$\sqrt{\frac{B_n}{n}}\, t = \frac{1}{n} \overline{A}_n = \frac{1}{n} \sum_{j=1}^{n} \sum_{k=2}^{\infty} \frac{\gamma_{kj}}{(k-1)!} h^{k-1},$$

where t is a given real number. Set:

$$\Gamma_{kn} = \frac{1}{n} \sum_{j=1}^{n} \gamma_{kj}.$$

Noting that $\Gamma_{2n} = B_n/n$, we have:

$$\sqrt{\Gamma_{2n}}\, t = \sum_{k=2}^{\infty} \frac{\Gamma_{kn}}{(k-1)!} h^{k-1}.$$

Inverting series (8) we obtain:

$$h = \frac{t}{\sqrt{\Gamma_{2n}}} - \frac{\Gamma_{3n}}{2\Gamma_{2n}^2} t^2 - \frac{\Gamma_{4n}\Gamma_{2n} - 3\Gamma_{3n}^2}{6\Gamma_{2n}^{7/2}} t^3 + \ldots, \tag{8}$$

and series (9) is dominated by a series convergent in some interval with coefficients independent of n. From (6) and the definition of R_j we have:

$$\frac{1}{n}\{h\overline{A}_n - \ln(R_1 \ldots R_n)\} = \sum_{k=2}^{\infty} \frac{(k-1)\Gamma_{kn}}{k!} h^k. \tag{9}$$

Substituting (9) into the last equation, we obtain

$$\frac{1}{n}\left\{\frac{1}{2}\frac{\overline{A}_n^2}{B_n} - h\overline{A}_n + \ln(R_1 \ldots R_n)\right\} = t^3\lambda_n(t), \tag{10}$$

where

$$\lambda_n(t) = \frac{1}{6}\frac{\Gamma_{3n}}{\Gamma_{2n}^{3/2}} + \frac{\Gamma_{4n}\Gamma_{2n} - 3\Gamma_{3n}^2}{24\Gamma_{2n}^3} t + \ldots$$

is a series convergent uniformly in all n for $|t|$ sufficiently small.

4. We put $\overline{F}_n(y) = \Phi(y) + Q_n(y)$ and show that for all n, h and y

$$|Q_n(y)| < \frac{L}{\sqrt{n}}, \tag{11}$$

where L is independent of n, h and y. By [4, p. 43], the last inequality is satisfied if there exist positive constants g and G such that

$$\frac{1}{n}\sum_{j=1}^{n} \overline{\sigma}_j^2 \geqslant g, \quad \frac{1}{n}\sum_{j=1}^{n}\int_{-\infty}^{\infty} |y - \overline{m}_j|^3 d\overline{V}_j(y) \leqslant G.$$

We note that from (6) and (4) it follows that $|\overline{m}_j| \le M$, $j = 1, 2, \cdots$. First we consider h to be positive, $0 < h < A_2$ and $M > 1$. Then

$$\frac{1}{n}\sum_{j=1}^{n} \overline{\sigma}_j^2 \geqslant \frac{D}{n}\sum_{j=1}^{n}\int_{-C}^{\infty} (y - \overline{m}_j)^2 dV_j(y), \tag{12}$$

where $C > 1$, but which otherwise is arbitrary for the time being, $D = e^{-A_2 C}/K$, and K is defined by condition 1. Since $EZ_j = 0$,

$$\left|\frac{2}{n}\sum_{j=1}^{n} \overline{m}_j \int_{-C}^{\infty} y\, dV_j(y)\right| \leqslant \frac{2M}{n}\sum_{j=1}^{n}\int_{-\infty}^{-C} y^2 dV_j(y).$$

By condition 3, given the number $\delta/4M$, where δ is determined by condition 2, one can find $C > 1$ such that

$$\frac{1}{n}\sum_{j=1}^{n}\int_{-\infty}^{-C} y^2 dV_j(y) < \frac{\delta}{4M}. \tag{13}$$

Assuming now that C satisfies this condition, we have

$$\frac{1}{n}\sum_{j=1}^{n} \overline{\sigma}_j^2 > D\left(\delta - \frac{\delta}{4M} - \frac{\delta}{2}\right) > \frac{D\delta}{4} = g > 0.$$

But if h is negative then, considering in equation (12) the region of integration

$(-\infty, C)$ instead of region $(-C, \infty)$ we arrive to the same result.[1)]

The existence of a number $G > 0$, with the property indicated above follows from condition 1.

5. Now let h be a real positive variable which tends to zero in such a way that $h\sqrt{n}$, is bounded from below for $n \to \infty$, by a positive number. From (6) and (7) we obtain:

$$h\sqrt{\overline{B}_n} = \frac{\overline{A}_n}{\sqrt{\overline{B}_n}} + O(h^2\sqrt{n}), \tag{14}$$

while $h\sqrt{\overline{B}_n} \geq h\sqrt{ng}$ is bounded from below by a positive number.

Replacing the integral in (5) by the sum of integrals with respect to $\Phi(y)$ and $Q_n(y)$ and utilizing (11) and (14) we obtain after simple calculations

$$\int_0^\infty e^{-h\sqrt{\overline{B}_n}\, y}\, d\overline{F}_n(y) = \left[1 - \Phi\left(\frac{\overline{A}_n}{\sqrt{\overline{B}_n}}\right)\right] e^{\frac{\overline{A}_n^2}{2\overline{B}_n}} [1 + O(h)]. \tag{15}$$

The equation $x = \overline{A}_n/\sqrt{\overline{B}_n}$, upon the substitution $t = x/\sqrt{n}$, becomes equation $\sqrt{n\overline{B}_n}\, t = \overline{A}_n$ which was investigated above. For t sufficiently small this equation has a single positive root $h \to 0$ as $t \to 0$. But $t = x/\sqrt{n}$ tends to zero since $x = o(\sqrt{n})$. And in this case, as was established above, equation (10) is valid. The product $h\sqrt{n}$ is indeed bounded from below by a positive number since $h \sim t/\sqrt{\Gamma_{2n}} = x/\sqrt{n\Gamma_{2n}}$, while $x > 1$ and $\Gamma_{2n} < G$. Therefore, taking h in (15) to be equal to the root of the equation $x = \overline{A}_n/\sqrt{\overline{B}_n}$ and taking (5) and (10) into account we obtain (2).

The relationship (3) is proved in an analogous manner.

In some problems it may turn out to be useful to consider the case when $1 < x < \epsilon_0\sqrt{n}$, where ϵ_0 is a fixed arbitrarily small number. In this case one cannot apply Theorem 1 directly; however, an examination of the proof shows its validity for the case $x < \epsilon_0\sqrt{n}$ as well.

Namely, the following theorem holds:

Theorem 2. *Let conditions* 1–3 *be satisfied and let* $1 < x < \epsilon_0\sqrt{n}$ *where* ϵ_0 *is a fixed arbitrarily small number. Then*

$$\frac{1 - F_n(x)}{1 - \Phi(x)} = e^{\frac{x^3}{\sqrt{n}}\lambda_n\left(\frac{x}{\sqrt{n}}\right)}[1 + r_1\varepsilon_0],$$

$$\frac{F_n(-x)}{\Phi(-x)} = e^{-\frac{x^3}{\sqrt{n}}\lambda_n\left(-\frac{x}{\sqrt{n}}\right)}[1 + r_2\varepsilon_0],$$

where $|r_1| \leq R$, $|r_2| \leq R$ *and* R *is an absolute constant.*

1) One must consider the case $h > 0$ for the proof of (3). Therefore condition (13), as well as condition 3 of paper [3] should be supplemented by an analogous condition for (C, ∞).

In the particular case of identical distributions, conditions 1–3 are reduced to the condition of convergence of the integral $\int_{-\infty}^{\infty} e^{hy}\,dV(y)$ for $|h| < A$, where A is some positive number, and the series $\lambda_n(t)$ becomes the series $\lambda(t)$ with coefficients independent of n.

Corollaries follow from Theorem 1 which are analogous to the corollaries resulting from Cramér's theorem.

Theorem 3. *If conditions 1–3 are satisfied, then for $x > 1$ and $x = O(n^{1/6})$ we have:*

$$\frac{1-F_n(x)}{1-\Phi(x)} = e^{\frac{c_n x^3}{\sqrt{n}}}\left[1+O\left(\frac{x}{\sqrt{n}}\right)\right],$$

$$\frac{F_n(-x)}{\Phi(-x)} = e^{-\frac{c_n x^3}{\sqrt{n}}}\left[1+O\left(\frac{x}{\sqrt{n}}\right)\right],$$

where $c_n = \Gamma_{3n}/6\Gamma_{2n}^{3/2}$ is bounded for all n.

In particular, if $x = o(n^{1/6})$ then both ratios tend to 1 as $n \to \infty$.

Theorem 4. *If conditions 1–3 are satisfied, then for $x > 1$ and $x = O(n^{1/6})$ we have:*

$$1 - F_n(x) = [1-\Phi(x)]\, e^{\frac{c_n x^3}{\sqrt{n}}} + O\left(\frac{1}{\sqrt{n}} e^{-\frac{x^2}{2}}\right),$$

$$F_n(-x) = \Phi(-x)\, e^{\frac{-c_n x^3}{\sqrt{n}}} + O\left(\frac{1}{\sqrt{n}} e^{-\frac{x^2}{2}}\right),$$

where $c_n = \Gamma_{3n}/6\Gamma_{2n}^{3/2}$ is bounded for all n.

Theorem 5. *Let c be a positive constant. If conditions 1–3 are satisfied then for $x = o(\sqrt{n})$ the ratios*

$$\frac{F_n\left(x+\frac{c}{x}\right)-F_n(x)}{1-F_n(x)} \quad \text{and} \quad \frac{\Phi\left(x+\frac{c}{x}\right)-\Phi(x)}{1-\Phi(x)}$$

tend to $1-e^{-c}$ as $n \to \infty$.

BIBLIOGRAPHY

[1] H. Cramér, *Sur un nouveau théorème-limite de la théorie des probabilités*, Actualités Sci. Ind., Paris, 1938; Russian transl., Uspehi Mat. Nauk 10 (1944), 166–178.

[2] W. Feller, *Generalization of a probability limit theorem of Cramér*, Trans. Amer. Math. Soc. 54 (1943), 361–372. MR 5, 125.

[3] V. V. Petrov, *Extension of Cramér's limit theorem to nonidentically distributed independent quantities*, Vestnik Leningrad. Univ. 1953, no. 8, 13–25. MR 17, 979.

[4] C. G. Esseen, *Fourier analysis of distribution functions. A mathematical study of the Laplace-Gaussian law*, Acta Math. 77 (1945), 1–125. MR 7, 312.

Translated by:
S. Kotz

ADDITIVE FUNCTIONALS OF MARKOV PROCESSES AND EXCESSIVE FUNCTIONS

M. G. ŠUR

Conditions are investigated for representing excessive functions in the form of potentials of additive functionals of Markov processes. For excessive functions a decomposition is constructed close to the one for superharmonic functions obtained by Riesz.

Investigations of additive functionals of Markov processes and excessive functions feature prominently in the contemporary theory of Markov processes. V. A. Volkonskiĭ showed in [2] interesting connections between additive functionals and excessive functions and gave sufficient conditions for representing an excessive function in the form of the potential of an additive functional.[1] This result of V. A. Volkonskiĭ has been developed by P.-A. Meyer [9] and the author [6,7].

The present article contains detailed proofs of the results published in the notes [6,7]. The basic terminology is collected in §1, together with certain results required subsequently. Conditions are examined in §2 and §3 for an excessive function to be the potential of a continuous additive functional. In §4 and §5 a decomposition is constructed for excessive functions which is a generalization of the Riesz decomposition for classical superharmonic functions.

§1. Preliminary remarks

We shall use principally the terminology and notations of monographs [3,4].

1. We recall certain definitions. Suppose that in a measurable space (E, B) a homogeneous Markov process $X = (x_t, \zeta, \mathfrak{M}_t, P_x)$ is given with a space of elementary events Ω. The transition function of this process is denoted by $P(t, x, \Gamma)$, where $t \geq 0$, $x \in E$, $\Gamma \in B$. The symbol $\mathfrak{N}_t$ denotes the σ-algebra in the space $\Omega_t = \{\zeta > t\}$ generated by the events $\{x_u \in \Gamma, \zeta > t\}$, where $t > u \geq 0$, $\Gamma \in E$. By $\mathfrak{N}^0$ we denote the σ-algebra in Ω generated by the events $\{x_u \in \Gamma\}$, where $u \geq 0$, $\Gamma \in E$. We write $A \in \mathfrak{N}_{t+0}$ if for every $u > t$ the event $\{A, \zeta > u\}$ belongs to $\mathfrak{N}_u$ and $A \in \overline{\mathfrak{N}}_{t+0}$ if, for every finite measure μ given on the σ-algebra B, sets A_1 and A_2 from $\mathfrak{N}_{t+0}$ can be found such that $A_1 \subseteq A \subseteq A_2$ and

$$\int_E P_x\{A_2 \setminus A_1\}\, \mu(dx) = 0.$$

[1] This result is contained in the monograph [4].

The σ-algebras $\overline{\mathfrak{N}}_t$ and $\overline{\mathfrak{N}}^0$ are defined similarly to the σ-algebra $\overline{\mathfrak{N}}_{t+0}$.

The process X is called standard if the following conditions are satisfied:

A. A topology $\mathcal{C}$ is given in the space E, the topological space $(E, \mathcal{C})$ being a locally compact Hausdorff space with a countable base and the σ-algebra B consisting of Borel sets of this topological space.

B. The process X is a right-continuous strictly Markov process.

C. For any nondecreasing sequence of Markov moments τ_n and any $x \in E$ the sequence $x(\tau_n)$ tends to $x(\tau)$ as $n \uparrow \infty$ (almost certain $\widetilde{\Omega}$, P_x),[1)] where

$$\tau(\omega) = \lim_{n\to\infty} \tau_n(\omega), \qquad \widetilde{\Omega} = \{\omega : \tau(\omega) < \zeta(\omega)\}.$$

D. The σ-algebra $\mathfrak{M}_t$ contains within itself the σ-algebra $\overline{\mathfrak{N}}_{t+0}$ for all $t \geq 0$.

The term "standard processes" was suggested by E. B. Dynkin; noncutoff standard processes were first studied by Hunt [5]. Apart from standard processes we shall have to deal with processes satisfying the following condition:

$(\mathcal{H})$. The process X and the noncutoff process X_e, obtained from X by introducing the additional condition e, are standard.

A precise definition of X_e is given in the Appendix. The process X_e can be intuitively described as a noncutoff Markov process whose paths $x_e(t)$ coincide with the paths of the original process X up to the moment of cutoff ζ of the process X and for which $x_e(t)$ is identically equal to e (where $e \notin E$) for values of the parameter t $(t < \infty)$ exceeding ζ.

Unless the contrary is indicated the rest of §1 deals exclusively with standard processes.

If a finite measure ν is given on the σ-algebra B, for $A \in \mathfrak{N}^0$ we put

$$P_\nu\{A\} = \int_E P_x\{A\}\, \nu(dx).$$

But if $A \in \overline{\mathfrak{N}}^0$, we choose A_1 and A_2 from $\mathfrak{N}^0$ so that $A_1 \subseteq A \subseteq A_2$ and $P_\nu\{A_2 \setminus A_1\} = 0$ and put

$$P_\nu\{A\} = P_\nu\{A_1\}.$$

A function of the set $P_\nu\{A\}$ is a measure on $\overline{\mathfrak{N}}^0$ and the measure ν itself is interpreted as the initial distribution of the process.

2. The random function $\phi_t(\omega)$, given over the interval $0 \leq t < \zeta(\omega)$ and assuming nonnegative values including, possibly, the value $+\infty$, is called a homogeneous additive functional of the process X if:

E_1. For any nonnegative t the function $\phi_t(\omega)$ is measurable with respect

[1)] The notation "almost certain $\widetilde{\Omega}$, P_x" denotes "almost certain on $\widetilde{\Omega}$ relative to the measure P_x".

to $\bar{\mathfrak{N}}_{t+0}$.

E_2. For any $x \in E$ and any $s \geq 0$ and $t \geq 0$ the relation

$$\phi_s + \theta_s \phi_t = \phi_{t+s} \quad (\text{almost certain } \Omega_1, P_x)$$

holds, where $\Omega_1 = \{\omega: t + s < \zeta(\omega)\}$ and θ_s is the shift operator of events.[1)]

We shall consider only homogeneous additive functionals for which the following condition is satisfied:

E_3. Considered as a function of t, $\phi_t(\omega)$ does not decrease in the interval $0 \leq t < \zeta(\omega)$ (almost certain P_x) for any $x \in E$.

This definition differs somewhat from the definition of a homogeneous additive functional adopted in [4]. In place of the term "additive functional" we shall generally use the shorter form "functional". For $t \geq \zeta$ we shall define the functional ϕ_t as

$$\phi_t = \lim_{t \uparrow \zeta} \phi_u.$$

The function $M_x \phi_\zeta$ is called the potential of the functional ϕ_t.

Two functionals ϕ_t and ϕ_t' are called equivalent if $P_x\{\phi_t \neq \phi_t'\} = 0$ for all $x \in E$ and $t \geq 0$. The functional ϕ_t is called right-continuous if for any $x \in E$ the function $\phi_t(\omega)$ is right-continuous with respect to t $(0 \leq t < \zeta(\omega))$ for P_x-almost all ω. We shall define a continuous functional similarly.

Following P. A. Meyer (see [9]) we shall say that the functional ϕ_t belongs to class (U) if for any $x \in E$ the functions $x_t(\omega)$ and $\phi_t(\omega)$, as functions of the argument t, do not have common points of discontinuity for P_x-almost all $\omega \in \Omega$ $(0 \leq t < \zeta(\omega))$.

3. Let us consider a function $f(x)$, given on E and assuming numerical values including, possibly, the values $-\infty$ and $+\infty$. We shall call $f(x)$ almost Borel[2)] if for every initial distribution ν two Borel-measurable functions $f_1(x)$ and $f_2(x)$ can be found such that $f_1(x) \leq f(x) \leq f_2(x)$ everywhere on E and $f_1(x_t) = f_2(x_t)$ for all t $(0 \leq t < \zeta)$ for P_x-almost all $\omega \in \Omega$. The set F $(F \subseteq E)$ will be called almost Borel if its characteristic function is almost Borel.

For any almost Borel set V we put

$$\tau_V = \begin{cases} \inf(t:\ t \geq 0,\ x_t \bar{\in} V), & \text{if } x_t \bar{\in} V \text{ for some } t < \zeta; \\ \zeta, & \text{otherwise.} \end{cases}$$

---

1) Following E. B. Dynkin, we define θ_s as the operator transforming events of the form $\{x_t \in \Gamma\}$ into events of the form $\{x_{t+s} \in \Gamma\}$ and retaining all set-theoretic operations.

2) All definitions and results in §1.3 and §1.4 except for Lemma A, belong to Hunt (see [5]).

The random variable τ_V is called the moment of the first exit from the set V. Then the probability $P_x\{\tau_V = 0\}$ is either zero or unity. In the first case the point x is called regular for V.

We shall need the following lemma.

Lemma A. *Let the process X satisfy condition* $(\mathcal{H})$. *If the functionals ϕ_t and ψ_t belong to the class (U) and if their potentials are equal and finite on some almost Borel set A, then for any $t \geq 0$ and any point $x \in A$ regular for A the relation*

$$P_x\{\phi_t \neq \psi_t,\ t < \tau_A\} = 0$$

holds.

In the case in which the potentials of the functionals of the lemma are equal and finite everywhere in E, the lemma reduces to a known result of P.-A. Meyer (see [9], p. 193). The case $A \subset E$ is considered basically in the same way as Meyer's result.[1)]

4. The measures $P(t, x, \Gamma)$ are extended uniquely onto the σ-algebra of almost Borel functions. For an almost Borel function $f(x)$ we put

$$T_t f(x) = \int_E f(y)\, P(t, x, dy),$$

if this integral has sense.

A nonnegative almost Borel function $f(x)$ is called an excessive function if for any $t \geq 0$ and $x \in E$

$$f(x) \geq T_t f(x)$$

and

$$f(x) = \lim_{t \downarrow 0} T_t f(x).$$

As an example of an excessive function we can take the potential of any homogeneous right-continuous functional with finite values. We shall consider certain properties of excessive functions as follows:

(a) for any Markov moment ξ the inequality

$$M_x f(x_\xi) \leq f(x)$$

is satisfied everywhere in E;

(b) for any $x \in E$

---

1) The only difference being that functions with compact supports contained in A are chosen as the functions a, b, a_i appearing in Meyer's proof (see [9], p. 192) and in this connection small modifications have to be made to Meyer's arguments.

$$f(x) = \lim_{t \downarrow 0} f(x_t) \quad \text{(almost certain } \Omega, P_x);$$

(c) for any $x \in E$ the following limit exists:

$$\lim_{t \uparrow \zeta} f(x_t) \quad \text{(almost certain } \Omega, P_x);$$

(d) if functions f and g are excessive, the function $\min(f, g)$ is also excessive.

The following assertion will be needed subsequently.

Lemma B. *Let an almost Borel function $f \geq 0$ satisfy for all $t \geq 0$ the inequality $f \geq T_t f$. Then the function $f' = \lim_{t \downarrow 0} T_t f$ is excessive.*

For the proof see [5], p. 121.

§2. Potentials of continuous functionals

In §2 and §3 we consider the standard process $X = (x_t, \zeta, \mathfrak{M}_t, P_x)$ in the measurable space (E, B).

Definition. The almost Borel function $f(x)$ satisfies condition $(\mathfrak{A})$ if for any nondecreasing sequence of Markov moments τ_n and any $x \in E$

$$\lim_{n \to \infty} M_x f(x(\tau_n)) = M_x f(x(\tau)), \tag{$\mathcal{A}$}$$

where τ denotes $\lim_{n \to \infty} \tau_n$.

The role of condition $(\mathfrak{A})$ is seen from the following theorem.

Theorem 1. *A necessary and sufficient condition for the bounded excessive function $f(x)$ to be the potential of a certain continuous functional $\phi_t(\omega)$ of the process X is that it should satisfy condition $(\mathfrak{A})$.*

Sufficient conditions for the representation of an excessive function f in the form of a potential of a continuous functional were first given, as has already been mentioned, by V. A. Volkonskiĭ (see [2], §1). These conditions consist in that the function $T_t f$ tends uniformly with respect to $x \in E$ to the function f as $t \downarrow 0$. Subsequently, conditions for representing an excessive function in the form of a potential of a continuous functional were studied by P.-A. Meyer and the author, the class of processes considered by P.-A. Meyer being somewhat narrower than the class of standard processes.

Proof of necessity. Let $f(x) = M_x \phi_\zeta$, where ϕ_t is a continuous functional. Then for any Markov moment ξ we shall have

$$M_x f(x_\xi) = M_x \{\phi_\zeta - \phi_\xi\}.$$

Therefore, on considering the nondecreasing sequence of Markov moments τ_n, we obtain:

$$\lim_{n\to\infty} M_x f(x(\tau_n)) = M_x\{\varphi_\zeta - \varphi_\tau\}, \qquad (1)$$

where $\tau = \lim_{n\to\infty}\tau_n$. But the right side of (1) equals $M_x f(x(\tau))$ and thus $f(x)$ satisfied ($\mathfrak{A}$).

Proof of sufficiency. This occupies the principal part of §2 and §3 and is considerably more complicated. We shall first establish a few auxiliary results. The function $f(x)$ will be assumed to be bounded and excessive and to satisfy ($\mathfrak{A}$). It can easily be verified that for any $t \geq 0$ the function $T_t f(x)$ also possesses these properties.

We denote by $\tau(h)$ the moment of first attainment of the set

$$B_h = \{x: |f(x) - T_h f(x)| > \epsilon\},$$

where $\epsilon > 0$ is fixed.

Lemma 1. *For any* $x \in E$

$$\lim_{h\to 0} P_x\{\tau(h) = \zeta\} = 1.$$

Proof. It is obvious that for $h \leq l$ the set B_h is contained in the set B_l and $\tau(h) \geq \tau(l)$. We put

$$\Omega_n = \{\omega : \tau(n^{-1}) < \zeta\}, \qquad y_n = x(\tau(n^{-1}))$$

and suppose that $k > n$. Then

$$f(y_k) - T_{n^{-1}} f(y_k) \geqslant f(y_k) - T_{k^{-1}} f(y_k). \qquad (2)$$

Since the functions f and $T_t f$ satisfy ($\mathfrak{A}$) and since, by virtue of (b),

$$f(y_k) - T_{k^{-1}} f(y_k) \geq \epsilon \quad (\text{almost certain } \Omega_k,\ P_x),$$

as $k \to \infty$ inequality (2) gives:

$$M_x f(x(\tau)) - M_x\{T_{n^{-1}} f(x(\tau))\} \geqslant \varepsilon \lim_{k\to\infty} P_x\{\tau(k^{-1}) < \zeta\},$$

where $\tau = \lim_{n\to\infty}\tau(n^{-1})$. As $n \to \infty$ the left side of the inequality tends to zero and therefore

$$\lim_{k\to\infty} P_x\{\tau(k^{-1}) < \zeta\} = 0.$$

Remark 1. Let us consider the set B_h on which $f(x)$ differs from

$$h^{-1}\int_0^h T_s f(x)\, ds$$

by more than ϵ and denote the moment of first attainment of B_h' by $\tau'(h)$. Clearly,

$$f(x) - h^{-1}\int_0^h T_s f(x)\, ds \leqslant f(x) - T_h f(x),$$

and therefore $\tau'(h) \geq \tau(h)$. Hence

$$\lim_{h \to 0} P_x\{\tau'(h) = \zeta\} = 1. \tag{3}$$

We shall use this result below.

Now let

$$\sup_{x \in E} f(x) = C$$

and

$$f_h(x) = h^{-1}[f(x) - T_h f(x)].$$

For $u \geq \zeta$ we put

$$f_h(x_u) = 0.$$

Lemma 2. *The following estimates are valid:*

$$M_x \int_s^t f_h(x_u)\,du \leq C$$

and

$$M_x\left\{\left[\int_s^t f_h(x_u)\,du\right]^2\right\} \leqslant 2C^2.$$

Proof. The first estimate easily follows from

$$M_x \int_s^t f_h(x_u)\,du = \frac{1}{h}\left[M_x \int_s^{s+h} f(x_u)\,du - M_x \int_t^{t+h} f(x_u)\,du\right]. \tag{4}$$

We now pass on to the second estimate. It is clear that

$$M_x\left\{\left[\int_s^t f_h(x_u)\,du\right]^2\right\} = 2M_x \int_s^t f_h(x_u)\left[\int_u^t f_h(x_v)\,dv\right] du, \tag{5}$$

and therefore

$$M_x\left\{\left[\int_s^t f_h(x_u)\,du\right]^2\right\} = 2M_x \int_s^t f_h(x_u)\, M_{x_u}\left[\int_0^{t-u} f_h(x_v)\,dv\right] du,$$

which, by virtue of the first estimate, does not exceed $2C^2$.

Theorem 2. *For any $s \geq 0$ and $t \geq 0$ $(s \leq t)$ there exists the limit*

$$\lim_{h \downarrow 0} \int_s^t f_h(x_u)\,du$$

in the sense of mean square convergence relative to any measure P_x.

Proof. The difference of the functions $f_h(x)$ and $f_l(x)$ will be denoted by $g(x)$ and their sum by $g_1(x)$. Let n be a positive integer and

$$s_k = s + \frac{k(t-s)}{n} \quad (n \geqslant k \geqslant 0).$$

We define the function $w(u)$ on the segment $[s, t]$ and suppose that $w(u) = s_k$ for $s_{k-1} < u \leq s_k$.

We have

$$M_x\left\{\left[\int_s^t g(x_u)\,du\right]^2\right\} = 2M_x\int_s^t g(x_u)\left[\int_u^t g(x_v)\,dv\right]du$$

(cf. [5]). We estimate the integral

$$I_1 = M_x\int_s^t g(x_u)\left[\int_{w(u)}^t g(x_v)\,dv\right]du = \sum_{k=1}^n M_x\int_{s_{k-1}}^{s_k} g(x_u)\left[\int_{s_k}^t g(x_v)\,dv\right]du.$$

Using the Markov property we obtain

$$I_1 = \sum_{k=1}^n M_x\int_{s_{k-1}}^{s_k} g(x_u)\,M_{x(s_k)}\int_0^{\delta_{n-k}} g(x_v)\,dv, \tag{6}$$

where $\delta_{n-k} = (n-k)(t-s)/n$. It easily follows from (4) that the modulus of the function

$$F_k(x) = M_x\int_0^{\delta_{n-k}} g(x_v)\,dv$$

does not exceed $2C$.

We denote by χ the characteristic function of the event $\bar{A}$ which is that during the time from s to t the path at least once falls in a set where the module of at least one of the functions F_k exceeds ϵ. From (6) and the inequality $|F_k| < 2C$ we obtain

$$|I_1| = \varepsilon M_x\int_s^t g_1(x_u)\,du + 2CM_x\left[\chi\int_s^t g_1(x_u)\,du\right].$$

On account of Lemma 2 the first term of this sum does not exceed $2C\epsilon$ and by virtue of the Cauchy-Bunjakovskiĭ inequality the second term does not exceed

$$2C\cdot\left(P_x\{\bar{A}\}\cdot M_x\left\{\left[\int_s^t g_1(x_u)\,du\right]^2\right\}\right)^{\frac{1}{2}}.$$

The second factor inside the round brackets, in accordance with Lemma 2, is uniformly bounded with respect to h and l. The first factor tends to zero for fixed n as $h \to 0$, $l \to 0$. In fact, the function F_k is equal to

$$\left\{\frac{1}{h}\int_0^h T_s f\,ds - \frac{1}{l}\int_0^l T_s f\,ds\right\} + \left\{\frac{1}{l}\int_0^l T_s\left[T_{\delta_{n-k}} f\right]ds - \frac{1}{h}\int_0^h T_s\left[T_{\delta_{n-k}} f\right]ds\right\}$$

and our assertion follows from Remark 1.

Thus

$$\lim_{h,l\to 0} I_1 = 0.$$

It remains for us to estimate the integral

$$I_2 = M_x \int_s^t g(x_u) \left[\int_u^{w(u)} g(x_v)\, dv \right] du$$

$$= M_x \int_s^t g(x_u) \left[M_{x(u)} \int_0^{\delta} g(x_v)\, dv \right] du \qquad (\delta = w(u) - u).$$

From (4) we see that the function

$$F(x) = M_x \int_0^{\delta} g(x_v)\, dv$$

is equal to the expression

$$\left\{ h^{-1} \int_0^h T_s f\, ds - h^{-1} \int_0^h T_{s+\delta} f\, ds \right\} + \left\{ l^{-1} \int_0^l T_{s+\delta} f\, ds - l^{-1} \int_0^l T_s f\, ds \right\},$$

which does not exceed

$$\left| h^{-1} \int_0^h T_s f\, ds - h^{-1} \int_0^h T_{s+\gamma} f\, ds \right| + \left| l^{-1} \int_0^l T_s f\, ds - l^{-1} \int_0^l T_{s+\gamma} f\, ds \right|, \qquad (7)$$

where $\gamma = (t - s)/n$. Let us consider the events:

$\Lambda = \{$the path of x_t falls at least once in the set on which $|T_\gamma f - f| \geq \epsilon/2\}$;

$\Lambda_1 = \{$the path of x_t falls at least once in the set on which $|h^{-1}\int_0^h T_s f ds - f| + |l^{-1}\int_0^l T_s f ds - f| \geq \epsilon/2\}$;

$\Lambda_2 = \{$the path of x_t falls at least once in the set on which $|h^{-1}\int_0^h T_{s+\gamma} f ds - T_\gamma f| + |l^{-1}\int_0^l T_{s+\gamma} f ds - T_\gamma f| \geq \epsilon/2\}$.

We choose n so that $P_x\{\Lambda\} < \epsilon/2$ and then choose h and l so small that $P_x\{\Lambda_1 \cup \Lambda_2\} < \epsilon/2$. Then the P_x-probability that the paths of x_t attain the set on which $F(x) \geq \epsilon$ does not exceed ϵ.

Taking this into account and repeating the arguments used in estimating I_1, we see that, if n is sufficiently large (but fixed) and h and l are sufficiently small, I_2 does not exceed 2ϵ. Theorem 2 is thereby proved.

Remark 2. Similar arguments show that for any fixed $x \in E$ and $\epsilon > 0$ it is possible to choose $t > 0$ so small that $M_x[\int_0^t f_h(x_u)\, du]^2$ does not exceed ϵ for all sufficiently small h.

§3. Potentials of additive functionals (continued)

We proceed to the completion of the proof of Theorem 1.

1. For each $x \in E$ we choose a sequence h_k (depending on x) such that $h_k \downarrow 0$ as $k \to \infty$ and for all rational $t \geq 0$ the limit

$$\phi_t(x, \omega) = \lim_{k\to\infty} \int_0^t f_{h_k}(x(v))\,dv \qquad (\text{almost certain } \Omega, P_x)$$

exists.[1] The function $\phi_t(x, \omega)$ is defined for the time being only for rational $t \geq 0$. It does not decrease with respect to t (almost certain Ω, P_x). We denote by Ω_x the aggregate of $\omega \in \Omega$ for which this function is uniformly continuous with respect to t when t runs through the rational values of any finite interval.

Lemma 3. *The probability* $P_x\{\Omega_x\}$ *is identically equal to unity.*

Proof. It is sufficient to show that, for any positive integer m,

$$J_n = M_x \sum_{k=0}^{2^n m} [\varphi_{2^{-n(k+1)}}(x, \omega) - \varphi_{2^{-n}k}(x, \omega)]^2$$

tends to zero as $n \to \infty$ (cf. [2], p. 148). It is clear that

$$J_n = M_x \sum_{k=0}^{2^n m} \lim_{h\to 0} \left[\int_{2^{-n}k}^{2^{+n(k+1)}} f_h(x_u)\,du\right]^2,$$

and, by Fatou's lemma,

$$J_n \leqslant \overline{\lim_{h\to 0}}\, M_x \sum_{k=0}^{2^n m} \left[\int_{2^{-n}k}^{2^{n}(k+1)} f_h(x_u)\,du\right]^2.$$

The expression governed by the upper limit sign does not exceed (cf. [5])

$$2M_x \sum_{k=0}^{2^n m} \left[\int_{2^{-n}k}^{2^{-n}(k+1)} f_h(x_u)\,du \left\{M_{x(u)} \int_0^{2^{-n}} f_h(x_v)\,dv\right\} du\right.$$

$$= 2M_x \int_0^m f_h(x_u)\left[M_{x(u)} \int_0^{2^{-n}} f_h(x_v)\,dv\right] du. \tag{8}$$

The last expression is estimated in the same way as I_2 above for it is easy to show that, if n is sufficiently large (and fixed) and h is sufficiently small, (8) does not exceed any preassigned number. Hence our lemma follows.

For irrational t and $\omega \in \Omega_x$ we put $\phi_t(x, \omega)$ equal to the limit of $\phi_r(x, \omega)$ when r runs through the rational numbers and $r \to t$ (but if t is irrational and $\omega \bar{\in} \Omega_x$, then $\phi_t(x, \omega)$ is assumed equal to an arbitrary nonnegative number α).

It is easy to see that

$$\varphi_t(x, \omega) = \lim_{h\to 0} \int_0^t f_h(x_u)\,du, \tag{9}$$

where the limit is understood in the sense of mean square convergence relative to the measure P_x. For rational t this assertion follows from the definition of

---

[1] For those values of ω for which this limit does not exist we assume $\phi_t(x, \omega)$ to be equal to an arbitrary nonnegative number α.

ϕ_t. Let t be irrational, $x \in E$ and $\epsilon > 0$. By virtue of (4) and condition ($\mathfrak{A}$) there exists $\delta > 0$ such that, for $t - \delta < s < t$ and $h < \delta$, $M_x \int_0^t f_h(x_u)\, du$ does not exceed ϵ. Furthermore,

$$M_x \left| \int_0^t f_h(x_u)\, du - \varphi_t(x, \omega) \right| \leqslant M_x \left| \int_0^s f_h(x_u)\, du - \varphi_s(x, \omega) \right|$$

$$+ M_x \int_0^t f_h(x_u)\, du + M_x \left| \varphi_s(x, \omega) - \varphi_t(x, \omega) \right|, \tag{10}$$

and if s is rational and fixed so that the last term of the sum in (10) is less than ϵ and $t - \delta < s < t$, then for all sufficiently small h the sum in (10) does not exceed 3ϵ. From this and from Theorem 2 follows (9).

We are now in a position to prove the following important lemma.

Lemma 4. *There exists a right-continuous functional* $\phi_t(\omega)$ *such that*

$$P_x\{\phi_t(\omega) = \phi_t(x, \omega)\} = 1$$

for all $x \in E$ *and* $t \geq 0$.

Proof. We shall use the ideas put forward in the proof of Theorem 5.3 in [8]. We put

$$\alpha_n(t) = \exp\left\{-\int_0^t f_{n^{-1}}(x_u)\, du\right\}.$$

For fixed n the family of random variables $\alpha_n(t)$ forms a multiplicative functional of the process X.[1] We consider the subprocess X_n of the process X corresponding to the functional $\alpha_n(t)$ and denote the transition function of the process X_n by $P_n(t, x, \Gamma)$. In accordance with Theorem 3.11 of [3],

$$P_n(t, x, \Gamma) = M_x[\chi_\Gamma(x_t)\,\alpha_n(t)],$$

where χ_Γ is the characteristic function of the set Γ. Therefore, for any $\Gamma \in B$

$$| P_n(t, x, \Gamma) - P_m(t, x, \Gamma) | \leqslant M_x[\chi_\Gamma(x_t)\, | \alpha_n(t) - \alpha_m(t) |], \tag{11}$$

and by Theorem 2 the limit

$$Q(t, x, \Gamma) = \lim_{n \to \infty} P_n(t, x, \Gamma)$$

exists.

The function $Q(t, x, \Gamma)$, as a function of the set $\Gamma \in B$, is a measure. In fact, if the sets Γ_i do not intersect and $-\Gamma = \bigcup_{i=1}^\infty \Gamma_i$, the series

$$P_n(t, x, \Gamma) = \sum_{i=1}^\infty P_n(t, x, \Gamma_i)$$

---

[1] We say that the family of nonnegative random variables $\beta(t, \omega)$ $(\omega \in \Omega,\ t \geq 0)$ forms a multiplicative functional if the family $-\log \beta(t, \omega)$ forms an additive functional. The properties of multiplicative functionals are described in [3], Chapter 3.

for any fixed $t \geq 0$ and $x \in E$ are majorized by the series

$$\sum_{i=1}^{\infty} P(t, x, \Gamma_i)$$

and are therefore uniformly convergent with respect to n. Therefore in the expression

$$\sum_{i=1}^{\infty} P_n(t, x, \Gamma_i)$$

we allow a term-by-term passage to the limit and hence the given assertion follows.

We now prove that $Q(t, x, \Gamma)$ is a homogeneous transition function. In accordance with (11), for any Borel-measurable function h not exceeding unity in modulus, the following estimate[1)] holds:

$$\left| \int_E h(y) P_n(t, x, dy) - \int_E h(y) P_m(t, x, dy) \right| \leqslant M_x \left| \alpha_n(t) - \alpha_m(t) \right|.$$

Therefore on fixing $\epsilon > 0$ and $x \in E$, we obtain from the Chapman-Kolmogorov relation

$$P_n(t+s, x, \Gamma) = \int_E P_n(t, x, dy) P_n(s, y, \Gamma)$$

the estimate

$$\left| P_n(t+s, x, \Gamma) - \int_E Q(t, x, dy) P_n(s, y, \Gamma) \right| < \varepsilon$$

for all sufficiently large n. If in this estimate we take the limit as $n \to \infty$ and then let ϵ tend to zero we deduce the Chapman-Kolmogorov relation for $Q(t, x, \Gamma)$.

Thus $Q(t, x, \Gamma)$ is a homogeneous transition function. It follows from Remark 2 that

$$\lim_{t \downarrow 0} Q(t, x, E) = 1.$$

Further, $Q(t, x, \Gamma)$ is majorized by the function $P(t, x, \Gamma)$. These properties of the function $Q(t, x, \Gamma)$ ensure the existence, as was shown by P.-A. Meyer, of a right-continuous multiplicative functional $\alpha_t(\omega)$ of X to which corresponds a certain subprocess $\widetilde{X} = (\widetilde{x}_t, \widetilde{\zeta}, \widetilde{\mathfrak{M}}_t, \widetilde{P}_x)$ of the process X with transition function $Q(t, x, \Gamma)$ (see [9], p. 138, Theorem 2.2).

The additive functional

$$\phi_t(\omega) = -\log \alpha_t(\omega)$$

is, as we shall now see, the required functional. It will be sufficient for us to verify that, for any $t \geq 0$ and $x \in E$

---

1) This estimate is trivial if h assumes only a finite set of values. For other functions h the estimate is obtained by an obvious passage to the limit.

$$P_x\{\alpha_t(\omega) = \beta_t(x, \omega)\} = 1, \tag{12}$$

where $\beta_t(x, \omega) = \exp\{-\phi_t(x, \omega)\}$. We fix $x \in E$ and $t \geq 0$. By Lemma 3.1 of the monograph [3]

$$\widetilde{P}_x\{x(s_1) \in \Gamma_1, \ldots, x(s_n) \in \Gamma_n, \widetilde{x}(u) \in E\} = M_x\{\chi_A \alpha_u\},$$

where $s_1 < s_2 < \cdots < s_n < u$, $\Gamma_i \in B$ and χ_A is the characteristic function of the event $\{x(s_1) \in \Gamma_1, \cdots, x(s_n) \in \Gamma_n\}$. On the other hand, the expression on the left side of this relation is equal to

$$\int_{x_1 \in \Gamma_1} Q(s_1, x, dx_1) \int_{x_2 \in \Gamma_2} Q(s_2 - s_1, x_1, dx_2) \ldots \int_{x_n \in \Gamma_n} Q(u - s_n, x_n, E). \tag{13}$$

By using the method of mathematical induction and arguing in the same way as in the proof of the Chapman-Kolmogorov relation for the function $Q(t, x, \Gamma)$ we show that (13) is equal to the limit (as $m \to \infty$) of expressions obtained from (13) on replacing Q by P_m. Therefore

$$M_x\{\chi_A \alpha_u\} = \lim_{n \to \infty} M_x\{\chi_A \alpha_n(u)\} = M_x\{\chi_A \beta_u(x, \omega)\}$$

and for any $u > t$

$$M_x\{\alpha_u | \mathfrak{N}_t\} = M_x\{\beta_u(x, \omega) | \mathfrak{N}_t\} \quad (\text{almost certain } \{\zeta > t\}, P_x).$$

Hence we have:

$$M_x\{\alpha_u | \overline{\mathfrak{N}}_t\} = M_x\{\beta_u(x, \omega) | \overline{\mathfrak{N}}_t\} \quad (\text{almost certain } \{\zeta > t\}, P_x).$$

By virtue of the properties of conditional expectation, $\overline{\mathfrak{N}}_t$ can be replaced in the last relation by $\overline{\mathfrak{N}}_{t+0}$. If we now consider the arbitrary sequence $u_i \downarrow t$ and take into account the right-continuity of the functional α_t and the continuity with respect to u for P_x-almost all ω of the function $\beta_u(x, \omega)$ we obtain from what has been said:

$$\alpha_t = P_x\{\alpha_t | \overline{\mathfrak{N}}_{t+0}\} = P_x\{\beta_t(x, \omega) | \overline{\mathfrak{N}}_{t+0}\} = \beta_t(x, \omega) \quad (\text{almost certain } \{\zeta > t\}, P_x),$$

which completes the proof of Lemma 4.

We note that the functional $\phi_t(\omega)$ introduced in the proof of the last lemma is continuous (this can be proved in exactly the same way as Lemma 3). Therefore Theorem 1 will be completely proved if it can be established that

$$f(x) = M_x \phi_\infty. \tag{14}$$

But from (9) we have $(t < \infty)$

$$M_x \phi_t = f(x) - T_t f(x).$$

Here if we make t tend to ∞ and recall that f possesses property $(\mathfrak{A})$ and that therefore

$$\lim_{t \to \infty} T_t f(x) = 0,$$

we obtain (14).

Thus the proof of Theorem 1 is complete. In subsequent work the following remark is important:

Remark 3. Let the bounded excessive function $f(x)$ satisfy ($\mathfrak{A}$) for any nondecreasing sequence of Markov moments τ_n and for all points $x \in E$ except for points of a certain almost Borel set S. Suppose that, in addition, this set S is unattainable from all $x \bar{\in} S$ (i.e. for all $x \bar{\in} S$ the path of $x_t(\omega)$ does not fall in S for any $t \geq 0$ (almost certain P_x)). Then all the assertions of §2 remain valid for $x \bar{\in} S$. Defining $\phi_t(x, \omega)$ by means of (9) for $x \bar{\in} S$ and $x_0(\omega) \bar{\in} S$ and putting $\phi_t(n, \omega) = \infty$ otherwise, we see that Lemma 4 remains valid. It is therefore easy to deduce the existence of a continuous functional ϕ_t such that for $x \bar{\in} S$

$$f(x) = M_x \varphi_\zeta \text{ and } P_x\{\varphi_t = \varphi_t(x, \omega)\} \equiv 1,$$

and for $x \in S$

$$P_x\{\phi_t = \infty\} \equiv 1.$$

2. We shall now show how we may dispense with the requirement in Theorem 1 for the excessive function $f(x)$ to be bounded.

Definition. An almost Borel function f satisfies the condition ($\mathfrak{A}'$) if for any nondecreasing sequence of Markov moments τ_n and any point $x \in E$ at which f has a finite value the equality ($\mathfrak{A}$) is satisfied.

Theorem 1′. *A necessary and sufficient condition for the excessive function f to be a potential of some continuous functional ϕ_t of the process X is that it should satisfy condition* ($\mathfrak{A}'$).

Proof. The necessity is proved as in Theorem 1. We therefore restrict ourselves to proving the sufficiency.

Let the excessive function $f(x)$ be bounded and satisfy ($\mathfrak{A}'$). We put

$$f_n(x) = M_x f(x(\xi_n)),$$

where ξ_n is the moment of the first exit from $B_n\{x: f(x) \geq n\}$. These functions are excessive (see, for example, [5], §6), do not decrease as n increases and $f = f_n$ on the set $C_n = E \setminus B_n$. Further, the set $S = \{x: f(x) = \infty\}$, in accordance with assertion (a), is unattainable from $x \bar{\in} S$. Hence it follows that each of the functions $f_n(x)$ satisfies the conditions considered in Remark 3.

By virtue of this remark, there exist continuous additive functionals $\phi_t^{(n)}$ for which, for $x \bar{\in} S$,

$$f_n(x) = M_x \phi_\zeta^{(n)},$$

$\phi_t^{(n)}(\omega)$ being the mean square limit (relative to any measure P_x, $x \bar{\in} S$) of the integrals

$$\frac{1}{h}\int_0^t [f_n(x_u) - T_h f_n(x_u)]\, du \qquad (h \to 0)$$

and $\phi_t^{(n)}(\omega) = \infty$ (almost certain P_x) if $x \in S$. From the equality $f_n = f_m$, which is satisfied everywhere on the set C_n for $m > n$, and from the inequality $T_h f_n \leq T_h f_m$, which also holds for $m > n$, we deduce that

$$f_n(x_u) - T_h f_n(x_u) \geqslant f_m(x_u) - T_h f_m(x_u)$$

and therefore, for $m > n$

$$\phi_t^{(n)}(\omega) \geq \phi_t^{(m)}(\omega) \tag{15}$$

P_x-almost everywhere on the set $\{\omega: t < \tau_{C_n}\}$, where $x \in S$. Since $\tau_{C_n} \uparrow \zeta$ (almost certain P_x), for all $t \geq 0$ the limit

$$\phi_t = \lim_{n \to \infty} \phi_t^{(n)}(\omega) \quad \text{(almost certain } P_x)$$

exists, equal to P_x-almost everywhere $+\infty$ if $x \in S$. It follows from (15) that ϕ_t is a continuous functional.

It remains for us to verify that

$$f(x) = M_x \phi_\zeta.$$

It is obviously satisfied for $x \in S$. We now fix $x \bar{\in} S$ and $\epsilon > 0$ and choose n so that $M_x f(x(\tau_{C_n})) < \epsilon$ (such a choice of n is possible since $\lim_{n\to\infty} M_x f(x(\tau_{C_n})) = 0$ by virtue of condition ($\mathfrak{A}'$)). Then for $m > n$

$$f_m(x) = M_x \varphi_\zeta^{(m)} = M_x \varphi_{\tau_n}^{(m)} + M_x \{\varphi_\zeta^{(m)} - \varphi_{\tau_n}^{(m)}\}$$
$$= M_x \varphi_{\tau_n}^{(m)} + M_x f_m(x(\tau_n)),$$

where we have put $\tau_n = \tau_{C_n}$. Recalling that $f_m \leq f$ and taking into account the choice of the number n, we obtain:

$$|f_m(x) - M_x \phi_{\tau_n}^{(m)}| < \epsilon.$$

Passing here to the limit as $m \to \infty$ we obtain:

$$|f(x) - M_x \phi_{\tau_n}| \leq \epsilon.$$

One further passage to the limit as $n \to \infty$ gives the estimate

$$|f(x) - M_x \phi_\zeta| \leq \epsilon,$$

from which our assertion follows.

§4. Some classes of excessive functions

1. In §2 and §3 we considered potentials of continuous functionals. In §4, preparatory to the final §5, we isolate out certain classes of excessive functions which are not potentials. In addition a condition is given here for representing an excessive function as the potential of a right-continuous functional. All processes considered in §4 are standard.

Definition. An excessive function $f(x)$ is called harmonic if for any compactum $F \subseteq E$

$$f(x) = M_x f(x_{\tau_c(F)}), \tag{16}$$

where c is a nonnegative constant $(c < \infty)$ and

$$\tau_c(F) = \min(c, \tau_F). \tag{16a}$$

The following criteria for harmonicity are often useful.

Lemma 5. *A necessary and sufficient condition for an excessive function $f(x)$ to be harmonic is that for any fixed sequence of compacta $F_n \uparrow E$ the identities*

$$f(x) \equiv M_x f(x_{\tau_n(F_n)})$$

are satisfied.

Proof. The necessity of the condition for harmonicity is obvious. We shall prove its sufficiency. Let $f(x)$ be harmonic, c be a nonnegative constant $(c < \infty)$ and F a certain compactum. We choose n such that $c \leq n$, $F \subseteq F_n$. Then $\tau_c(F) \leq \tau_n(F_n)$ and, using property (a) of excessive functions and the strictly Markov property of the process X, we obtain

$$M_x f(x_{\tau_c(F)}) \geqslant M_x \{g(x_{\tau_c(F)})\} = M_x \{\theta_{\tau_c(F)} f(x_{\tau_n(F_n)})\} = M_x f(x_{\tau_n(F_n)}),$$

where $g(x) = M_x f(x_{\tau_n(F_n)})$. The right-hand term of this chain of relations is equal to $f(x)$ and the left-hand term does not exceed $f(x)$. Therefore the left-hand term also equals $f(x)$, as was to be proved.

The identity of our definition of harmonicity with that of P.-A. Meyer (see [9], p. 167) follows from this lemma.

The following is a useful criterion for harmonicity:

Lemma 6. *Suppose that the process X is such that for any compactum F the probability $P_x\{\tau_F < \infty\}$ is identically equal to unity. Under this condition the excessive function $f(x)$, bounded on each compactum, will be harmonic if and only if for any compactum F the identity*

$$f(x) \equiv M_x f(x(\tau_F)) \tag{17}$$

holds.

Proof. If $f(x)$ is excessive and (17) holds for any compactum, (16) also holds (cf. the chain of formulas in the proof of the preceding lemma); hence $f(x)$ is harmonic. It therefore remains for us to prove that if the excessive function $f(x)$ is harmonic and bounded on any compactum $F \subseteq E$, condition (17) is valid. But in this case for any compactum F we have:

$$M_x f(x(\tau_F)) = \lim_{n\to\infty} M_x f(x(\tau_n(F))),$$

since the expressions governed by the signs of mathematical expectation do not exceed $\sup_{x\in F} f(x)$ and since for all points of the set $\{\omega: \tau_F < \zeta\}$ the second of these expressions coincides with the first for all sufficiently large n. The lemma is thereby proved.

We note that the condition of this lemma is a sufficient condition for harmonicity of any excessive function $f(x)$ connected with any standard process.

2. Suppose that E is not a compactum. We divide by Ω^* the aggregate $\omega \in \Omega$ for which at least one of the two following assertions is valid:

(1) $\tau_F(\omega) < \zeta$ for any compactum F;

(2) $\zeta(\omega) = \infty$.

If E is a compactum, we put

$$\Omega^* = \{\omega: \zeta(\omega) = \infty\}.$$

For any excessive function $f(x)$ we put

$$\xi(f) = \begin{cases} \lim\limits_{t\uparrow\zeta} f(x_t), & \text{if } \omega \in \Omega^*, \\ 0, & \text{if } \omega \overline{\in} \Omega^* \end{cases}$$

(the random variable $\xi(f)$ is thus defined only for P_x-almost all ω). It is easily verified that, if the function $f(x)$ is excessive, the function $M_x\xi(f)$ is harmonic.

Definition. The harmonic function $f(x)$ is called regular if $f(x) = M_x\xi(f)$.

The following lemma will be useful in what follows. Different variants of the lemma are well known (see, for example [9], p. 169).

Lemma 7. *Let $f(x)$ be an arbitrary excessive function. Then the function $g(x) = M_x\xi(f)$ is a regular harmonic function such that $\xi(f) = \xi(g)$ (almost certain P_x).*

Proof. The excessiveness of the function $g(x)$ is easily established. Further, if the function f is unbounded and $f_n = \min(n, f)$, then $\xi(f_n) \uparrow \xi(f)$ and $M_x\xi(f_n) \uparrow M_x\xi(f)$. Therefore, in order to prove the lemma it is sufficient to verify that $\xi(f) = \xi(g)$ (almost certain P_x), assuming f to be bounded. With this aim we consider the sequence of compacta $F_n \uparrow E$ and put $\tau_n = \tau_n(F_n)$ (see (16a)). We denote by $\mathfrak{M}_{\tau_n}$ the σ-algebra of events A such that $A \subseteq \{\tau_n < \zeta\}$ and $\{A, \tau_n \le t < \zeta\} \in \mathfrak{M}_t$ for any $t > 0$. Then for any $A \in \mathfrak{M}_{\tau_n}$ we shall have:

$$M_x\{\chi_A \xi(f)\} = M_x\{\chi_A \theta_{\tau_n} \xi(f)\} = M_x\{\chi_A g(x(\tau_n))\},$$

where χ_A is the characteristic function of A. Hence

$$M_x\{\xi(f) \mid \mathcal{M}_{\tau_n}\} = M_x\{g(x(\tau_n)) \mid \mathcal{M}_{\tau_n}\} \text{ (almost certain } P_x)$$

and the required relation therefore follows.

The following definition plays an important part subsequently:

Definition. An excessive function $f(x)$ is called singular if for every point x at which $f(x)$ takes a finite value the following two conditions are satisfied:

(C_1) there exists a nonincreasing sequence of sets G_n such that

$$\lim_{n\to\infty} P_x\{\tau_n = \zeta\} = 1 \text{ and } f(x) \equiv M_x f(x(\tau_n)),$$

where τ_n is the moment of first attainment of the set G_n;

(C_2) the following equality is valid:

$$\xi(f) = 0 \quad \text{(almost certain } \Omega, P_x).$$

It follows from (C_1) that $f_0 \equiv 0$ is a unique bounded singular function. By virtue of (C_2) $f_2(x) \equiv 0$ is a unique regular harmonic finite-valued function not exceeding the singular function $f_1(x)$. In the case of a Wiener process in Euclidean space of dimension $m \geq 3$ an example of a singular function is $r^{2-m}(x, y)$ for fixed y, where $r(x, y)$ is the Euclidean distance between the points x and y.

3. Let us consider the conditions for an excessive function to be presented in the form of the potential of a right-continuous process. The following lemma is taken from P.-A. Meyer (see [9], p. 184).

Lemma C. *Any bounded excessive function $f(x)$ for which $\xi(f) = 0$ (almost certain P_x) for all $x \in E$ can be put in the form*

$$f(x) = f_1(x) + f_2(x),$$

where the excessive function $f_2(x)$ satisfies condition ($\mathfrak{A}$) and the function $f_1(x)$ is the potential of the functional β_t equal to the sum of the jumps of the function $f(x_u)$ at those moments u $(0 \leq u < \min(t, \zeta))$ when the path of $x_u(\omega)$ has no discontinuities.

From this lemma we can easily deduce

Theorem 3. *A necessary and sufficient condition for the excessive function f to be the potential of a certain right-continuous functional of X is that, for every $x \in E$ for which $f(x) < \infty$ and for every sequence of Markov moments τ_n such that $\tau_n \uparrow \zeta$ (almost certain P_x),*

$$\lim_{n\to\infty} M_x f(x(\tau_n)) = 0$$

holds.

Proof. The necessity of this condition for representing an excessive function as a potential of a right-continuous functional is obvious. To prove the sufficiency of this condition we note that, if it is satisfied for $f(x)$, it is satisfied also for any function of the form $f_n = \min(n, f)$, where n is a nonnegative constant. This together with Lemma C and Theorem 1 enables us to establish the existence of a right-continuous functional $\phi^{(n)}(t)$ for which f_n is a potential. By repeating the argument at the end of the proof of Theorem 1′ the proof of the theorem is completed.

§5. Decomposition of excessive functions

In the measurable space (E, B) let us consider the standard process X satisfying the condition $(\mathcal{H})$.

The theorem given below was first obtained (in a somewhat different form and for a narrower class of processes) by P.-A. Meyer (see, for example, [9]).

Theorem 4. *Any excessive function $f(x)$ with finite values can be put in the form*

$$f(x) = f_1(x) + f_2(x) + f_3(x), \tag{18}$$

where $f_1(x)$ is a regular harmonic function, $f_2(x)$ is a singular function and $f_3(x)$ is the potential of a certain functional ϕ_t of class (U). The functions $f_i(x)$ in the decomposition (18) can be chosen uniquely and the functional $\phi_t \in (U)$ is defined uniquely by the function $f(x)$ with accuracy up to equivalence.

Proof, We show first that if the decomposition (18) exists then it is unique. From (18), condition (C_2) and the easily verifiable relation

$$\lim_{n\to\infty} f_3(x(\zeta_n)) = 0 \qquad (\zeta_n \equiv \tau_n(E_n)),$$

where E_n is a nondecreasing sequence of compacta amounting to the components of E, we conclude that $\xi(f_1) = \xi(f)$ (almost certain P_x) and therefore the component f_1 is defined uniquely (see the definition of a regular harmonic function). Considering the sequence of sets C_n satisfying the condition (C_1) for the function f_2 and the function $F = f - f_1$, we have:

$$f_2(x) = \lim_{n\to\infty} M_x f_2(x(\tau_{E\setminus G_n}))$$

$$= \lim_{n\to\infty} M_x [F(x(\tau_{E\setminus G_n})) - f_3(x(\tau_{E\setminus G_n}))] = \lim_{n\to\infty} M_x F(x(\tau_{E\setminus G_n})).$$

Therefore the component f_2 (and hence f_3 also) is also defined uniquely. The fact that the functional $\phi_t \in (U)$ is defined uniquely with respect to the function f_3 with an accuracy up to equivalence follows from Lemma A.

We now prove the existence of the decomposition (18). For bounded excessive functions f the existence of the decomposition (18) is a corollary of

Theorem 1 and Lemma C, the second term of the decomposition (18) for such functions being everywhere zero and the first equal to $M_x\xi(f)$. We assume that f is unbounded and put

$$F_1(x) = M_x\xi(f), \qquad h_n(x) = \min\{n, f(x) - F_1(x)\}.$$

In accordance with Lemma 7, F_1 is a regular harmonic function. The functions h_n are excessive and bounded in $\xi(h_n) = 0$ (almost certain Ω, P_x). From what has already been proved for bounded excessive functions of the decomposition (18), it follows from Lemma C and Theorem 1 that each function h_n is the potential of a certain functional $\phi_t^{(n)} \in (U)$ equal to the sum of a certain continuous functional $\alpha_t^{(n)}$ and the number of jumps $\beta_t^{(n)}$ of the function $f(x_u)$ at those times u $(0 \leq u < \min(t, \zeta))$ when the function x_u is continuous.

Let us now consider the moment τ_n of first attainment of the set $G_n = \{x: f(x) \geq n\}$. Since the functions h_n and h_m coincide on the set $E\setminus G_m$ for $m \leq n$, by Lemma A the functions $\phi_t^{(n)}$ coincide with one another for all $h \geq m$ and all $t \geq 0$ for which $t < \tau_m$ (almost certain Ω, P_x). Here, from property (a) of excessive functions it follows that $\tau_n \uparrow \zeta$ (almost certain P_x) and therefore the limit

$$\phi_t = \lim_{n\to\infty} \phi_t^{(n)}$$

exists for all $t \geq 0$ (almost certain P_x). Where this limit does not exist we assume $\phi_t = 0$. The random function ϕ_t is a functional of class (U).

Further, from $\phi_t = \phi_t^{(n)}$ for $t \leq \tau_n$ (almost certain P_x) we deduce that:

$$h_n(x) = M_x\varphi_\infty^{(n)} = M_x\{\varphi_{\tau_n - 0}\} + M_x\{\varphi_\infty^{(n)} - \varphi_{\tau_n-0}^{(n)}\}. \tag{19}$$

Denoting the second term of the last sum by $H_n(x)$ and taking into account that $\phi_{\tau_n - 0} \uparrow \phi_\zeta$ as $n \to \infty$ (almost certain P_x), from (19) we obtain

$$\lim_{n\to\infty} H_n(x) = \lim_{n\to\infty} h_n(x) - \lim_{n\to\infty} M_x\{\varphi_{\tau_n-0}\} = f(x) - F_1(x) - M_x\varphi_\infty.$$

We shall show that $F_2 = \lim_{n\to\infty} H_n$ is a singular excessive function. The equality of the extreme terms of the last chain of relations allows us to assert that this function processes property (b) of excessive functions. Therefore, in order to prove its excessiveness it is sufficient to establish that

$$T_tF_2(x) \leq F_2(x). \tag{20}$$

For the verification of this we consider the quantity $\xi = \min(t, \tau_k)$. For $n \geq k$ we shall have:

$$M_xH_n(\xi) = M_x\theta_\xi[\varphi_\infty^{(n)} - \varphi_{\tau_n-0}^{(n)}] = M_x\theta_\xi[\alpha_\infty^{(n)} - \alpha_{\tau_n}^{(n)}] + M_x\theta_\xi[\beta_\infty^{(n)} - \beta_{\tau_n-0}^{(n)}].$$

The first of these terms equals

$$M_x\{[\alpha_\infty^{(n)} - \alpha_\xi^{(n)}] - [\alpha_{\tau_n}^{(n)} - \alpha_\xi^{(n)}]\} = M_x\{\alpha_\infty^{(n)} - \alpha_{\tau_n}^{(n)}\},$$

and the second term obviously does not exceed $M_x\{\beta_\infty^{(n)} - \beta_{\tau_n-0}^{(n)}\}$. Thus

$$M_xH_n(\xi) \leqslant M_x\{\alpha_\infty^{(n)} + \beta_\infty^{(n)} - \alpha_{\tau_n}^{(n)} - \beta_{\tau_n-0}^{(n)}\} = M_x[\varphi_\infty^{(n)} - \varphi_{\tau_n-0}^{(n)}] = H_n(x),$$

and hence we obtain

$$M_xH_n(\xi) \le H_n(x) \text{ and } M_xF_2(\xi) \le F_2(x)$$

by virtue of Fatou's lemma. The last inequality gives

$$\int_{t\le\tau_k} F_2(x_t)P_x\{d\omega\} \le F_2(x).$$

Applying Fatou's lemma again and letting $k \to \infty$, we obtain (20). The excessiveness of F_2 is thereby established.

We shall show that F_2 is singular. For this we shall prove that

$$F_2(x) = M_xF_2(x(\tau_m)) \quad (m \ge 1). \tag{21}$$

Arguing in the same way as in the proof of (20), we obtain:

$$M_x H_n(x(\tau_m)) = M_x \theta_{\tau_m}\{\varphi_\infty^{(n)} - \varphi_{\tau_n-0}^{(n)}\}$$

$$= M_x\{[\varphi_\infty^{(n)} - \varphi_{\tau_n-0}^{(n)}]\chi_{A_1}\} + M_x\{\theta_{\tau_m}[\varphi_\infty^{(n)} - \varphi_{\tau_n-0}^{(n)}]\chi_{A_2}\}, \tag{21a}$$

which can be written in the form:

$$M_x\{\phi_\infty^{(n)} - \phi_{\tau_n-0}^{(n)}\} + M_x\{\theta_{\tau_m}[\phi_\infty^{(n)} - \phi_{\tau_n-0}^{(n)}]\chi_{A_2}\} - M_x\{[\phi_\infty^{(n)} - \phi_{\tau_n-0}^{(n)}]\chi_{A_2}\}, \tag{22}$$

where $m \le n$ and χ_{A_1} and χ_{A_2} are characteristic functions of the events $A_1 = \{\omega\colon \tau_m < \tau_n\}$ and $A_2 = \{\omega\colon \tau_m = \tau_n\}$.

We denote the second of the mathematical expectations in (22) by I_1 and the third by I_2 and shall prove that for fixed m and $n \to \infty$ both these expressions tend to zero. This assertion, as far as I_1 is concerned, follows from the relations

$$I_1 = M_x[H_n(x(\tau_m))\chi_{A_2}] \leqslant \int_{\tau_n<\zeta} f(x(\tau_m))P_x\{d\omega\},$$

where the last integral tends to zero since the probability $P_x\{\tau_n = \zeta\}$ tends to 1. On considering I_2, we note that

$$I_2 \leqslant M_x[\varphi_\infty^{(n)}\chi_{A_2}] = \int_{\tau_m=\tau_n<\zeta} [\varphi_{\tau_m} + \theta_{\tau_m}\varphi_\infty^{(n)}]P_x\{d\omega\}$$

$$\leqslant \int_{\tau_n<\zeta} \varphi_\infty P_x\{d\omega\} + \int_{\tau_n<\zeta} H_n(x(\tau_m))P_x\{d\omega\},$$

and the last integral here does not exceed

$$\int_{\tau_n<\zeta} f(x(\tau_m))P_x\{d\omega\}.$$

From these estimates the assertion for I_2 also follows.

Thus as $n \to \infty$ the last two expressions in (22) tend to zero. On the other hand the first expression in (22), equal to $H_n(x)$, tends to $F_2(x)$. Therefore from (21a) and (22) we have:

$$F_2(x) = \lim_{n\to\infty} M_x H_n(x(\tau_m)).$$

Hence, on taking into account that $H_n \leq f$, we easily obtain (21).

Thus the function F_2 satisfies the first of the conditions entering into the definition of a singular function. It satisfies the second condition also by virtue of the estimate $f - F_1 \geq F_2$. Putting $f_1 = F_1$, $f_2 = F_2$ and $f_3 = M_x\phi_\infty$, we obtain the decomposition (18). Theorem 4 is therefore proved.

The decomposition (18) can also be extended to excessive functions taking the value $+\infty$. This is achieved by means of the following theorem.

Theorem 4′. *If the process X satisfies the condition* ($\mathcal{H}$), *decomposition* (18) *holds for any excessive function* $f(x)$.

Proof. Let us consider the function $F_1(x) = M_x\xi(f)$, which is by Lemma 7 regular and harmonic, and the function

$$h(x) = \begin{cases} f(x) - F_1(x) & \text{for } F_1(x) < \infty, \\ \infty & \text{for } F_1(x) = \infty. \end{cases}$$

We shall show that $h(x) \geq T_t h(x)$ for any $t \geq 0$. For points of the set $A = \{x: f(x) = \infty\}$ this is obvious. But if $x \bar{\in} A$ and $t \geq 0$, then

$$T_t f(x) + F_1(x) - T_t F_1(x)$$

$$= \int_{t<\zeta} f(x_t) P_x\{d\omega\} + \int_{t\geq\zeta} \chi \lim_{k\to\infty} f(x(\tau_k)) P_x\{d\omega\}, \tag{23}$$

where χ is the characteristic function of the set Ω^* introduced in §4 and $\tau_k \equiv \tau_k(F_k)$ for a certain sequence of compacta $F_k \uparrow E$ (see (16a)). The right-hand term of (23) does not exceed

$$\int_{t<\tau_k} f(x_t) P_x\{d\omega\} + \int_{\tau_k\leq t<\zeta} f(x_t) P_x\{d\omega\} + \int_{t\geq\tau_k} \chi \lim_{n\to\infty} f(x(\tau_n)) P_x\{d\omega\}. \tag{24}$$

The third term of this sum being equal to

$$\int_{t\geq\tau_k} g(x(\tau_k)) P_x\{d\omega\},$$

where $g(x) = M_x\xi(f)$ is majorized by

$$\int_{t\geq\tau_k} f(x(\tau_k)) P_x\{d\omega\}.$$

The sum of this expression and the first term of (24) is equal to $M_x f(x(\eta))$, where $\eta = \min(t, \tau_k)$, which does not exceed the value of $f(x)$ (by property (a) of excessive functions). Therefore the sum of the first and third terms in (24) does not exceed $f(x)$. In addition, as $k \to \infty$ the second term in (24) tends to zero. From this and (23) we obtain:

$$T_t f(x) + F_1(x) - T_1 F_1(x) \leq f(x),$$

which proves the inequality $h \geq T_t h$. The argument given here is taken from Meyer (see [9], p. 170).

It is also easy to see that $h(x) = \lim_{t \to 0} T_t h(x)$ if $x \bar{\in} A$. Therefore, by Lemma B, the function $h' = \lim_{t \to 0} T_t h$ is excessive and h coincides with h' on the set $E \setminus A$.

Let us consider the part X' of the process X on $E \setminus A$. Applying Theorem 3 to h' and X' [1] we see that the function h' can be put in the form

$$h'(x) = h_1(x) + h_2(x) + h_3(x) \qquad (x \in E \setminus A),$$

where $h_1(x)$ is a regular harmonic function connected with X', $h_2(x)$ is a singular function connected with X' and $h_3(x) = M_x \phi'_\infty$, where ϕ'_t is a functional of class (U) for X'. From the definitions of $h(x)$ and $h'(x)$ it follows that $h_1(x) = 0$ $(x \in E \setminus A)$.

Let

$$\varphi_t(\omega) = \begin{cases} \varphi'_t(\omega), & \text{if } x_0(\omega) \bar{\in} A, \\ +\infty, & \text{if } x_0(\omega) \in A. \end{cases}$$

Since the set A is attainable by the path $x_t(\omega)$ starting from the point $x \in E \setminus A$ only with P_x-probability zero, $\phi_t(\omega)$ is a functional of class (U) for the process X. It is clear that $M_x \phi_\infty = \infty$ if $x \in A$.

As far as the function h_2 is concerned it can be defined on the set A so that an excessive function connected with X is obtained. In fact, let $h'_2(x) = +\infty$ if $x \in A$, and $h'_2(x) = h_2(x)$ if $x \bar{\in} A$. The function $h'_2(x)$ satisfies the condition of Lemma B. Therefore, the function $H_2 = \lim_{t \to 0} T_t h'_2$ is excessive for X and coincides with h_2 outside A.

We shall show that the function H_2 is singular for X. For this it is necessary to verify that conditions (C_1) and (C_2) are satisfied for every point x at which $H_2(x) < \infty$. If $x \in E \setminus A$ these conditions are satisfied as is seen from the proof of the preceding theorem (both here and subsequently the sequence G_n is taken to be $G_n = \{x: H_2(n) \geq n\}$). Thus it remains for us to verify these conditions

---

[1] The process X' satisfies conditions B, C and D. Moreover, the set A is unattainable from points $x \in E \setminus A$. Therefore the proof of Theorem 3 retains its validity in the case of the process X'. In carrying over the definitions of §4 to the case of the process X' it is necessary to consider in these definitions compacta in E (and not in $E \setminus A$) and points $x \in E \setminus A$. The quantity $\xi(f)$ is defined, however, in exactly the same way as for the process X.

for x from the set $B = \{x: x \in A, H_2(x) < \infty\}$.

We fix $x \in B$. For any $t > 0$ the probability $P_x\{x_t \in A\}$ equals zero since otherwise we would have had

$$H_2(x) \geq T_t h_2'(x) = \infty$$

for some $t > 0$. From this both conditions (C_1) and (C_2) are easily deduced. Let us, for example, prove condition (C_1). Let $H_2(x) < n$. Putting

$$g(y) = M_y H_2(x(\tau_{E \setminus G_n}))$$

and taking into account that (C_1) is already proved for points of the set $E \setminus A$, we obtain

$$P_x\{g(x_t) \neq H_2(x_t)\} = 0.$$

Here,

$$H_2(x) = \lim_{t \to 0} H_2(x_t) \quad \text{(almost certain } P_x)$$

and

$$g(x) = \lim_{t \to 0} g(x_t) \quad \text{(almost certain } P_x),$$

since the function g is excessive for the part of the process X on the set $E \setminus G_n$. From what has been said condition (C_2) follows.

Appendix

Suppose that a homogeneous Markov process $X = (x_t, \zeta, \overline{N}_{t \to 0}, P_x)$ is given in the measurable space (E, B). If $P_n\{\zeta = \infty\} \not\equiv 1$, then in place of the process X there sometimes enters into consideration a new noncutoff process X_e in relation to which the original process is a part. This process is constructed in the following way.[1)]

Let e and ω' be certain points not belonging to E and Ω respectively. We put

$$E' = E \cup \{e\}, \quad \Omega' = \Omega \cup \{\omega'\}$$

and denote by B' the σ-algebra of subsets of the set E' whose intersections with E belong to B. Further, let

---

1) We assume that $\mathfrak{M}_t = \overline{\mathfrak{N}}_{t+0}$ for the original process. This results in no loss of generality since the condition can be satisfied in the case of a standard process by contracting the σ-algebras $\mathfrak{M}_t$. The numbering of the formulas and results in the Appendix has no relation to that of §§1–5.

$$x'_t \equiv x'_t(\omega) = \begin{cases} x_t(\omega), & \text{if} \quad \omega \in \Omega \text{ and } t < \zeta(\omega), \\ e, & \text{if} \quad \omega \in \Omega \text{ and } t \geq \zeta(\omega), \\ e, & \text{if} \quad \omega = \omega', \end{cases}$$

and let $\mathfrak{R}_t$, $\overline{\mathfrak{R}}_t$, $\mathfrak{R}_{t+0}$, $\overline{\mathfrak{R}}'_{t+0}$ be defined by means of the functions x'_t in the same way as the σ-algebras $\mathfrak{N}_t$, $\overline{\mathfrak{N}}_t$ etc. were defined by means of the functions x_t (only in place of ζ we take $+\infty$ in the definition of the new σ-algebras). We denote by $\mathfrak{R}^0$ the smallest σ-algebra containing all the $\mathfrak{R}_t$ and by $\overline{\mathfrak{R}}^0$ the smallest σ-algebra containing all the σ-algebras $\overline{\mathfrak{R}}_t$.

We now proceed to construct a system of measures P'_x on $\overline{\mathfrak{R}}^0$ $(x \in E')$. If the set A has the form

$$A = \{x'_s \in \Gamma\} \quad (s \geq 0, \quad \Gamma \in B'),$$

we put

$$P'_x\{A\} \equiv P'(s, x, \Gamma) = \begin{cases} P(s, x, \Gamma \cap E) + \chi_\Gamma(e) P_x\{\zeta > s\}, & \text{if} \quad x \neq e, \\ \chi_\Gamma(e), & \text{if} \quad x = e, \end{cases}$$

where χ_Γ is the characteristic function of the set Γ (cf. (4.6) in the book [3]). But if A has the form

$$A = \{x'_{s_1} \in \Gamma_1, \ldots, x'_{s_n} \in \Gamma_n\} \quad (s_{i+1} > s_i, \quad \Gamma_i \in B'), \tag{1}$$

then we equate $P'_x\{A\}$ to the integral

$$\int_{y_1 \in \Gamma_1} P'(s_1, x, dy_1) \int_{y_2 \in \Gamma_2} P'(s_2 - s_1, y_1, dy_2) \ldots \int_{y_{n-1} \in \Gamma_{n-1}} P'(s_n - s_{n-1}, y_{n-1}, \Gamma_n). \tag{2}$$

The following lemma is valid:

Lemma 1. *If the sets* A_n $(n \geq 0)$ *have the form* (1), $A_0 = \bigcup_{n=1}^{\infty} A_n$ *and the sets* $A_i \cap A_j$ *are empty for* $i \neq j$ $(i, j \geq 1)$, *then*

$$P'_x\{A_0\} = \sum_{n-1}^{\infty} P'_x\{A_n\}.$$

The assertion of this lemma becomes obvious if it is reformulated in terms of the process X.

By means of this lemma the function P'_x given on the class of sets (1) by formula (2) is uniquely extended onto $\mathfrak{R}^0$ (see Theorem 1.1 from [3]) and $\overline{\mathfrak{R}}^0$.

Now let $\mathfrak{R}^*$ be the minimal system of subsets of the set Ω' containing all sets of the form $\{x'_s \in \Gamma\}$ and closed under the operations of addition and intersection of any number of sets and under the operation of taking the complement. To each set $A \in \mathfrak{R}^*$ we can compare the set $\theta'_t A$ $(t \geq 0)$ taking the operator θ'_t as invariant under the operations of addition and intersection of any number of sets and assuming that

$$\theta_t'(A\setminus B) = \theta_t'A\setminus\theta_t'B$$

and

$$\theta_t'\{x_h' \in \Gamma\} = \{x_{t+h}' \in \Gamma\} \quad (h \geqslant 0, \quad \Gamma \in B').$$

Lemma 2. *If* $A \in \mathfrak{R}_t$, *the set* $A_t = A\Omega\{\zeta > t\}$ *lies in* $\mathfrak{N}_t$. *If* $A \in \mathfrak{R}_{t+0}$, *then* $A_t \in \mathfrak{N}_{t+0}$. *If* $A \in \overline{\mathfrak{R}}_{t+0}$ *then* $A_t \in \overline{\mathfrak{N}}_{t+0}$.

Proof. If A has the form (1) for $s_i \leq t$, then A belongs to $\mathfrak{R}_t$ and $A_t \in \mathfrak{N}_t$. Hence the first assertion of our lemma is deduced by means of Lemma 1.1 from [3]. Now let $A \in \mathfrak{R}_{t+0}$. Then $A_{t+u} \in \mathfrak{R}_{t+u}$ and therefore $A_{t+u} \in \mathfrak{N}_{t+u}$ for any $u > 0$. The second assertion of the lemma is thereby proved. The third assertion follows from the second and from the definition of the σ-algebras $\overline{\mathfrak{N}}_{t+0}$ and $\overline{\mathfrak{R}}_{t+0}$. The proof of the lemma is complete.

Theorem 1. *The collection* $X_e = (x_t', \infty, \overline{\mathfrak{R}}_{t+0} P_x')$ *defines a homogeneous Markov process.*

Proof. Of all the conditions defining a homogeneous Markov process (see [3], §2.8) we shall require in the verification only the following: for any $t \geq 0$, $A \in \mathfrak{R}^0$ and $x \in E'$,

$$P_x'\{\theta_t' A \mid \overline{\mathfrak{R}}_{t+0}\} = P_{x_t}'\{A\} \text{ (almost certain } P_x'). \tag{3}$$

Let A have the form (1) and $\Gamma_i \subseteq E$. Then $\theta_t' A = \theta_t A$, where θ_t is the shift operator corresponding to the process X and the sets A and $\theta_t' A$ lie in Ω, and (3) follows from the Markov property of the process X and Lemma 2. Now let A have the form (1), where $\Gamma_i \leq E$ for $i < n$ and $\Gamma_n = \{e\}$. We put

$$B_1 \doteq \{x_{s_1}' \in \Gamma_1, \ldots, x_{s_{n-1}}' \in \Gamma_{n-1}, x_{s_n}' \in E\},$$

$$B_2 = \{x_{s_1}' \in \Gamma_1, \ldots, x_{s_{n-1}}' \in \Gamma_{n-1}\}.$$

By what has already been proved,

$$P_x'\{\theta_t' B_i \mid \overline{\mathfrak{R}}_{t+0}\} = P_{x_t}'\{B_i\} \text{ (almost certain } P_x').$$

Hence

$$P_x'\{\theta_t' A \mid \overline{\mathfrak{R}}_{t+0}\} = P_x'\{\theta_t'(B_2\setminus B_1) \mid \overline{\mathfrak{R}}_{t+0}\}$$
$$= P_{x_t}'\{B_2\} - P_{x_t}'\{B_1\} = P_{x_t}'\{A\} \text{ (almost certain } P_x').$$

Therefore (3) is also valid in the case under consideration. Thus (3) is proved for the two particular forms of the set A. The validity of (3) for all A of the form (1) therefore follows and hence, by virtue of Lemma 1.1 from [3], for all $A \in \mathfrak{R}^0$.

Theorem 2. *Let the process* X *be strictly Markov. Then the process* X_e *is also strictly Markov.*

Proof. Let τ be a random variable not depending on the future for the process X_e. We denote by $\mathfrak{R}_\tau$ the σ-algebra in the space $\Omega_\tau = \{\tau < \infty\}$ composed of the sets

$A \subseteq \Omega_\tau$ for which $\{A, \tau \leq t\} \in \overline{\mathfrak{R}}_t$ for all $t \geq 0$. We must show that, for any nonnegative $\mathfrak{R}_\tau$-measurable function $\delta = \delta(\omega)$ and any $\Gamma \in B'$,

$$P'_x\{x'_{\tau+\delta} \in \Gamma \mid \mathscr{R}_\tau\} = P'(\delta, x'_\tau, \Gamma) \quad (\text{almost certain } \Omega_\tau, P'_x).$$

We note that if τ does not depend on the future for X_e, by virtue of Lemma 2 the quantity τ', induced by the quantity τ on Ω, does not depend on the future for X. From this and from Lemma 2 the proof follows for $\Gamma \subseteq E$. If, however, $\Gamma = \{e\}$, then

$$\begin{aligned} P'_x\{x'_{\tau+\delta} \in \Gamma \mid \mathscr{R}_\tau\} &= P'_x\{\tau < \infty \mid \mathscr{R}_\tau\} - P'_x\{x'_{\tau+\delta} \in E \mid \mathscr{R}_\tau\} \\ &= 1 - P'(\delta, x'_\tau, E) = P'(\delta, x'_\tau, \Gamma) \quad (\text{almost certain } \Omega_\tau, P'_x), \end{aligned}$$

which was required to be proved.

We conclude with a few words on the choice of topology in E'. If the space E is locally bicompact but not bicompact, we shall take to be open in E' all open sets from E and all sets of the form $S \cup \{e\}$, where S is a set in E with bicompact complement. If, however, E is bicompact we shall take to be open in E' any sets $S \subseteq E'$ whose intersections with E are open in E.

BIBLIOGRAPHY

[1] A. D. Ventcel′, *Non-negative additive functionals of Markov processes*, Dokl. Akad. Nauk SSSR 137 (1961), 17–20 = Soviet Math. Dokl. 2 (1961), 218–221. MR 22 #10018.

[2] V. A. Volkonskiĭ, *Additive functionals of Markov processes*, Trudy Moskov. Mat. Obšč. 9 (1960), 143–189. MR 25 #610.

[3] E. B. Dynkin, *Foundations of the theory of Markov processes*, Fizmatgiz, Moscow, 1959; English transl., *Theory of Markov processes*, Prentice-Hall, Englewood Cliffs, N. J. and Pergamon Press, Oxford, 1961. MR 24 #A1745; MR 24 #A1747.

[4] ———, *Markov processes*, Fizmatgiz, Moscow, 1963. (Russian)

[5] G. A. Hunt, *Markoff processes and potentials*. I, II, Illinois J. Math. 1 (1957), 44–93, 316–369; III, ibid. 2 (1958), 151–213; Russian transl., IL, Moscow, 1962. MR 19, 951; MR 21 #5824.

[6] M. G. Šur, *Continuous additive functionals of Markov processes and excessive functions*, Dokl. Akad. Nauk SSSR 137 (1961), 800–803 = Soviet Math. Dokl. 2 (1961), 365–368. MR 26 #7023.

[7] ———, *Kurtosis functions and additive functionals of Markov processes*, Dokl. Akad. Nauk SSSR 143 (1962), 293–296 = Soviet Math. Dokl. 3 (1962), 398. MR 28 #5507.

[8] R. M. Blumenthal, R. K. Getoor and H. P. McKean, Jr., *Markov processes with identical hitting distributions*, Illinois J. Math. 6 (1962), 402–420. MR 25 #5550.

[9] P.-A. Meyer, *Fonctionnelles multiplicatives et additives de Markov*, Ann. Inst. Fourier (Grenoble) 12 (1962), 125–230. MR 25 #3570.

Translated by:
T. Garrett

SOME REMARKS ON ASYMPTOTICALLY EFFECTIVE LINEAR ESTIMATES OF REGRESSION COEFFICIENTS

S. A. AĬVAZJAN AND Ju. A. ROZANOV

Suppose we are given a random sequence of the form

$$\eta(t) = \sum_{k=1}^{n} \alpha_k \cdot A_k(t) + \xi(t) \quad (t = 0, 1, 2, \ldots, T),$$

where $A_k(t)$ are known (nonrandom) linear independent functions, α_k are unknown numerical parameters (regression coefficients), and $\xi(t)$ is a stationary (in the wide sense) random sequence with $M\xi(t) = 0$ and $M\xi(t_i)\overline{\xi(t_j)} < \infty$. [Here M is the mathematical expectation.]

As is known (see, for example, [1]), linear efficient estimates $\hat{a}_k$ of the parameters α_k $(k = \overline{1, n})$ have the form

$$\hat{a} = (A^* B^{-1} A)^{-1} A^* B^{-1} \eta, \tag{1}$$

where $\hat{a}$ is the n-vector column of the efficient estimates, η is the $(T + 1)$-vector column of observed values of the given process $\eta(t)$, $B = (M\xi(t_i) \cdot \overline{\xi(t_j)})$ is the square $(T + 1) \times (T + 1)$-correlation matrix of the sequence $\xi(t)$ and $A = (A_j(t_i))$ is the $(T + 1) \times n$-matrix of values of the given regression functions $A_j(t)$ $(j = \overline{1, n})$. From this it is obvious that the finding of efficient estimates involves very difficult computations, (multiple inversion of matrices of high order, etc.), and even this is possible only in the case where the correlation function of the $\xi(t)$ process is known (which happens quite rarely in practical problems). Thus it is of interest to investigate the properties of much more available estimates, so-called least squares estimates $\tilde{a}_k$ $(k = \overline{1, n})$, i.e., those which minimize the quadratic form $\Sigma_{t=0}^{T} |\eta(t) - \Sigma_{k=1}^{n} a_k \cdot A_k(t)|^2$ and accordingly have the form

$$\tilde{a} = (A^* A)^{-1} A^* \eta. \tag{2}$$

The most definitive results in this direction were obtained by Grenander and Rosenblatt [1], who, in particular, set forth conditions (necessary and sufficient) under which least squares estimates are asymptotically efficient.

In the presentation to follow it will be implicit that the $\xi(t)$ process has a continuous spectral density $f(e^{-i\lambda})$ $(-\pi \leq \lambda \leq \pi)$ which nowhere goes to zero. This rather general restriction is not, however, necessary, but makes possible the simplification of certain steps of the proofs. In particular, without loss of generality we can relax this condition, requiring that the spectral density $f(e^{-i\lambda})$ be strictly positive only at the elements of the corresponding regression spectrum S (see [3]). To clarify the meaning of this concept we introduce an asymptotic characteristic connecting the given regression functions:

$$R_{ij}(h)=\lim_{T\to\infty}\frac{\sum_{t=0}^{T} A_i(t+h)\cdot\overline{A_j(t)}}{\sqrt{\sum_{t=0}^{T}|A_i(t+h)|^2\,|A_j(t)|^2}}$$

and let $R_h=(R_{ij}(h))$ and $M(\Delta\lambda)=(M_{ij}(\Delta\lambda))$, where $M_{ij}(\lambda)$ is the spectral function of the regression vectors $\overline{A}_i=\{A_i(0), A_i(1), \cdots\}$ and $\overline{A}_j=\{A_j(0), A_j(1), \cdots\}$, i.e., $R_h=\int_{-\pi}^{\pi} e^{i\lambda h}\,dM(\lambda)$. The collection of all points λ such that for any interval $[\lambda_1, \lambda_2]$ containing this point $M(\lambda_2)-M(\lambda_1)>0$, is called the *regression spectrum* S. The sets $\mathcal{E}_1, \mathcal{E}_2, \cdots$ are called *elements of the regression spectrum* if $N(\mathcal{E}_i)>0$, $\Sigma_i N(\mathcal{E}_i)=E$ and $N(\mathcal{E}_i)\cdot N(\mathcal{E}_j)=\delta_{ij}\cdot N(\mathcal{E}_i)$, where $N=M^{-1/2}\cdot M\cdot M^{-1/2}$, E is the identity matrix and δ_{ij} is the Kronecker symbol.

Remark 1. Here we are concerned with the formulation of conditions under which least squares estimates are consistent (obviously, if consistency of the estimates is violated, even if they are efficient, there is no practical interest). The following example describes a situation in which least squares estimates (in the given case, efficient) are not consistent. Let $\eta(t)=\alpha_1 A_1(t)+\alpha_2 A_2(t)+\xi(t)$, where

$$A_1(t)=\begin{cases}A_{1.0}, & \text{for } t=0\\ A, & \text{for } t>0\end{cases}\quad\text{and}\quad A_2(t)=\begin{cases}A_{2.0}, & \text{for } t=0\\ -A, & \text{for } t>0\end{cases}$$

(here $A_{1\cdot 0}\neq -A_{2\cdot 0}$), and $\xi(t)$ is a sequence of uncorrelated variables ($t=0, 1, 2, \cdots, T$). In this case, as is known, least squares estimates are exactly efficient, i.e., $\tilde{a}_k=\hat{a}_k$ ($k=\overline{1, n}$). Computation according to (2) yields $\tilde{a}_k=\Sigma_{t=0}^{T} c_k(t)\cdot\eta(t)$, where

$$c_k(t)=\begin{cases}(A_{1.0}+A_{2.0})^{-1}, & \text{for } t=0\\ \dfrac{(-1)^k A_{k.0}}{A(A_{1.0}+A_{2.0})T}, & \text{for } t>0\ (k=1, 2),\end{cases}$$

so that, taking into account the unbiasedness of $\tilde{a}_k$ and the uncorrelatedness of $\xi(t)$, it is easy to obtain $\lim_{T\to\infty}\mathrm{M}|\tilde{a}_k(T)-\alpha_k|^2=\mathrm{M}|\xi(t)|^2/|A_{1\cdot 0}+A_{2\cdot 0}|>0$, which also implies inconsistency of the estimates $\tilde{a}_k$.

Grenander and Rosenblatt discussed certain conditions guaranteeing consistency of least squares esimates. We verify these arguments in the following theorem which we prove below.

Theorem 1. *If the functions* $A_k(t)$ *are such that*

(I) *the angles* γ_k *between the vectors* $\overline{A}_k=\{A_k(0), A_k(1), \cdots, A(T)\}$ *and their projections on the linear span* H_k *of the vectors* $\overline{A}_j$ ($j=\overline{1, n}$ *but* $j\neq k$) *are uniformly* (*in* T) *greater than zero, or equivalently,* $R_0=(R_{ij}(0))$ *is nondegenerate;*

(II) *the quantities* $\Sigma_{t=0}^{T} |A_k(t)|^2$ *increase unboundedly as* T *increases, where*

$$\lim_{T\to\infty} \frac{\sum_{t=0}^{T-N} |A_k(t)|^2}{\sum_{t=0}^{T} |A_k(t)|^2} = 1$$

for each $k = \overline{1, n}$ *and for each fixed* N; *then the least squares estimates are consistent.*

It is easy to see that if condition (I) is not satisfied, the functions $A_k(t)$, being linearly independent on the interval $[0, T]$ for each fixed T, are asymptotically (in $T \to \infty$) linearly dependent; condition (II) imposes restrictions on the growth of the functions $\Sigma_{t=0}^{T} |A_k(t)|^2$ at infinity (incidentally, condition (I) is not satisfied in our example (p. 2)).

Remark 2. Here we formulate the definition of *asymptotically efficient estimates* in a fashion more general and natural (in our opinion) than that of [1]. According to the definition of Grenander-Rosenblatt, the estimates $\tilde{a}_k$ $(k = \overline{1, n})$ are said to be asymptotically efficient if

$$\lim_{T\to\infty} Q_T \cdot \mathbf{M}\{(\tilde{a}(T) - \alpha)(\tilde{a}(T) - \alpha)^*\} \cdot Q_T$$
$$= \lim_{T\to\infty} Q_T \cdot \mathbf{M}\{(\hat{a}(T) - \alpha)(\hat{a}(T) - \alpha)^*\} \cdot Q_T, \tag{3}$$

where the normalizing factor Q_T is a diagonal matrix with main diagonal elements $d_{kk}^{1/2} = (\Sigma_{t=0}^{T} A_k(t) \cdot \overline{A_k(t)})^{1/2}$, and $a(T)$ and α are the column vectors of estimates and estimated parameters respectively.

Our definition, not involving the matrix Q_T, is constructed by a natural analogy with the one-dimensional case, namely: the estimates $\tilde{a}_k$ $(k = \overline{1, n})$ are said to be asymptotically efficient if

$$\lim_{T\to\infty} \frac{\mathbf{M}\,|\tilde{a}_k(T) - a_k|^2}{\mathbf{M}\,|\hat{a}_k(T) - a_k|^2} = 1 \tag{3'}$$

for all $k = \overline{1, n}$. Here we prove

Theorem 2. *If the functions* $A_k(t)$ *satisfy conditions* (I) *and* (II), *then definitions* (3) *and* (3′) *are equivalent.*

Remark 3. Here we formulate a sufficient condition for asymptotic efficiency of least squares estimates. We note that the necessary and sufficient condition for asymptotic efficiency formulated in [1] uses a spectral representation of the correlation matrix R_0 of the regression functions $A_k(t)$ and moreover assumes information about the form of the spectral density of the $\xi(t)$ process.

However, in practice, as a rule the function $f(e^{-e\lambda})$ is unknown, which excludes the possibility of practical verification of the condition set forth by

Grenander and Rosenblatt. At the same time, if the spectral density is known, then in many cases it is possible to compute exact efficient estimates (see [2], and also formula (1) and Lemma 8 of the present work) and accordingly, the urgency for investigation of properties of other estimates (including also least squares estimates) is diminished. This brings to light the interesting

Theorem 3. *If the functions* $A_k(t)$ *have a certain regularity of growth, namely*

$$\text{(III)}\quad A_k(t+h)-A_k(t)=\beta_k(h)\cdot A_k(t)+\delta_k(t,\ h),$$

where $\sum_{t,h=0}^{T}|\delta_k(t,h)|^2 = o(\sum_{t=0}^{T}|A_k(t)|^2)$ *for* $T\to\infty$ $(k=\overline{1,n})$, *then the least squares estimates* $\tilde{a}_k$ *are asymptotically efficient.*

Proof of Theorems 1–3 is based on a series of lemmas given below.

Lemma 1. *Suppose* $x_1, x_2, \cdots, x_m$ *are vectors in some Euclidean space and*

$$x=\sum_{i=1}^{m} x_i.$$

Suppose γ_k *is the angle between the vector* x_k *and its projection on the linear span* H_k *of all the remaining vectors* x_j $(j\neq k)$. *The following inequalities hold:*

$$\|x_k\|\leqslant\frac{1}{\sin\gamma_k}\cdot\|x\|\quad (k=\overline{1,m}),$$

$$\frac{\sin\gamma}{m}\sum_{i=1}^{m}\|x_i\|\leqslant\|x\|\leqslant\sum_{i=1}^{m}\|x_i\|\quad \left(\gamma=\min_k\gamma_k\right). \tag{4}$$

Proof. The vector

$$x=\sum_{j=1}^{m}x_j=\sum_{j\neq k}x_j+\hat{x}_k+(x_k-\hat{x}_k)$$

is the sum of the elements $y_1=\sum_{j\neq k}x_j+\hat{x}_k$ and $y_2=x_k-\hat{x}_k$, of which the first belongs to the subspace H_k, while the other is orthogonal to it. Therefore

$$\|x\|^2=\|y_1\|^2+\|y_2\|^2\geqslant\|y_2\|^2.$$

But obviously

$$\|y_2\|=\|x_k\|\cdot\sin\gamma_k,$$

and consequently

$$\|x\|\geqslant\|x_k\|\cdot\sin\gamma_k,$$

$$\frac{\sin\gamma}{m}\sum_{i=1}^{m}\|x_i\|\leqslant\|x\|\leqslant\sum_{i=1}^{m}\|x_i\|.$$

The lemma is proved.

Lemma 2. *Let*

$$D=\det(d_{kj}),\quad d_{kj}=(x_k,\ x_j)$$

be the Gram determinant of the vectors $x_1, x_2, \cdots, x_m$ *satisfying the conditions of Lemma* 1. *The following relations hold*:

$$(\sin\gamma)^{2m}\cdot\prod_{k=1}^{m} d_{kk}\leqslant D\leqslant\frac{(m!)^2}{(\sin\gamma)^{2m}}\cdot\prod_{k=1}^{m} d_{kk}. \tag{5}$$

Proof. Let D_{kj} be the algebraic cofactor of the element d_{kj}. Since the matrix (d_{kj}) corresponding to the Gram determinant is positive definite,

$$|d_{kj}|\leqslant d_{kk}^{\frac{1}{2}} d_{jj}^{\frac{1}{2}},$$

$$|D_{kj}|\leqslant D_{kk}^{\frac{1}{2}} D_{jj}^{\frac{1}{2}}.$$

The quantity $D^{1/2}$, as is known, represents the area of the parallelepiped formed by the vectors x_k $(k=\overline{1, m})$. Correspondingly, $D_{kk}^{1/2}$ is the area of the parallelepiped formed by the vectors x_j $(j=\overline{1, m}$, but $j\neq k)$.

If $\hat{x}_k$ is the projection of the vector x_k on the subspace H_k spanned by the elements x_j $(j\neq k)$, then obviously

$$D^{\frac{1}{2}}=\|x_k-\hat{x}_k\|\cdot D_{kk}^{\frac{1}{2}}=\|x_k\|\cdot D_{kk}^{\frac{1}{2}}\cdot\sin\gamma_k.$$

But

$$d_{kk}=\|x_k\|^2,\quad \sin\gamma_k\geqslant\sin\gamma$$

and therefore

$$D^{\frac{1}{2}}\geqslant d_{kk}^{\frac{1}{2}}\cdot D_{kk}^{\frac{1}{2}}\cdot\sin\gamma,$$

$$D\leqslant\sum_{j=1}^{m}|d_{kj}\cdot D_{kj}|\leqslant d_{kk}^{\frac{1}{2}}\cdot D_{kk}^{\frac{1}{2}}\cdot\sum_{j=1}^{m} d_{jj}^{\frac{1}{2}}\cdot D_{jj}^{\frac{1}{2}}\leqslant D^{\frac{1}{2}}\cdot\frac{m}{\sin\gamma}\cdot d_{kk}^{\frac{1}{2}}\cdot D_{kk}^{\frac{1}{2}},$$

whence

$$D^{\frac{1}{2}}\leqslant\frac{m}{\sin\gamma}\cdot d_{kk}^{\frac{1}{2}}\cdot D_{kk}^{\frac{1}{2}}.$$

Using the obtained inequalities for the principal minors of the determinant D of order $m-1, m-2, \cdots$, at the mth step we arrive at relations (5). The lemma is proved.

We say that the variables u and v, depending on some parameter T $(T=0, 1, 2, \cdots)$, are equivalent $(u\sim v)$ if

$$\lim_{T\to\infty}\frac{u}{v}=1,$$

and roughly equivalent $(u\asymp v)$ if

$$0<\varliminf_{T\to\infty}\frac{u}{v}\leqslant\varlimsup_{T\to\infty}\frac{u}{v}<\infty.$$

Suppose there are certain functions $A'_k(t)$ of the form

$$A'_k(t) = \beta'_k(t) \cdot A_k(t) + \delta'_k(t), \tag{6}$$

where $\beta'_k(t)$ are uniformly bounded numbers, $\beta'_k(t) \equiv 1$ for $N \leq t \leq T - N$, N fixed, and the functions $\delta'_k(t)$ are such that

$$\sum_{t=0}^{T} |\delta'_k(t)|^2 = o\left[\sum_{t=0}^{T} |A_k(t)|^2\right] \tag{7}$$

as

$$T \to \infty \quad (k = \overline{1, n}).$$

Let

$$D' = \det(d'_{kj}) \tag{8}$$

$$d'_{kj} = \sum_{t=0}^{T} \overline{A'_k(t)} \cdot A'_j(t) \quad (k, j = \overline{1, n})$$

and let D'_{kj} be the algebraic cofactor of the element d'_{kj}.

Lemma 3. *Given the conditions and notation* (6)–(8), *the following equivalence relations hold:*

$$D' \sim D$$

and

$$D'_{kj} \sim D_{kj},$$

if

$$D'_{kj} \asymp D_{kk}^{\frac{1}{2}} \cdot D_{jj}^{\frac{1}{2}}. \tag{9}$$

Proof. Each of the $n!$ terms of the determinant D' is the product $\Pi d'_{kj}$ of n of its elements taken from various rows and columns. On the strength of conditions (II) and (7), the elements of the determinant D' are equivalent to the elements d_{kk} $(k = \overline{1, n})$ of the determinant D and therefore

$$\prod_{k=1}^{n} d'_{kk} \sim \prod_{k=1}^{n} d_{kk}.$$

Further, on the strength of the same conditions (II) and (7),

$$|d'_{kj}| \leqslant C \cdot d_{kk}^{\frac{1}{2}} \cdot d_{jj}^{\frac{1}{2}}$$

for all $k, j = \overline{1, n}$, whence it follows that each of the terms of the determinant D' satisfies the inequality

$$\left|\prod d'_{kj}\right| \leqslant C^n \cdot \prod_{k=1}^{n} d_{kk}$$

and consequently

$$|D'| \leqslant n! \cdot C^n \cdot \prod_{k=1}^{n} d_{kk}.$$

We see that the principal terms of the determinant D' are only those which are roughly equivalent to the product $\Pi_{k=1}^{n} d_{kk}$. Let $\Pi d'_{kj}$ be such a term. It is easy to see that $\Pi d'_{kj} \asymp \Pi_{k=1}^{n} d_{kk}$ if and only if each of the elements d'_{kj} involved in the the expression $\Pi d'_{kj}$ is roughly equivalent to $d_{kk}^{1/2} \cdot d_{jj}^{1/2}$. From the conditions (II) and (7) it follows that if $d'_{kj} \asymp d_{kk}^{1/2} \cdot d_{jj}^{1/2}$, then $d'_{kj} \asymp d_{kj}$ and, moreover, $d'_{kj} \sim d_{kj}$.

Thus the principal terms $\Pi d'_{kj}$ and Πd_{kj} of the determinants D' and D, i.e., having order $\Pi_{k=1}^{n} d_{kk}$, are equivalent to each other, $\Pi d'_{kj} \sim \Pi d_{kj}$. Since (see Lemma 2) the ratio

$$\frac{|\prod d_{kj}|}{D} \leqslant (\sin \gamma)^{-2n},$$

then from the equivalence of the principal terms follows the equivalence of the determinants D' and D themselves. Exactly analogously it is shown that $D'_{kj} \sim D_{kj}$ if $D'_{kj} \asymp D_{kk}^{1/2} \cdot d_{jj}^{1/2}$. This completes the proof of the lemma.

Now we consider estimates $\tilde{a}'_k$ of the parameters α_k $(k = \overline{1, n})$ of the form

$$\tilde{a}'_k = \sum_{t=0}^{T} \left[\sum_{j=1}^{n} C'_{kj} \cdot \overline{A'_j(t)} \right] \cdot \eta(t), \tag{10}$$

where C'_{kj} are constants and the functions $A'_j(t)$ satisfy conditions (6). In order for these estimates to satisfy the requirement of unbiasedness, we need to let

$$C'_{kj} = \frac{D'_{kj}}{D'}. \tag{11}$$

Lemma 4. *Suppose the stationary sequence $\xi(t)$ has spectral density $f(e^{-i\lambda})$ such that*

$$0 < c_1 \leqslant f(e^{-i\lambda}) \leqslant c_2 < \infty. \tag{12}$$

Then any estimates $\tilde{a}'_k$ of the form (10) are equivalent to the corresponding estimates $\tilde{a}_k$ by the method of least squares, i.e.,

$$\mathbf{M}\, |\tilde{a}'_k - \alpha_k|^2 \sim \mathbf{M}\, |\tilde{a}_k - \alpha_k|^2, \tag{13}$$

where

$$\mathbf{M}\, |\tilde{a}_k - \alpha_k|^2 \asymp d_{kk}^{-1} \quad (k = \overline{1, n}). \tag{14}$$

Proof. We have

$$\tilde{a}_k - \alpha_k = \sum_{j=1}^{n} \frac{D_{kj}}{D} \sum_{t=0}^{T} \overline{A_j(t)} \cdot \xi(t)$$

and

$$\mathbf{M}\,|\tilde{\alpha}_k-\alpha_k|^2=\int_{-\pi}^{\pi}\left|\sum_{t=0}^{T}e^{i\lambda t}\left[\sum_{j=1}^{n}\frac{D_{kj}}{D}\cdot\overline{A_j(t)}\right]\right|^2 f(e^{-i\lambda})\,d\lambda$$
$$\asymp\sum_{t=0}^{T}\left|\sum_{j=1}^{n}\frac{D_{kj}}{D}\cdot\overline{A_j(t)}\right|^2.$$

Applying Lemma 1 to the vectors $x_j=D_{kj}A_j/D$ of the space R_{T+1}, we obtain that

$$\left[\sum_{t=0}^{T}\left|\sum_{j=1}^{n}\frac{D_{kj}}{D}\overline{A_j(t)}\right|^2\right]^{\frac{1}{2}}\asymp\sum_{j=1}^{n}\frac{|D_{kj}|}{D}d_{jj}^{\frac{1}{2}}.$$

Since always

$$|D_{kj}|\leqslant D_{kk}^{\frac{1}{2}}D_{jj}^{\frac{1}{2}},$$

and on the strength of inequalities (5) of Lemma 2

$$D\asymp\prod_{i=1}^{n}d_{ii};\quad D_{kk}\asymp\prod_{i\neq k}d_{ii};\quad D_{jj}\asymp\prod_{i\neq j}d_{ii},$$

then

$$\sum_{j=1}^{n}\frac{|D_{kj}|}{D}d_{jj}^{\frac{1}{2}}\asymp d_{kk}^{-\frac{1}{2}}$$

and consequently

$$\mathbf{M}\,|\tilde{\alpha}_k-\alpha_k|^2\asymp d_{kk}^{-1}.$$

Further,

$$\tilde{\alpha}'_k-\alpha_k=\sum_{j=1}^{n}\frac{D'_{kj}}{D'}\sum_{t=0}^{T}\overline{A'_j(t)}\cdot\xi(t).$$

In this expression either $D'_{kj}\asymp D_{kk}^{1/2}\cdot D_{jj}^{1/2}$, or $D'_{kj}=o[D_{kk}^{1/2}\cdot D_{jj}^{1/2}]$. In the first case $D'_{kj}\sim D_{kj}$ (see Lemma 3), and therefore

$$\mathbf{M}\left|\frac{D'_{kj}}{D'}\cdot\sum_{t=0}^{T}\overline{A'_j(t)}\cdot\xi(t)-\frac{D_{kj}}{D}\sum_{t=0}^{T}\overline{A_j(t)}\cdot\xi(t)\right|^2$$
$$\asymp\mathbf{M}\left|\frac{D'_{kj}}{D'}\cdot\sum_{t=N}^{T-N}\overline{A'_j(t)}\cdot\xi(t)-\frac{D_{kj}}{D}\sum_{t=N}^{T-N}\overline{A_j(t)}\,\xi(t)\right|^2$$
$$\asymp d_{jj}\left|\frac{D'_{kj}}{D'}-\frac{D_{kj}}{D}\right|^2\asymp\frac{|D_{kj}|^2 d_{jj}}{D^2}\cdot\left|\frac{D'_{kj}}{D_{kj}}\cdot\frac{D}{D'}-1\right|=o(d_{kk}^{-1});$$

in the second case

$$\mathbf{M}\left|\frac{D'_{kj}}{D'}\sum_{t=0}^{T}\overline{A'_j(t)}\cdot\xi(t)\right|^2 \asymp \frac{|D_{kj}|^2}{D^2}\,d_{jj}=o(d_{kk}^{-1}).$$

Hence it is easy to deduce that

$$\mathbf{M}\,|\tilde{a}'_k-\tilde{a}_k|^2=o(d_{kk}^{-1})$$

and consequently

$$\mathbf{M}\,|\tilde{a}'_k-\alpha_k|^2\sim\mathbf{M}\,|\tilde{a}_k-\alpha_k|^2.$$

The lemma is proved.

Lemma 5. *If the functions $A_k(t)$ satisfy conditions* (I) *and* (II), *then in order for the estimates $\tilde{a}_k$ to be asymptotically efficient in the sense of Grenander-Rosenblatt* (*i.e., to satisfy relation* (3)), *it is necessary and sufficient that relation* (3′) *be satisfied.*

Proof.

(a) **Necessity.** By Lemmas 3 and 4 the limits from the left and the right in the matrix relation (3) exist and are positive, so that instead of (3) we may write

$$\lim_{T\to\infty}\frac{Q_T\cdot\mathbf{M}\,(\hat{a}-\alpha)\,(\hat{a}-\alpha)^*\cdot Q_T}{Q_T\cdot\mathbf{M}\,(\tilde{a}-\alpha)\,(\tilde{a}-\alpha)^*\cdot Q_T}=E \tag{3^0}$$

(E is the identity matrix) whence, in particular, it also follows that relations (3′) are satisfied (they coincide with relation (3^0) for its diagonal elements).

(b) **Sufficiency.** For simplicity let $\alpha_k\equiv 0$ $(k=\overline{1,n})$ and let $\mathrm{M}a_k^2=||a_k||^2$. It is easy to see that

$$\|\tilde{a}_k\|^2=\|\hat{a}_k\|^2+\|\tilde{a}_k-a_k\|^2.$$

Thus

$$Q_T\cdot\mathbf{M}\,(\tilde{a}-\alpha)\,(\tilde{a}-\alpha)^*\cdot Q_T=(z_{ik}), \tag{15}$$

where $z_{ij}=d_{ii}^{1/2}\,\mathrm{M}(\tilde{a}_i\tilde{a}_k)\cdot d_{kk}^{1/2}$. But since by (3′)

$$\lim_{T\to\infty}\frac{\|\tilde{a}_k\|^2}{\|\hat{a}_k:}=1,\quad\text{or}\quad\lim_{T\to\infty}\frac{\|\hat{a}_k\|^2+\|\tilde{a}_k-\hat{a}_k\|^2}{\|\hat{a}_k\|^2}$$

$$=\lim_{T\to\infty}\frac{\|\hat{a}_k\|^2}{\|\hat{a}_k\|^2}+\lim_{T\to\infty}\frac{\|\tilde{a}_k-\hat{a}_k\|^2}{\|\hat{a}_k\|^2}=1+\lim_{T\to\infty}\frac{\|\tilde{a}_k-\hat{a}_k\|^2}{\|\hat{a}_k\|^2}=1,$$

consequently $\lim_{T\to\infty}\|\tilde{a}_k-\hat{a}_k\|^2/\|\hat{a}_k\|^2=0$, and thus also $\lim_{T\to\infty}\|\tilde{a}_k-\hat{a}_k\|\cdot d_{kk}=0$, and, returning to (15), we obtain

$$\lim_{T\to\infty} z_{kk}(T) = \lim_{T\to\infty} d_{kk}^{\frac{1}{2}} \| \tilde{a}_k \|^2 d_{kk}^{\frac{1}{2}} = \lim_{T\to\infty} d_{kk}^{\frac{1}{2}} \| \tilde{a}_k - \hat{a}_k \|^2 d_{kk}^{\frac{1}{2}}$$

$$+ \lim_{T\to\infty} d_{kk}^{\frac{1}{2}} \| \hat{a}_{kk} \|^2 d_{kk}^{\frac{1}{2}} = \lim_{T\to\infty} d_{kk}^{\frac{1}{2}} \| \hat{a}_k \| d_{kk}^{\frac{1}{2}},$$

i.e., we have equality of diagonal elements in the matrix relation (3).

Further:

$$\mathbf{M}\tilde{a}_i\tilde{a}_k = \mathbf{M}\hat{a}_i\hat{a}_k + \mathbf{M}[\tilde{a}_k(\tilde{a}_i - \hat{a}_i)] + \mathbf{M}[\hat{a}_i(\tilde{a}_k - \hat{a}_k)].$$

But using the Cauchy inequality, we have

$$\mathbf{M}[\tilde{a}_k(\tilde{a}_i - \hat{a}_i)] \leqslant \| \tilde{a}_i - \hat{a}_i \| \cdot \| \tilde{a}_k \|$$

and

$$\mathbf{M}[\hat{a}_i(\tilde{a}_k - \hat{a}_k)] \leqslant \| \tilde{a}_k - \hat{a}_k \| \cdot \| \hat{a}_i \|,$$

while since $d_{ii}^{1/2} \| \tilde{a}_i - \hat{a}_i \| \cdot \| a_k \| d_{kk}^{1/2} \longrightarrow 0$ as $T \longrightarrow \infty$, then, returning to (15), we obtain

$$\lim_{T\to\infty} z_{ik}(T) = \lim_{T\to\infty} d_{ii}^{\frac{1}{2}} \cdot \mathbf{M}(\tilde{a}_i\tilde{a}_k) \cdot d_{kk}^{\frac{1}{2}} = \lim_{T\to\infty} d_{ii}^{\frac{1}{2}} \cdot \mathbf{M}(\hat{a}_i\hat{a}_k) \cdot d_{kk}^{\frac{1}{2}},$$

which is what we wished to prove.

Lemma 6. *Suppose the linear estimates* $\tilde{a}_k$ *of the parameters* α_k *are asymptotically efficient for a certain class of spectral densities. Then they are also asymptotically efficient for the spectral densities* $f(e^{-i\lambda})$ *for which there is a sequence of functions* $f_N(e^{-i\lambda})$, $N = 1, 2, \cdots$ *from this class such that* $f(e^{-i\lambda})/f_N(e^{-i\lambda}) \longrightarrow 1$ *as* $N \longrightarrow \infty$ *uniformly in* λ, $-\pi \leq \lambda < \pi$.

Proof. Efficiency of the estimates $\tilde{a}_k$ is obviously equivalent to the following. For any ϵ there exists T_0 such that

$$\frac{\mathbf{M}\,|\,\tilde{a}_k - \alpha_k\,|^2}{\mathbf{M}\,|\,\hat{a}_k - \alpha_k\,|^2} \leqslant 1 + \varepsilon \quad \text{for all} \quad T \geqslant T_0.$$

Let $f_N(e^{-i\lambda})$ be such that

$$1 - \varepsilon' \leqslant \frac{f(e^{-i\lambda})}{f_N(e^{-i\lambda})} \leqslant 1 + \varepsilon'$$

for all λ. Further, choose T_0 such that

$$\frac{\mathbf{M}_N\,|\,\tilde{a}_k - \alpha_k\,|^2}{\mathbf{M}_N\,|\,\hat{a}_k - \alpha_k\,|^2} \leqslant 1 + \varepsilon'$$

for $T \geq T_0$; here the index N indicates that the variables correspond to the spectral density $f_N(e^{-i\lambda})$.

We have:

$$\mathbf{M}_N \,|\, \tilde{\alpha}_k - \alpha_k \,|^2 = \int_{-\pi}^{\pi} \left| \sum_{t=0}^{T} c_k(t) \cdot e^{i\lambda t} \right|^2 f_N(e^{-i\lambda})\, d\lambda,$$

$$\mathbf{M} \,|\, \tilde{\alpha}_k - \alpha_k \,|^2 = \int_{-\pi}^{\pi} \left| \sum_{t=0}^{T} c_k(t) \cdot e^{i\lambda t} \right|^2 f(e^{-i\lambda})\, d\lambda$$

and by the choice of the function $f_N(e^{-i\lambda})$

$$\frac{\mathbf{M} \,|\, \tilde{\alpha}_k - \alpha_k \,|^2}{\mathbf{M}_N \,|\, \tilde{\alpha}_k - \alpha_k \,|^2} \leqslant 1 + \varepsilon'$$

for all T. Analogously

$$\frac{\mathbf{M}_N \,|\, \hat{\alpha}_k - \alpha_k \,|^2}{\mathbf{M} \,|\, \hat{\alpha}_k - \alpha_k \,|^2} \leqslant \frac{1}{1-\varepsilon'} .$$

Therefore

$$\frac{\mathbf{M} \,|\, \tilde{\alpha}_k - \alpha_k \,|^2}{\mathbf{M} \,|\, \hat{\alpha}_k - \alpha_k \,|^2} = \frac{\mathbf{M}_N \,|\, \hat{\alpha}_k - \alpha_k \,|^2}{\mathbf{M} \,|\, \hat{\alpha}_k - \alpha_k \,|^2} \cdot \frac{\mathbf{M} \,|\, \tilde{\alpha}_k - \alpha_k \,|^2}{\mathbf{M}_N \,|\, \tilde{\alpha}_k - \alpha_k \,|^2} \cdot \frac{\mathbf{M}_N \,|\, \tilde{\alpha}_k - \alpha_k \,|^2}{\mathbf{M}_N \,|\, \hat{\alpha}_k - \alpha_k \,|^2}$$

$$\leqslant \frac{1}{1-\varepsilon'} (1+\varepsilon')^2 \leqslant 1 + \varepsilon$$

for $T \geq T_0$ if ϵ' is taken sufficiently small from the start. The lemma is proved.

Lemma 7. *Suppose the continuous function* $f(e^{-i\lambda})$ *is positive everywhere on the interval* $[-\pi, \pi]$. *Then there exists a sequence of polynomials*

$$P_N(z) = \sum_{s=0}^{N} p_s \cdot z^s$$

such that $f(e^{-i\lambda})$ *is the uniform (in* λ*) limit of the functions* $|P_N(e^{-i\lambda})|^{-2}$, *i.e.,*

$$f(e^{-i\lambda}) = \lim_{N\to\infty} \frac{1}{|P_N(e^{-i\lambda})|^2} . \tag{16}$$

Here if $\beta_k(s)$ *are the functions involved in the asymptotic expansion* (III), *then these polynomials* $P_N(z)$ *can be chosen such that*

$$\sum_{s=0}^{t_1} \sum_{u=0}^{t_2} p_s \cdot p_u \cdot \overline{\alpha_k(s-u)} \neq 0, \quad 0 \leqslant t_1 \leqslant N, \quad 0 \leqslant t_2 \leqslant N. \tag{17}$$

Proof. Obviously if there exists a sequence of polynomials $P_N(z)$ for which relation (16) holds, then it is possible to "adjust" the coefficients p_s of these polynomials in such a way that relation (16) remains valid and together with this conditions (17) are satisfied. Further, the function $f^{-1/2}(z)$ is continuous on the unit circle and consequently there exists a sequence of polynomials $P_N(z)$ converging uniformly to it on the unit circle:

$$f^{-\frac{1}{2}}(z) = \lim_{N\to\infty} P_N(z).$$

Hence it follows that

$$f^{\frac{1}{2}}(z)=\lim_{N\to\infty}\frac{1}{P_N(z)}$$

and

$$f(z)=\lim_{N\to\infty}\frac{1}{|P_N(z)|^2}.$$

The lemma is proved.

Lemma 8. *Suppose the stationary sequence* $\xi(t)$ *has spectral density* $f(e^{-i\lambda})$ *of the form*

$$f(e^{-i\lambda})=\frac{1}{|P_N(e^{-i\lambda})|^2}, \tag{18}$$

where the polynomial

$$P_N(z)=\sum_{s=0}^{N}p_s\cdot z^s$$

has no zero inside the unit circle. Then the coefficients* $c_k(t)$ *in the expansion of the efficient estimates* $\hat{a}_k=\sum_{t=0}^{T}c_k(t)\cdot\eta(t)$ *of the parameters* α_k $(0\le t\le T;\ N<T)$ *can be found by the formulas*

$$c_k(t)=\begin{cases}\frac{1}{2\pi}\sum_{s=0}^{N}\sum_{u=0}^{t}p_s\cdot p_u\cdot\sum_{j=1}^{n}C_{kj}\cdot\overline{A_j(t+s-u)} & \text{for } 0\leqslant t\leqslant N-1\\ \frac{1}{2\pi}\sum_{s=0}^{N}\sum_{u=0}^{N}p_s\cdot p_u\cdot\sum_{j=1}^{n}C_{kj}\cdot\overline{A_j(t+s-u)} & \text{for } N\leqslant t\leqslant T-N\\ \frac{1}{2\pi}\sum_{s=0}^{T-t}\sum_{u=0}^{N}p_s\cdot p_u\cdot\sum_{j=1}^{n}C_{kj}\cdot\overline{A_j(t+s-u)} & \text{for } T-N+1\leqslant t\leqslant T,\end{cases} \tag{19}$$

where the constants C_{kj} *must be determined from the condition of unbiasedness of the estimates* $\hat{a}_k$.

Proof. The system of equations for the coefficients $c_k(t)$ can be represented in the form

$$-i\int_{|z|=1}z^{t-1}\sum_{t=0}^{T}c_k(t)\cdot z^{-t}\frac{dz}{P(z)\cdot P(z^{-1})} \tag{20}$$

$$=\sum_{j=1}^{n}C_{kj}\cdot\overline{A_j(t)},\quad 0\leqslant t\leqslant T;\ k=\overline{1,n}.$$

*If the spectral density has the form (18), then the polynomial $P(z)$ can always be chosen so that it has no zero inside the unit circle.

We define the functions $B_k(t)$ for all t as the right parts of equalities (20), and both parts of these equalities apply the operator $P(\Delta)P(\Delta^{-1})$, where Δ is the shift operator. As a result of this we obtain that

$$c_k(t) = \frac{1}{2\pi i} \int\limits_{|z|=1} z^{t-1} \sum_{s=0}^{T} c_k(s)\, z^{-s} dz = \frac{1}{2\pi} P(\Delta) \cdot P(\Delta^{-1})\, B_k(t).$$

From the analyticity of the functions $1/P(z)$ and $1/P(z^{-1})$ inside and outside the unit circle, respectively,

$$P(\Delta) \cdot B_k(t) = -i \int\limits_{|z|=1} z^{t-1} \sum_{s=0}^{T} c_k(s)\, z^{-s} \frac{dz}{P(z^{-1})} = 0 \quad \text{for} \quad t < 0,$$

$$P(\Delta^{-1}) \cdot B_k(t) = -i \int\limits_{|z|=1} z^{t-1} \sum_{s=0}^{T} c_k(s) \cdot z^{-s} \frac{dz}{P(z)} = 0 \quad \text{for} \quad t > T$$

for all $k = \overline{1, n}$. Using the permutability of the operators $P(\Delta)$ and $P(\Delta^{-1})$, it is not difficult to obtain relations (19) for the coefficients $c_k(t)$. The lemma is proved.

Lemma 9. *Suppose the spectral density* $f(e^{-i\lambda})$ *of the stationary sequence* $\xi(t)$ *has the form* (18), *and condition* (17) *is satisfied. Then the least squares estimates* $\tilde{a}_k$ *are asymptotically efficient.*

Proof. It is easy to see that if the coefficients p_s of the polynomial $P_N(z)$ satisfy condition (17), then, making use of the representation (III) of the functions $A_k(t)$ and formulas (19), efficient estimates can be represented in the form (10). Therefore, by Lemma 4, least squares estimates are in this case asymptotically efficient. The lemma is proved.

In conclusion, we note that the conditions of Theorems 1–3 are satisfied by a rather wide class of functions; in particular, by functions of the form $A_k(t) = t^k e^{i\omega_k t}$, where ω_k are known numbers.

We also note that the methods of proof used for Lemmas 1–9 can be useful in further investigation of the problem of asymptotically efficient estimates of regression coefficients. For example, least squares estimates can be considered as a "zeroth approximation" to exact efficient estimates when in their computation the true correlation function of the process $B^{\xi}(\tau)$ is replaced by its "zeroth approximation" $B_0(\tau)$, where $B_0(0) = B^{\xi}(0)$ and $B_0(\tau) \equiv 0$ for all $\tau > 0$. It is natural to expect that exact efficient estimates, computed for the correlation function $B_s(\tau)$ where $B_s(k) = B^{\xi}(k)$ for all $0 \le k \le s$, are closer to exact efficient estimates of the process $\xi(t)$ than estimates of the preceding approximation, and, in particular, than least squares estimates (i.e., computed for $B^{\xi}(\tau) \approx B_0(\tau)$). Here the correlation functions $B_s(\tau)$, with whose help the true correlation function $B^{\xi}(\tau)$ is "approximated," are naturally chosen in the most convenient way possible for computation. Thus on the one hand we simplify the algorithm for finding estimates of the regression coefficients (here, especially, is where the use of least squares estimates makes sense), while on the other hand we improve their efficiency properties. Apparently solutions of problems thus posed should also be sought with a reasonable combination of these two considerations (degree of simplification of computation and amount of loss of properties of efficient estimates).

We consider such an example from this point of view. We estimated the mean value m of the process $\eta(t)$, $\eta(t) = m + \xi(t)$, where $\xi(t)$ is a stationary process in the broad sense with spectral density $f_N^{\xi}(e^{-i\lambda}) = 1/|P_N(e^{-i\lambda})|^2$ and corresponding correlation function $B_N^{\xi}(\tau)$ (here, as before, $P_N(z) = p_0 + p_1 z + \cdots + p_{N-1} z^{N-1} + p_N z^N$). We investigated the improvement of efficiency of estimates when instead of the "zeroth approximation" (i.e., instead of least squares estimates $\hat{m}_0 = 1/(T+1)\Sigma_{t=0}^{T} \eta(t)$) we consider the sth approximation by estimates of so-called "Markov type" $\hat{m}_s$ ($s \ll N$), i.e., in the computations, instead of the spectral density $f_N^{\xi}(e^{-i\lambda})$, we use the function $f_s(e^{-i\lambda}) = 1/|\hat{P}_s(e^{-i\lambda})|^2$, where the coefficients $\hat{p}_i$ ($i = \overline{0, s}$) of the polynomial $\hat{P}_s$ are chosen such that $B_s(k) = B_N^{\xi}(k)$ for all $0 \leq k \leq s$. Computations performed on the basis of the results of Lemma 8 showed that the considerable complication of the computations caused by passage from the estimate $\hat{m}_0$ to the estimates $\hat{m}_s$ (especially for $s \geq 2$) is very weakly compensated for by the improvement of the efficiency properties of the estimates.

In particular, it turned out that if we represent the variance of the estimates $\hat{m}_s$ in the form

$$D\hat{m}_k = \sigma^2 + \Delta_k^2 \quad (k = \overline{0, s}),$$

where σ^2 is the variance of the exact efficient estimate of the parameter m, then $\Delta_k^2 \approx C_0(N-k)^2/T^2$, to within $O(1/T^3)$, so that $\Delta_0^2/\Delta_s^2 \approx (N/(N-s))^2$. Quite apart from the fact that we did not succeed in increasing the rate of convergence of asymptotically efficient estimates to the exact efficient estimates (both Δ_0^2 and Δ_s^2 are quantities of order T^{-2}), even the gain in the coefficients for T^{-2} turns out to be very modest. This example confirms the desirable properties of least squares estimates.

BIBLIOGRAHHY

[1] U. Grenander and M. Rosenblatt, *Statistical analysis of stationary time series*, Wiley, New York and Almqvist & Wiksell, Stokholm, 1957. MR 18, 959.

[2] V. F. Pisarenko and Ju. A. Rozanov, *On certain problems for stationary processes leading to integral equations related to the Wiener-Hopf equation*, Vol. 14, Izv. Akad. Nauk SSSR, Moscow, 1963.

[3] Chiang Tse-pei, *On the estimation of regression coefficients of a continuous parameter time series with a stationary residual*, Teor. Verojatnost. i Primenen. 4(1959, 405–423. MR 22 #12670.

Translated by L. Rosenblatt

THE PROBLEM OF OPTIMAL SEARCH

V. I. ARKIN

Suppose we have a countable number of cells, indexed by the numbers 1, 2,⋯ ⋯, n, ⋯. In one of them is a point which we wish to find. The probability that the point is in the ith cell is $p_i \geq 0$, where $p_1 \geq p_2 \geq \cdots \geq p_m \geq \cdots$ and $\Sigma_{i=1}^{\infty} p_i = 1$.

We assume that the search effort can be directed simultaneously into an arbitrary number of cells. Also, if the search is made simultaneously in k cells, then the probability of detecting the point by time t in any one of them, under the condition that the point is in that cell, equals $1 - e^{-(a/k)t}$, and equals 0 if the point is not in that cell.

A search strategy is a law by which, at any time t, we establish in which cells the search must be made. In our problem, search strategy will mean a collection of functions $\boldsymbol{\phi} = \{\phi_1(t), \phi_2(t), \cdots, \phi_n(t), \cdots\}$ such that $\phi_i(t) = \phi_i(t+0)$ and

$$\phi_i(t) = \begin{cases} 1 & \text{if the search is made in the } i\text{th cell at time } t, \\ 0 & \text{otherwise.} \end{cases}$$

The strategy $\boldsymbol{\phi}_T$ is said to be truncated at T if $\phi_i(t) \equiv 0$ for $t > T$. From the strategy $\boldsymbol{\phi}$, for each $T > 0$ we can form the strategy $\boldsymbol{\phi}_T$ truncated at T, setting

$$\varphi_T = \begin{cases} \varphi, & t \leqslant T, \\ 0, & t > T. \end{cases}$$

With each strategy ϕ_T truncated at T we associate a functional $\mathbf{P}(\phi_T)$—the probability of detecting the point by time T. We say that the strategy $\tilde{\boldsymbol{\phi}}_T$ is optimal if $\mathbf{P}(\tilde{\boldsymbol{\phi}}_T) = \sup_{\boldsymbol{\phi}_T} \mathbf{P}(\boldsymbol{\phi}_T)$. The strategy $\boldsymbol{\phi}^*$ is called uniformly optimal if for each T its truncation at T is optimal, i.e., $\mathbf{P}(\boldsymbol{\phi}_T^*) = \mathbf{P}(\tilde{\boldsymbol{\phi}}_T)$ for each $0 \leq T < \infty$.

In the present article we find the strategy $\boldsymbol{\phi}^*$ and prove its uniqueness. This work clearly relates to the general scheme of control by Markov processes "on incomplete data" proposed in [1], and generalizes results obtained in [2],[3].

Theorem. *The strategy* $\boldsymbol{\phi}^* = \{\phi_1^*(t), \phi_2^*(t), \cdots, \phi_n^*(t), \cdots\}$, *where*

$$\varphi_j^*(t) = 1,\ 1 \leqslant j \leqslant k;\ \varphi_i^*(t) = 0,\ i \geqslant k+1;\ \sum_{j=0}^{k-1} t_j < t \leqslant \sum_{j=0}^{k} t_j \tag{1}$$

and

$$t_j = \frac{j}{a} \ln \frac{p_j}{p_{j+1}},\ t_0 = 0 \tag{2}$$

is the unique uniformly optimal strategy.

Proof. Consider an arbitrary strategy ϕ_T truncated at T. It is easy to show that

$$\mathbf{P}(\varphi_T) = \sum_{i=1}^{\infty} p_i \left\{ 1 - \exp\left[-\alpha \sum_{k=1}^{\infty} \frac{1}{k} \int_{\theta_k} \varphi_i(t)\, dt \right] \right\}, \tag{3}$$

where $\theta_k = \{t\colon \phi i_1(t) = \phi i_2(t) = \cdots = \phi i_k(t) = 1,\ t \leq T\}$; $i_1, i_2, \cdots, i_k$ are k arbitrary natural numbers.

It is obvious that

$$\sum_{k=1}^{\infty} \int_{\theta_k} dt = T. \tag{4}$$

We shall show that for each T

$$\mathbf{P}(\varphi_T^*) = \sup_{\varphi_T} \mathbf{P}(\varphi_T). \tag{5}$$

Let

$$\tau_i = \sum_{k=1}^{\infty} \frac{1}{k} \int_{\theta_k} \varphi_i(t)\, dt. \tag{6}$$

Then we have the equalities:

$$\sum_{i=1}^{\infty} \tau_i = \sum_{i=1}^{\infty} \sum_{k=1}^{\infty} \frac{1}{k} \int_{\theta_k} \varphi_i(t)\, dt = \sum_{k=1}^{\infty} \frac{1}{k} \int_{\theta_k} \sum_{i=1}^{\infty} \varphi_i(t)\, dt = T.$$

Therefore (3) and (4) can be written in the form:

$$\mathbf{P}(\varphi_T) = \sum_{i=1}^{\infty} p_i (1 - e^{-\alpha \tau_i}), \tag{7}$$

$$\sum_{i=1}^{\infty} \tau_i = T. \tag{8}$$

Thus with each strategy ϕ_T truncated at T we can associate a collection of variables $\{\tau_i\}$, $1 \leq i < \infty$. Assume $\tau_i = \tau_i^*$ for the strategy ϕ_T^*. Using (1) and (2), we have:

$$\begin{aligned} &\tau_i^* = \frac{1}{\alpha} \ln p_i + \lambda; \ i = 1, 2, \ldots, m, \\ &\tau_\nu^* = 0, \ \nu > m, \end{aligned} \tag{9}$$

where the constant λ is found from the condition of normalization (8) and m satisfies the inequalities:

$$\sum_{i=1}^{m} \ln p_i - m \ln p_m < \alpha T \leqslant \sum_{i=1}^{m} \ln p_i - m \ln p_{m+1}. \tag{10}$$

To complete the proof of the theorem we need the following

Lemma. *The maximum of the function* $P(\tau_1, \tau_2, \cdots) = \Sigma_{i=1}^{\infty} p_i(1 - e^{-\alpha\tau_i})$, *given* $\Sigma_{i=1}^{\infty} \tau_i = T$, $\tau_i \geq 0$, *is attained only for* $\tau_i = \tau_i^*$.

Proof. Together with the collection $\{\tau_i^*\}$, consider the collection $\{\tau_i' = \tau_i^* + \Delta_i\}$, $\tau_i' \geq 0$ satisfying the condition $\Sigma_{i=1}^{\infty} \tau_i' = T$, i.e., $\Sigma_{i=1}^{\infty} \Delta_i = 0$ and at least one $\Delta_i \neq 0$. It is easy to show that $\delta P \leq 0$ for $\tau_i = \tau_i^*$, while $\delta^2 P < 0$. Uniqueness of the collection $\{\tau_i^*\}$ follows from the convexity of the function $P(\tau_1, \tau_2, \cdots)$. The lemma is proved.

Thus we have shown that an arbitrary strategy ϕ_T^*, the truncation of the strategy ϕ^* at T, is optimal. We shall show uniqueness of the uniformly optimal strategy. Suppose there exists another uniformly optimal strategy $\hat{\phi} \neq \phi^*$. Let $T \geq 0$ be the first time these strategies become distinct. This implies that there exist k and $\delta > 0$ such that $|\hat{\phi}_k(t) - \phi_k^*(t)| = 1$ and sign $[\hat{\phi}_k(t) - \phi_k^*(t)] = \text{const}$ for $T < t \leq T + \epsilon$. Obviously for given k and ϵ we have $\tau_k^* \neq \hat{\tau}_k$, whence follows $P(\phi_{T+\epsilon}^*) > P(\hat{\phi}_{T+\epsilon})$ on the strength of the lemma. The theorem is proved.

Corollary 1. *With each search strategy* ϕ *is related a random variable* τ_ϕ, *namely the time of detection of the point, if the search is carried out according to the strategy* ϕ. *The distribution function of* τ_{ϕ^*}, *namely the detection time with use of the uniformly optimal strategy, has the form:*

$$\mathbf{P}(\tau_{\varphi^*} \leqslant t) = \sum_{i=1}^{m} p_i - m \exp\left[\frac{1}{m}\left(\sum_{i=1}^{m} \ln p_i - \alpha t\right)\right], \tag{11}$$

where m *satisfies inequalities* (10). *Formula* (11) *is obtained by substituting* (9) *into* (7).

Corollary 2. $\mathbf{M}\tau_{\phi^*} = \inf \mathbf{M}\tau_\phi/\phi$ and, as can be shown:

$$\mathbf{M}\tau_{\varphi^*} = \frac{1}{\alpha}\sum_{m=1}^{\infty}\left\{m \ln \frac{p_m}{p_{m+1}} \sum_{i=m+1}^{\infty} p_i + m^2 [p_m - p_{m+1}]\right\}.$$

Proof of this assertion follows from the inequality

$$\mathbf{M}\tau_{\varphi} = \int_0^{\infty} [1 - \mathbf{P}(\varphi_T)]\, dT \geqslant \int_0^{\infty} [1 - \mathbf{P}(\varphi_T^*)]\, dT = \mathbf{M}\tau_{\varphi^*},$$

from formula (11), and the theorem.

Remark 1. Instead of pure strategies, we can consider arbitrary randomized strategies represented as probability measures π concentrated on piecewise-constant pure strategies. The criterion of optimality for randomized strategies is the functional $\mathbf{M}^{\pi}\mathbf{P}(\phi_T)$. It is easy to establish that the pure strategy ϕ^* is the uniformly optimal strategy in this class.

Remark 2. In many practical problems the probability of detecting the point by time t in each of the k cells under search, given that the point is in that cell, is equal to

$$1 - e^{-\frac{1}{k}\int_0^t \alpha(t)dt}$$

$\alpha(t) \geq 0$. This case reduces to the one considered above by replacing the time:

$$s = \int_0^t \alpha(t)\, dt.$$

In conclusion the author expresses his deep gratitude to Ju. V. Prohorov for his constant interest in this work, and also to D. V. Anosov and A. N. Širjaev for many valuable comments.

BIBLIOGRAPHY

[1] A. N. Širjaev, *On the theory of resolving functions and control of processes of observation with incomplete data,* Transactions of the Third Prague Conference on Information Theory, Statistical Decision Functions, Random Processes, 1962, Publ. House Czechoslovak Acad. Sci., Prague, 1964. MR **28** #3910.

[2] O. V. Staroverov, *On a search problem,* Teor. Verojatnost. i Primenen 8 (1963), 196–201. MR 27 #845.

[3] A. Charnes and W. Cooper, *Theory of search: optimum distribution of search effort,* Management Sci. 5 (1958), 44–50. MR 20 #5088.

Translated by:
L. Rosenblatt

ON FINDING OPTIMAL CONTROLS [1)]

V. I. ARKIN, V. A. KOLEMAEV AND A. N. ŠIRJAEV

1. Let $\xi(t)$ be a one-dimensional diffusion process satisfying the stochastic differential equation

$$d\xi(t) = b(t, \xi)\,dt + a(t, \xi)\,d\eta, \quad \xi(0) = 0, \tag{1}$$

where $\eta(t)$ is a Wiener process with $\mathbf{M}\Delta\eta = 0$, $\mathbf{D}\Delta\eta = \Delta t$, $\eta(0) = 0$. [Here $\mathbf{M}$ is the expectation and $\mathbf{D}$ is the variance.] In equation (1)

$$b(t, x) = B[t, x, u(t, x)], \; a(t, x) = A[t, x, v(t, x)],$$

where A and B are given functions; the measurable functions $u = \{u(t, x)\}$, $v = \{v(t, x)\}$ define controls [2)] and belong to some given class of admissible functions.

We assume that the process $\xi(t)$ for a chosen control $\delta = (u, v)$ is determined by the given equation (1) in some simply-connected region G with boundary Γ. Let τ denote the moment of first exit of the trajectory of the process $\xi(t)$ on Γ, and let

$$V^{\delta}(t, x) = M_x^{\delta}\left\{\varphi[\tau, \xi(\tau)] + \int_t^{\tau} F[t, \xi(t), u(t, \xi(t)), v(t, \xi(t))]\,dt\right\} \tag{2}$$

be a functional related to the behavior of the process $\xi(t)$ for the chosen control δ up to the moment of exit on Γ under the condition that $\xi(t) = x$. In formula (2) M^{δ} is the mathematical expectation computed with respect to the probability distribution of the process $\xi(t)$, induced by the control δ. The quantity $V_{\delta}(0, x)$ defines the total gain from use of the control δ under the condition $\xi(0) = x$. Let

$$\tilde{V}(t, x) = \sup V^{\delta}(t, x),$$

where the sum is taken over all admissible controls $\delta = (u, v)$. If there exists a control $\tilde{\delta} = (\tilde{u}, \tilde{v})$ for which $\tilde{V} = V^{\tilde{\delta}}(t, x)$, we say it is optimal. This note has as its aim, under the assumption of existence of an optimal control, [3)] establishment of certain methods (§§2, 3, 4) which make possible efficient determination of the optimal control $(\tilde{u}, \tilde{v})$ in many cases.

---

1) The results of this work were published by the authors in the report "On finding optimal controlling influences" in the VII All-Union Conference on the theory of probability and mathematical statistics, Tbilisi, 1963.

2) Below we restrict consideration to so-called Markov controls characterized by the fact that at each moment of time t the value of the controls is not determined by the entire past course of the process $\xi(\cdot)$, but only by the values at the preceding moment of time. For optimality of Markov controls see, for example, [1].

3) Questions of existence in such problems were investigated in [2, 3].

2. Suppose a control δ has been chosen; then $V^{\delta}(t, x)$ satisfies the backward Kolmogorov equation [4]

$$-\frac{\partial V^{\delta}}{\partial t}=\frac{1}{2}a(t, x)\frac{\partial^2 V^{\delta}}{\partial x^2}+b(t, x)\frac{\partial V^{\delta}}{\partial x}+f(t, x), \tag{3}$$

where $f(t, x)=F[t, x, u(t, x), v(t, x)]$; $V^{\delta}(t, x)|_{\Gamma}=\phi(t, x)$.

In turn, for the optimal gain, the following nonlinear equation ([2, 3]) is satisfied:

$$-\frac{\partial \tilde{V}}{\partial t}=\sup_{\{u, v\}}\left\{\frac{1}{2}A(t, x, v)\frac{\partial^2 \tilde{V}}{\partial x^2}+B(t, x, u)\frac{\partial \tilde{V}}{\partial x}+F(t, x, u, v)\right\}. \tag{4}$$

If it is possible to find the optimal gain $V(t, x)$, then in principle the optimal control itself is also found (see, for example, equation (7) of [2], equation (14) of [3]). But finding $\tilde{V}(t, x)$ is equivalent to solving the nonlinear equation (4), which is generally difficult. Therefore it is of interest to investigate those cases where, bypassing the direct finding of $\tilde{V}$, it is possible to find $\tilde{u}$, $\tilde{v}$.

Assume that the functions A, B, F, Φ do not explicitly depend on t, nor $A(x, v)$ on v; $A(x, v)=a(x)$. In this case 1) (4) may be written as

$$-\frac{1}{2}a(x)\frac{d^2\tilde{V}}{dx^2}=\sup_{\{u\}}\left[B(x, u)\frac{d\tilde{V}}{dx}+F(x, u)\right]. \tag{5}$$

Consider the right part of (5). In those cases when the optimal control is found at an interior point of the set of admissible values (this question must be investigated separately each time), the optimal control $\tilde{u}(x)$ can be found in the following way. Since V does not depend on u,

$$\frac{\partial B}{\partial u}\cdot\frac{d\tilde{V}}{dx}+\frac{\partial F}{\partial u}=0,$$

i.e.,

$$\frac{d\tilde{V}}{dx}=-\frac{\partial F}{\partial u}\Big/\frac{\partial B}{\partial u} \tag{6}$$

and

$$\frac{d^2\tilde{V}}{dx^2}=-\frac{\left(\frac{d}{dx}\frac{\partial F}{\partial u}\right)\frac{\partial B}{\partial u}-\left(\frac{d}{dx}\frac{\partial B}{\partial u}\right)\frac{\partial F}{\partial u}}{\left(\frac{\partial B}{\partial u}\right)^2}. \tag{7}$$

Since for the optimal control $u=\tilde{u}(x)$

---

1) Here for simplicity we restrict consideration to one parameter only.

$$\sup_{\{u\}}\left[B(x,\,u)\frac{\partial\tilde V}{\partial x}+F(x,\,u)\right]=B(x,\,\tilde u)\frac{d\tilde V}{dx}+F(x,\,\tilde u),$$

then, substituting (6) into (7) and (5), after simple transformations we find

$$\frac{du}{dx}=\frac{2}{a(x)}\frac{\frac{\partial B}{\partial u}\left[F\frac{\partial B}{\partial u}-B\frac{\partial F}{\partial u}\right]-\left[\frac{\partial^2F}{\partial x\partial u}\cdot\frac{\partial B}{\partial u}-\frac{\partial^2B}{\partial x\partial u}\cdot\frac{\partial F}{\partial u}\right]}{\left[\frac{\partial^2F}{\partial u^2}\cdot\frac{\partial B}{\partial u}-\frac{\partial^2B}{\partial u^2}\cdot\frac{\partial F}{\partial u}\right]}. \tag{8}$$

If F and B do not depend explicitly on x, we obtain the simpler equation

$$\frac{du}{dx}=\frac{2}{a(x)}\cdot\frac{dB}{du}\cdot\frac{F\frac{\partial B}{\partial u}-B\frac{\partial F}{\partial u}}{\frac{d^2F}{du^2}\cdot\frac{dB}{du}-\frac{d^2B}{du^2}\cdot\frac{dF}{du}}, \tag{9}$$

defining the optimal control $\tilde u(x)$.

We treat an interesting case considered in [5], where $G=(-a,\,a)\times[0,\,\infty)$, $a(x)=1$, $F(u)=1-C(u)$, $B(u)=u$. In this case

$$\tilde V(x)=\sup_{\{\delta\}}M_x^\delta\left[\tau-\int_0^\tau C(u_t)\,dt\right], \tag{10}$$

where $C(u)$ can be interpreted as the payoff after use of the local deflection of u with the aim of keeping the process longer within the limits $(-a,\,a)$. It is natural to assume the function $C(u)$ even and nondecreasing, with $C(0)=0$. For $C''>0$ from (9) we obtain $(u(0)=0)$

$$\frac{du}{dx}=2\frac{C(u)-1-uC'(u)}{C''(u)}. \tag{11}$$

The solution of this equation for $|a|\le\pi/2$ and $C(u)=u^2$ yields the following value for the optimal control:

$$\tilde u(x)=-\mathrm{tg}\,x. \tag{12}$$

3. Now we consider the case where control consists of the fact that local deflection of $u(x,\,t)$ takes on one of two values: u_1, u_2. Often one can, by a general argument, prove that there exists a line (of switching) $x=\tilde x(t)$ which partitions the half-plane $R\times[0,\,\infty)$ into two regions G_1 and G_2 such that in the first only the control u_1 is applied, and in the other, the control u_2. Then the problem of finding an optimal control reduces essentially to the problem of finding an optimal switching line $\tilde x(t)$. To find it one can use the method which is the fundamental principle of the work of V. S. Mihalevič [7, 8] (see also the more recent [9–12]). For example, applied to the problem involving equation (4) where $u-u_1$, u_2, A, B, and F do not depend on v, this method consists of the fact that

in the region G_i the equation ($V_i = \widetilde{V}$ on G_i)

$$-\frac{\partial V_i}{\partial t}=\frac{1}{2}A(t, x)\frac{\partial^2 V_i}{\partial x^2}+B(t, x, u_i)\frac{\partial V_i}{\partial x}+F(t, x, u_i) \tag{13}$$

holds, and on the unknown boundary $x = \widetilde{x}(t)$ the following condition is satisfied[1])

$$\frac{\partial V_1}{\partial x}=\frac{\partial V_2}{\partial x}, \quad \frac{\partial^2 V_1}{\partial x^2}=\frac{\partial^2 V_2}{\partial x^2}. \tag{14}$$

It also makes it possible to find the unknown optimal switching line $\widetilde{x}(t)$. This method was used in [6] to solve the problem involving the functional (10) in the case of a finite number of values of the control $u = u_1, \cdots, u_k$. As the number k increases, a passage to the limit can be used to obtain (11) once again.

4. In an example of this problem involving the functional (10), under the conditions imposed on the coefficients in §2, we show how to obtain equation (11) for the optimal control using Pontrjagin's maximum principle [13].

If a control δ has been chosen, then for it $V(x) = V^{\delta}(x)$ satisfies the equation

$$-\frac{1}{2}\frac{d^2 V}{dx^2}=u\cdot\frac{dV}{dx}+1-C(u), \tag{15}$$

which is equivalent, if we let $\pi_1(x) = V(x)$, $\pi_2(x) = V'(x)$, to the system

$$\frac{d\pi_1}{dx}=\pi_2, \quad \frac{d\pi_2}{dx}=2[-u\pi_2-1+C(u)]. \tag{16}$$

We represent $V(x)$ as

$$V(x)=\int_{-a}^{x} V'(x)\,dx. \tag{17}$$

Then the problem of finding an optimal control is equivalent to the problem of finding, for each x, a control which maximizes the functional (17). In accordance with [13] we let $\pi_0(x) = V(x)$, and to system (16) adjoin the equation

$$\frac{d\pi_0}{dx}=\pi_2. \tag{18}$$

The system adjoint to system (16), (18) has the form

$$\frac{d\psi_0}{dx}=0, \quad \frac{d\psi_1}{dx}=0, \quad \frac{d\psi_2}{dx}=-\psi_0-\psi_1+2u\psi_2.$$

The optimal control, if it exists, must by [13] yield a maximum of the functional

---

1) In the derivation of formula (14) it is assumed that $\widetilde{x}(t)$ has a finite derivative and the second derivatives of $V_{i;\ i=1,2}$ are bounded.

$$H = \psi_0\pi_2 + \psi_1\pi_2 + \psi_2\{2[-u\pi_2 + C(u) - 1]\}.$$

From the conditions of transversality at the point x it follows that $\psi_1 \equiv 0$, $\psi_2 \neq 0$. Here $\psi_0 < 0$. Consequently the optimal control $u = \tilde{u}(x)$ must be found from the condition

$$\max_u [-u\pi_2(x) - 1 + C(u)] = -\tilde{u}\pi_2(x) - 1 + C(\tilde{u}). \qquad (19)$$

Assuming the function $C(u)$ to be twice differentiable, from (19) we have

$$C'(u) = \pi_2(x), \quad C'' \frac{du}{dx} = \frac{d\pi_2}{dx},$$

which together with the second equation of (16) yields

$$\frac{du}{dx} = \frac{2[C(u) - 1 - uC'(u)]}{C''(u)}.$$

This equation for the optimal control coincides with equation (11) obtained earlier by another method.

BIBLIOGRAPHY

[1] I. V. Girsanov, *Some minimum problems in the theory of controlled Markov processes*, Proceedings of the Fourth All-Union Mathematical Congress, 1961.

[2] ———, *Minimax problems in the theory of differential processes*, Dokl. Akad. Nauk SSSR 136 (1961), 761–764 = Soviet Math. Dokl. 2 (1961), 118–121. MR 31 #509.

[3] W. H. Fleming, *Some Markovian optimization problems*, J. Math. Mech. 12 (1963), 131–140. MR 26 #1636.

[4] R. Z. Has′minskiĭ, *Probability distribution of functionals of trajectories of a random process of diffusion type*, Dokl. Akad. Nauk SSSR 104 (1955), 22–25. MR 17, 501.

[5] V. A. Kolemaev, *On optimal control of a Wiener process*, Litovsk. Mat. Sb. 3 (1963), no. 1, 175–180. MR 29 #3314.

[6] ———, *On statistical control of the dimensions of mass-produced articles*, Teknicheskaya kibernetika, No. 4, 1963.

[7] V. S. Mihalevič, *Consecutive Bayes' solutions and optimal methods of statistical acceptance control*, Dissertation, Moscow State University, 1956. Cf. MR 19, 694.

[8] ———, *Bayes' rule for choosing between two hypotheses for the mean value of a normal process*, Visnik Kievskogo Univ., No. 1 (1958), 101–104.

[9] H. Chernoff, *Sequential tests for the mean of a normal distribution*, Proc. 4th Berkeley Sympos. Math. Stat. and Prob., Vol. I, pp. 79–91, Univ. of California Press, Berkeley, Calif., 1961. MR **24** #A1788.

[10] R. L. Stratonovič, *Some extremal problems of mathematical statistics and conditional Markov processes*, Teor. Verojatnost. i Primenen 7 (1962), 226–229.

[11] A. N. Širjaev, *On the theory of resolving functions and control of processes of observation with incomplete data*, Transactions of the Third Prague Conference on Information Theory, Statistical Decision Functions, Random Processes, 1962, Publ. House Czechoslovak Acad. Sci., Prague, 1964. MR **28** #3910.

[12] J. A. Bather, *Bayes procedures for deciding the sign of a normal mean*, Proc. Cambridge Philos. Soc. **58** (1962), 599–620. MR 27 #869.

[13] L. S. Pontrjagin, V. G. Boltjanskiĭ, R. V. Gamkrelidze and E. F. Miščenko, *Mathematical theory of optimal processes*, Fizmatgiz, Moscow, 1961; English transl., Wiley, New York, 1962; Macmillan, New York, 1964. MR 29 #3316a; MR 29 #3316b.

Translated by:
L. Rosenblatt

ON WAITING TIMES IN A MIXED SYSTEM OF MASS SERVICE

O. V. VISKOV

1°. **Introduction.** Assume that at moments of time $t_1 = \xi_1$, $t_2 = \xi_1 + \xi_2, \ldots$, calls which have a Poisson distribution with parameter λ enter a service structure consisting of one device. If at the moment of arrival of a given call the device is free, then this entering call is immediately serviced. A call finding the device busy is refused service with probability q $(0 \leq q < 1)$ and is considered lost, and with probability $p = 1 - q$ remains in the service structure and is serviced immediately after all previous calls which have arrived and are already in line. Let η_n be the time needed to service the nth call. We assume that the sequence $\{\eta_n\}$ consists of independent (of each other and of the incoming flow) identically distributed random variables with common distribution function $B(x)$. We also assume that the event of "getting in line" does not depend on the other events.

Let $W(t)$ be the waiting time for the beginning of service if a call arrived at time t and entered the line. In particular, $W_n = W(t_n - 0)$ is the waiting time for the beginning of service by the nth call if it stands in line.

The described mass service system was studied in detail by L. Takács in [1], where in particular it was shown that the distribution of the random variable $W(t)$ tends to become stationary if $\rho = \lambda p \int_0^\infty x dB(x) < 1$. As ρ approaches unity this tendency becomes slower and slower. The aim of the present article is to estimate the "time of entering" of the described system into a stationary state. For example, one of the results (see e.g. formula (10)), applied to the case where the random variables η_n have exponential distribution with mean one, takes the form:

If $t \to \infty$ and $\delta = 1 - \lambda p \to 0$ with $t\delta^2 \to \tau$, $0 < \tau < \infty$, then

$$\mathbf{M}e^{-\delta s W(t)} \to s e^{s(s+1)\tau} \int_\tau^\infty e^{-s(s+1)u} f(u)\, du,$$

where

$$f(u) = 1 + \frac{1}{2\sqrt{\pi}} \int_u^\infty t^{-3/2} e^{-t/4} dt.$$

In particular, as $t\delta^2 \to \infty$

$$\mathbf{M}e^{-\delta s W(t)} \to (1+s)^{-1},$$

i.e.,

$$\mathbf{P}\left\{W(t) \leqslant \frac{v}{\delta}\right\} \to 1 - e^{-v}.$$

An analogous study of the behavior of the random variables W_n in the case $p = 1$ was made by E. G. Samandarov [2]. The behavior of the stationary distribution of the random variables $W_n (p = 1)$ was investigated by Kingman [3], [4] and I. N. Kovalenko. The most complete analysis of passage phenomena arising in a one-line mass service system under "heavy load" conditions $(p = 1)$ can be found in the article by Ju. V. Prohorov [5].

2°. **Auxiliary assertions.** Everywhere below we will assume that at the initial moment of time $t = 0$ the device is free, and consequently the first call is immediately serviced.

Without loss of generality we can assume that $\beta = \int_0^\infty x dB(x) = 1$, and let $\lambda - (1 - \delta)/p$, i.e., $\delta - 1 - \lambda p$. We let

$$\psi(s) = \mathbf{M}e^{-s\eta_n};\ F(t, x) = \mathbf{P}\{W(t) \leqslant x\};\ \Omega(t, s) = \mathbf{M}e^{-sW(t)};$$

$$P_0(t) = \mathbf{P}\{W(t) = 0\} \text{ and } \gamma(s) = \int_0^\infty e^{-st} P_0(t)\, dt.$$

Before going on to the precise formulation of the aim of this study, we cite some of the assertions of L. Takács [1] which are needed below.

Theorem 1. *If $W(0) = 0$ with probability one, then the Laplace-Stieltjes transform $\Omega(t, s) = \int_0^\infty e^{-sx} d_x F(t, x)$ is given by the formula*

$$\Omega(t, s) = e^{\{s-(1-\delta)[1-\psi(s)]\}t} \Bigg[1 - \left\{s + \frac{q(1-\delta)}{p}[1 - \psi(s)]\right\} \times \int_0^t e^{-\{s-(1-\delta)[1-\psi(s)]\}u} P_0(u)\, du\Bigg]. \qquad (1)$$

Theorem 2. *If $\rho = \lambda p \beta < 1$, then the limiting distribution $F(x) = \lim_{t\to\infty} F(t, x)$ exists and does not depend on the initial distribution $F(0, x)$; moreover,*

$$\Omega^*(s) = \lim_{t\to\infty} \Omega(t, s) = \frac{\delta\left\{s + \frac{q(1-\delta)}{p}[1 - \psi(s)]\right\}}{\left[1 + \frac{q(1-\delta)}{p}\right]\{s - (1-\delta)[1 - \psi(s)]\}}. \qquad (2)$$

If $\rho \geq 1$, then $\lim_{t\to\infty} F(t, x) = 0$ for all x.

We note further that the working time of the service structure consists of alternating periods of employment of the device (i.e., continuous work) and free periods, i.e., those time intervals during which the device is free of service. Clearly under the assumptions made, the durations of the successive periods of freedom and employment are independent so that identically distributed random variables, and also (since the incoming flow of calls is simple) the periods of freedom, have exponential distribution.

Let $G(x)$ be the distribution function of the employment period of the device and $x(s) = \int_0^\infty e^{-sx} dG(x)$. Now it is possible for us to formulate other results of

[1] which are needed below.

Theorem 3. *The Laplace-Stieltjes transform $x(s)$ of the distribution of the employment period is the least in absolute value root of the equation*

$$x(s)=\psi\{s+(1-\delta)[1-x(s)]\} \tag{3}$$

for $\operatorname{Re}(s)\geq 0$. *Moreover,*

$$\lim_{x\to\infty} G(x)=\omega,$$

where ω *is the root nearest to* 0 *of the equation*

$$\omega=\psi\{(1-\delta)(1-\omega)\}.$$

If $\rho\leq 1$, *then* $\omega=1$ *and* $G(x)$ *is a proper distribution function; but if* $\rho>1$, *then* $\omega<1$, *i.e., in this case the employment period is infinite with probability* $1-\omega$.

The Laplace transform $\gamma(s)=\int_0^\infty e^{-st}P_0(t)\,dt$ of the probability $P_0(t)$ that the device is free at time t is easily expressed in terms of $x(s)$. Indeed, we have

Theorem 4.

$$\gamma(s)=\int_0^\infty e^{-st}P_0(t)\,dt=\frac{p}{ps+(1-\delta)[1-x(s)]}, \tag{4}$$

where $x(s)$ *is defined by the relation* (3).

We omit the proofs of these results, directing the interested reader to [1]. We are interested in passage phenomena arising with the dual limit passage $t\to\infty$ and $\delta\to 0$. We shall show, for example, that if here $t\delta^2\to\tau$, $0<\tau<\infty$, then

$$\Omega(t,\ \delta s)\to\Omega_\tau(s)=se^{s(\sigma^2 s+1)\tau}\int_\tau^\infty e^{-s(\sigma^2 s+1)u}f(u)\,du,$$

where $2\sigma^2=\psi''(0)$ and

$$f(u)=1+\frac{\sigma}{2\sqrt{\pi}}\int_u^\infty v^{-3/2}e^{-v/4\sigma^2}\,dv.$$

From this relation it follows, in particular, that if $t\to\infty$, $\delta\to 0$ in such a way that $t\delta^2\to\infty$, then

$$\mathbf{P}\{\delta W(t)\leqslant x\}\to 1-e^{-x/\sigma^2}.$$

In the following analysis, a more careful study of the distribution of the employment period of the given system will play a basic role. We now proceed with it.

3°. **Asymptotic behavior of the moments of the distribution of the employment period.** Below we assume that all moments of the distribution $B(x)$ of the service time are finite. In other words, let

$$\psi(s)=\sum_{n=0}^\infty b_n s^n,$$

where $b_0 = 1$; $b_1 = -1$ and $b_2 = \sigma^2$.

It then follows from relation (3) that all moments of the distribution $G(x)$ of the employment period are also finite. Let $a_n^{(\delta)}$ be the nth coefficient of the expansion of the function $x(s)$ into a Taylor series in a neighborhood of $s = 0$, i.e. $a_n^{(\delta)} = (-1)^n x^{(n)}(0)/n!$ and $x(s) = \sum_{n=0}^{\infty} a_n^{(\delta)} s^n$. We note that the functions $x(s)$, $y(s)$, etc., depend on δ, but we shall not mention this dependence explicitly if it does not clarify the presentation.

Using Theorem 3 and relation (3), it is easy to see that

$$a_0^{(\delta)} = 1;\ a_1^{(\delta)} = -\delta^{-1};\ a_2^{(\delta)} = \sigma^2\delta^{-3}.$$

We show that, in general, $a_n^{(\delta)}$ has order δ^{-2n+1} as $\delta \to 0$.

Theorem 5. *As* $\delta \to 0$ *there exist finite limits*

$$\bar{a}_n = \lim_{\delta\to 0} \delta^{2n-1} a_n^{(\delta)} \quad (n = 1, 2, \ldots).$$

Proof. We start with relation (3). Using the so-called Bruno formula (see, for example, [6]) for the nth derivative of a composite function, we have

$$a_n^{(\delta)} = \sum \frac{k!\, b_k}{k_1! \ldots k_n!} \left(\frac{1}{\delta}\right)^{k_1} (a_2^{(\delta)})^{k_2} \ldots (a_n^{(\delta)})^{k_n} [-(1-\delta)]^{k_2+\ldots+k_n},$$

where $k = k_1 + k_2 + \cdots + k_n$ and the summation is taken over all solutions in nonnegative integers of the equation

$$k_1 + 2k_2 + \ldots + nk_n = n.$$

We note that k_n is nonzero in only one case, namely, when $k_n = 1$ and $k_1 = k_2 = \cdots = k_{n-1} = 0$. Consequently

$$a_n^{(\delta)} = \sum{}' \frac{k!\, b_k}{k_1! \ldots k_{n-1}!} \left(\frac{1}{\delta}\right)^{k_1} (a_2^{(\delta)})^{k_2} \ldots (a_{n-1}^{(\delta)})^{k_{n-1}}$$

$$\times [-(1-\delta)]^{k_2+\ldots+k_{n-1}} - b_1 (1-\delta)\, a_n^{(\delta)}, \tag{5}$$

where $k = k_1 + k_2 + \cdots + k_{n-1}$ and the symbol Σ' means summation over all nonnegative integer solutions of the equation

$$k_1 + 2k_2 + \ldots + (n-1)\, k_{n-1} = n. \tag{6}$$

Since $b_1 = -1$, then (5) implies

$$\delta a_n^{(\delta)} = \sum{}' \frac{k!\, b_k}{k_1! \ldots k_{n-1}!} \left(\frac{1}{\delta}\right)^{k_1} (a_2^{(\delta)})^{k_2} \ldots (a_{n-1}^{(\delta)})^{k_{n-1}} [-(1-\delta)]^{k_2+\ldots+k_{n-1}}. \tag{7}$$

On the basis of this recursion relation, the theorem can be proved by induction. Indeed, for $m = 1$ the assertion of the theorem is obvious. Suppose the limits $\lim_{\delta\to 0} \delta^{2m-1} a_m^{(\delta)} = \bar{a}_m$ exist for $m = \overline{1, n-1}$. Then from (7) we have

$$\delta^{2n-1}a_n^{(\delta)} = \sum{}' \frac{k!\,b_k}{k_1!\ldots k_{n-1}!}\left(\frac{1}{\delta}\right)^{k_1}\left(a_2^{(\delta)}\delta^3\right)^{k_2}\ldots\left(a_{n-1}^{(\delta)}\delta^{2n-3}\right)^{k_{n-1}}$$
$$\times\frac{\delta^{2n-2}}{\delta^{3k_2+\ldots+(2n-3)k_{n-1}}}\left[-(1-\delta)\right]^{k_2+\ldots+k_{n-1}}. \qquad (8)$$

For existence of the finite limit $\bar{a}_n = \lim_{\delta\to 0}\delta^{2n-1}a_n^{(\delta)}$ it is necessary that the exponent of δ in the denominator not exceed the exponent of δ in the numerator, i.e.,

$$2n-2 \geqslant k_1+3k_2+\ldots+(2n-3)k_{n-1}.$$

But this is actually the case, since from (6)

$$k_1+3k_2+\ldots+(2n-3)k_{n-1} = 2n-(k_1+k_2+\ldots+k_{n-1})$$

and obviously $k_1+k_2+\cdots+k_{n-1}\geq 2$.

Remark. It is obvious that on the right-hand side of relation (8) in the limit the only nonzero terms are those for which

$$k_1+k_2+\ldots+k_{n-1}=2.$$

Consequently, in order to express $\bar{a}_n$ in terms of the preceding, it is necessary to solve, in nonnegative integers, the system

$$\begin{cases} k_1+k_2+\ldots+k_{n-1}=2, \\ k_1+2k_2+\ldots+(n-1)k_{n-1}=n. \end{cases} \qquad (9)$$

It is clear that if $n = 2m$, then the only solutions of interest of the system are

$$k_j = k_{2m-j} = 1\ (j=\overline{1,\ m-1});\ k_i=0 \quad \text{for } i\neq j;\ 2m-j$$

and

$$k_m=2;\ k_i=0 \quad \text{for } i\neq m;$$

but if $n = 2m+1$, then the solutions of (9) have the form

$$k_j=k_{2m+1-j}=1\ (j=\overline{1,\ m});\ k_i=0 \quad \text{for } i\neq j;\ 2m+1-j.$$

This remark and equality (8) make it possible to write the recursion relations satisfied by the quantities $\bar{a}_n$.

4°. **Asymptotic behavior of the random variable $W(t)$.** First of all we note that the stationary waiting time $W=\lim_{t\to\infty}W(t)$ for small δ has asymptotically an exponential distribution. More precisely,

$$\lim_{\delta\to 0}\mathbf{P}\{\delta W\leqslant x\} = 1-e^{-\frac{x}{\sigma^2}},$$

where, as above, $\sigma^2=\psi''(0)/2$.

This conclusion follows easily from (2) if we use L'Hospital's rule twice in computing the limit of $\Omega^*(\delta s)$ as $\delta\to 0$. A stronger assertion can be made. Namely, we shall prove that if $t\to\infty$ and $\delta\to 0$ in such a way that $t\delta^2\to\infty$, then the random variable $\delta W(t)$ is asymptotically exponential. This will be a simple consequence of the following theorem.

Theorem 6. *If* $t \to \infty$ *and* $\delta \to 0$ *in such a way that* $t\delta^2 \to \tau$, $0 < \tau < \infty$, *then*

$$\Omega(t, \delta s) = \mathbf{M}e^{-\delta s W(t)} \to s e^{s(\sigma^2 s+1)\tau} \int_\tau^\infty e^{-s(\sigma^2 s+1)u} f(u)\, du, \tag{10}$$

where $\sigma^2 = \psi''(0)/2$ *and*

$$f(u) = 1 + \frac{\sigma}{2\sqrt{\pi}} \int_u^\infty v^{-3/2} e^{-v/4\sigma^2} dv.$$

Proof. Using (1), we have

$$\Omega\left(\frac{\tau}{\delta^2}, \delta s\right) = e^{\frac{\tau}{\delta^2}\left\{\delta s - [1-\psi(\delta s)](1-\delta)\right\}} \left[1 - \left\{\delta s + \frac{(1-\delta)q}{p}[1-\psi(\delta s)]\right\} \times \int_0^{\tau/\delta^2} e^{-u\{\delta s - (1-\delta)[1-\psi(\delta s)]\}} P_0(u)\, du\right]. \tag{11}$$

First of all we note that, as is easily verified,

$$\lim_{\delta \to 0} \frac{\tau\{\delta s - [1-\psi(\delta s)](1-\delta)\}}{\delta^2} = \tau s(\sigma^2 s + 1) \tag{12}$$

and

$$\delta s + \frac{(1-\delta)q}{p}[1-\psi(\delta s)] = \frac{s\delta}{p} + o(\delta). \tag{13}$$

Now we study the asymptotic behavior of the integral

$$I_\delta(\tau, s) = \frac{\delta}{p} \int_0^{\tau/\delta^2} e^{-u\{\delta s - (1-\delta)[1-\psi(\delta s)]\}} P_0(u)\, du. \tag{14}$$

Making the substitution $v = \delta^2 u$, we obtain

$$I_\delta(\tau, s) = \frac{1}{\delta p} \int_0^\tau e^{-\frac{v}{\delta^2}\{\delta s - (1-\delta)[1-\psi(\delta s)]\}} P_0\left(\frac{v}{\delta^2}\right) dv. \tag{15}$$

But by (12),

$$e^{-\frac{v}{\delta^2}\{\delta s - (1-\delta)[1-\psi(\delta s)]\}} = e^{-vs(\sigma^2 s+1)}(1 + o(1)). \tag{16}$$

Moreover,

$$\gamma_1(s) = \frac{1}{\delta p} \int_0^\infty e^{-sv} P_0\left(\frac{v}{\delta^2}\right) dv = \frac{\delta}{p} \int_0^\infty e^{-s\delta^2 u} P_0(u)\, du = \frac{\delta}{p}\gamma(s\delta^2),$$

where $\gamma(s)$ is defined by relation (4). Consequently

$$\frac{\delta}{p}\gamma(s\delta^2)=\frac{\delta}{p\delta^2 s+(1-\delta)\,[1-x\,(s\delta^2)]}\,. \tag{17}$$

Recall that in the notation used above

$$x\,(s)=\sum_{n=0}^{\infty} a_n^{(\delta)} s^n,$$

where, in particular, $a_0^{(\delta)} = 1$ and $a_1^{(\delta)} = -\delta^{-1}$. Therefore

$$\frac{\delta}{p}\gamma\,(s\delta^2)=\left\{s-q\delta s-(1-\delta)\sum_{2}^{\infty} a_n^{(\delta)}\delta^{2n-1}s^n\right\}^{-1}.$$

Applying Theorem 5, we obtain

$$\lim_{\delta\to 0}\frac{\delta}{p}\gamma(s\delta^2)=\left(s-\sum_{2}^{\infty}\bar{a}_n s^n\right)^{-1}. \tag{18}$$

Hence it follows that

$$\lim_{\delta\to 0}\frac{\delta}{p}P_0\left(\frac{v}{\delta^2}\right)=f\,(v)$$

exists, and

$$\bar{f}\,(s)=\int_0^{\infty} e^{-sv}f\,(v)\,dv=\left(s-\sum_{2}^{\infty}\bar{a}_n s^n\right)^{-1}. \tag{19}$$

Thus from (11)–(19) it follows that as $\delta\to 0$ and for each fixed τ, $0<\tau<\infty$, and s, $\mathrm{Re}\,(s)\geq 0$,

$$\Omega\left(\frac{\tau}{\delta^2},\ s\delta\right)\to\Omega_\tau\,(s)=e^{s(\sigma^2 s+1)\tau}\left[1-s\int_0^{\tau} e^{-s(\sigma^2 s+1)u}f\,(u)\,du\right]. \tag{20}$$

The form of the function $f(u)$ in relation (20) can be obtained by making use of the remark on Theorem 5. But we shall proceed otherwise. Since for any fixed τ, $0<\tau<\infty$ and s, $\mathrm{Re}\,(s)\geq 0$, $0\leq|\Omega_\tau(s)|\leq 1$, it is seen that as τ increases the expression in the square brackets in relation (20) must go to zero. Hence it follows immediately that

$$\int_0^{\infty} e^{-s(\sigma^2 s+1)u}\,f\,(u)\,du=s^{-1}.$$

Therefore, letting $s=(\sqrt{1+4\alpha\sigma^2}-1)/2\sigma^2$,

$$\bar{f}\,(\alpha)=\int_0^{\infty} e^{-\alpha u}f\,(u)\,du=\frac{1+2\sigma\sqrt{\alpha}\sqrt{1+\frac{1}{4\sigma^2\alpha}}}{2\alpha}\,.$$

Further, letting $2\sigma = c$ for brevity,

$$\tilde{f}(\alpha) = \frac{1}{2\alpha} + \frac{c}{2\sqrt{\alpha}}\left(1 + \frac{1}{c^2\alpha}\right)^{\frac{1}{2}} = \frac{1}{2\alpha} + \frac{c}{2}\sum_{k=0}^{\infty}\frac{\Gamma\left(\frac{3}{2}\right)}{\Gamma(k+1)\Gamma\left(\frac{3}{2}-k\right)}\cdot\frac{1}{c^{2k}\alpha^{k+\frac{1}{2}}}.$$

But, as is well known,

$$\frac{1}{\alpha^{\nu}} = \frac{1}{\Gamma(\nu)}\int_0^{\infty} e^{-\alpha t}t^{\nu-1}dt.$$

Therefore

$$f(t) = \frac{1}{2} + \frac{c}{2}\sum_{k=0}^{\infty}\frac{\Gamma\left(\frac{3}{2}\right)}{\Gamma(k+1)\Gamma\left(\frac{3}{2}-k\right)\Gamma\left(k+\frac{1}{2}\right)}\cdot\frac{t^{k-\frac{1}{2}}}{c^{2k}}. \tag{21}$$

Since (see, for example, [7])

$$\Gamma(k+1) = k!;\quad \Gamma\left(\frac{3}{2}\right) = \frac{\sqrt{\pi}}{2};\quad \Gamma\left(\frac{3}{2}-k\right)\Gamma\left(k+\frac{1}{2}\right) = (-1)^{k+1}\left(k-\frac{1}{2}\right)\frac{1}{\pi},$$

then from (21), after simple transformations, we obtain

$$\frac{df}{dt} = -\frac{ct^{-\frac{3}{2}}}{4\sqrt{\pi}}e^{-\frac{t}{c^2}}.$$

Thus

$$f(t) = 1 + \frac{c}{4\sqrt{\pi}}\int_t^{\infty} v^{-\frac{3}{2}}e^{-\frac{v}{c^2}}dv, \tag{22}$$

since, as is clear from (19),

$$\lim_{t\to\infty} f(t) = \lim_{s\to 0} s\tilde{f}(s) = 1.$$

This completes the proof of Theorem 6.

It is easy to verify that as $\tau\to\infty$

$$\Omega_{\tau}(s) \to (\sigma^2 s + 1)^{-1}.$$

Thus we have

Corollary 1. *If $\delta\to 0$ and $t\to\infty$ in such a way that $t\delta^2\to\infty$, then*

$$\mathbf{P}\{\delta W(t) \leqslant x\} \to 1 - e^{-\frac{x}{\sigma^2}}.$$

Now let $m(t) = MW(t)$. Then from Theorem 6 immediately follows

Corollary 2. *If $t\to\infty$ and $\delta\to 0$ in such a way that $t\delta^2\to\tau$, $0<\tau<\infty$, then*

$$\delta m(t) \to \mu(\tau) = \frac{\sigma}{2\sqrt{\pi}}\int_0^{\tau}\int_u^{\infty} v^{-\frac{3}{2}}e^{-\frac{v}{4\sigma^2}}dv\,du. \tag{23}$$

Moreover, for $\mu(\tau)$ we have the following asymptotic expansion:

$$\mu(\tau)=\sigma^2-\frac{\sigma^2}{\sqrt{\pi}}x^{-\frac{3}{2}}e^{-x}\left\{\sum_{k=0}^{n}\frac{(-1)^k(k+1)(2k+1)!!}{(2x)^k}+O\left(\frac{1}{x^{n+1}}\right)\right\}, \tag{24}$$

where $x=\tau/4\sigma^2$.

Proof. The validity of (23) is easily verified by direct differentiation of (10) at the point $s=0$ if we take (22) into account. Further, it is easy to see that $\mu(\infty)=\sigma^2$. Consequently,

$$\mu(\tau)=\sigma^2-\frac{\sigma}{2\sqrt{\pi}}\int_\tau^\infty\int_u^\infty v^{-\frac{3}{2}}e^{-\frac{v}{4\sigma^2}}\,dv\,du=\sigma^2-\frac{\sigma}{2\sqrt{\pi}}\int_\tau^\infty v^{-\frac{3}{2}}e^{-\frac{v}{4\sigma^2}}(v-\tau)\,dv. \tag{25}$$

But since

$$\int_\tau^\infty v^{-\frac{1}{2}}e^{-\frac{v}{4\sigma^2}}dv=2\sigma\Gamma\left(\frac{1}{2},\frac{\tau}{4\sigma^2}\right)\text{and}\int_\tau^\infty v^{-\frac{3}{2}}e^{-\frac{v}{4\sigma^2}}dv=\frac{1}{2\sigma}\Gamma\left(-\frac{1}{2},\frac{\tau}{4\sigma^2}\right),$$

where $\Gamma(\alpha,x)=\int_x^\infty e^{-v}v^{\alpha-1}dv$ is an incomplete Γ-function, equality (25) takes the form

$$\mu(\tau)=\sigma^2-\frac{\sigma^2}{\sqrt{\pi}}\left[\Gamma\left(\frac{1}{2},x\right)-x\Gamma\left(-\frac{1}{2},x\right)\right], \tag{26}$$

where $x=\tau/4\sigma^2$. Besides, using integration by parts it is easy to verify that

$$\Gamma\left(\frac{1}{2},x\right)=-\frac{1}{2}\Gamma\left(-\frac{1}{2},x\right)+x^{-\frac{1}{2}}e^{-x}. \tag{27}$$

Relation (24) is obtained after simple transformations from the corresponding asymptotic representation for the incomplete Γ-function

$$\Gamma(\alpha,x)=x^{\alpha-1}e^{-x}\left\{\sum_{k=0}^{n-1}\frac{(-1)^k\Gamma(1-\alpha+k)}{x^k\Gamma(1-\alpha)}+O(|x^{-n}|)\right\},$$

(see, for example, [7]), if we take into account (26) and (27).

Remark. The asymptotic expansion (24) yields a representation for the time it takes the given system to enter a stationary state, evaluated in terms of the behavior of the mean waiting time. It is possible to discuss the relative deviation of the quantity $\mu(\tau)$ from its stationary value σ^2 from the following table where, as before, $x=\tau/4\sigma^2$.

x	1	2	3	4	5	6	7	8	9
$1-\frac{\mu(\tau)}{\sigma^2}$	0,288	0,247	0,179	0,116	0,069	0,039	0,022	0,012	0,006

It is obvious that for sufficiently small δ an analogous situation holds for the quantity $\delta m(t) = M\delta W(t)$.

BIBLIOGRAPHY

[1] L. Takács, *The transient behavior of a single server queuing process with a Poisson input,* Proc. Fourth Berkeley Sympos. Math. Stat. and Prob., Vol. II, 535–567, Univ. of California Press, Berkley, Calif., 1961. MR 27 #4290.

[2] E. G. Samandarov, *Service systems in heavy traffic conditions,* Teor. Verojatnost. i Primenen. 8 (1963), 327–330. MR 27 #4292.

[3] J.F.C. Kingman, *The single server queue in heavy traffic,* Proc. Cambridge Philos. Soc. 57 (1961), 902–904. MR 24 #A1150.

[4] ———, *On a queue in heavy traffic,* J. Roy Statist. Soc. Ser. B 24 (1962), 383–392. MR 26 #5654.

[5] Ju. V. Prohorov, *Passage phenomena in mass service processes,* Litovsk. Mat. Sb., Vil'nyus, 1963.

[6] J. Riordan, *Introduction to combinatiorial analysis,* IL, Moscow, 1963.

[7] I. M. Ryžik and I. S. Gradšteĭn, *Tables of integrals, sums, series and products,* 3rd ed., GITTL, Moscow, 1951; English and German transl., 2nd. rev. ed., VEB Deutscher Verlag der Wissenschaften, Berlin and Plenum Press, New York, 1963. MR 14, 643; MR 28 #1326.

Translated by:
L. Rosenblatt

ON CONTROLS LEADING TO OPTIMAL STATIONARY STATES

O. V. VISKOV AND A. N. ŠIRJAEV

This paper deals with certain regulated systems whose statistical description can be given in terms of controlled chains.

The work contains several general results related to existence of homogeneous Markov controls in problems of minimization of functionals of the form (5), under the assumption that both the phase space of the controlled system and the space of values of the controlling influences consist of a finite number of points.

§1. Basic definitions and statement of the problem

1. Let $\mathfrak{X}$ be the (phase) space of states of a system, consisting of not more than a countable number of points. It is known [1] that a random process with discrete time $\xi = \{\xi_n, n = 0, 1, \cdots\}$, $\xi_i \in \mathfrak{X}$, forms a Markov (uncontrolled) chain if with probability one (with respect to the measure of this process)

$$P\{\xi_{n+1} = x_{n+1} \mid \xi_n, \ldots, \xi_0\} = P\{\xi_{n+1} = x_{n+1} \mid \xi_n\}. \tag{1}$$

For the definition of a *controlled* chain, we assume that we are given a space D, namely the space of solutions, or equivalently, the values of the controlling influences, and a collection of transition probabilities (of matrices of transition probabilities)

$$P(\xi_{n+1} = x_{n+1} \mid \xi_n, d_n);\ n = 0, 1, \ldots, \tag{2}$$

depending on the solutions $d_n \in D$.

We assume that at each moment of time $n = 0, 1, \cdots$ the choice of concrete value of the parameter d_n can be effected on the basis of the results of the preceding observations $x_0, \cdots, x_n$; $d_n = d_n(x_0, \cdots, x_n)$. Every such function, defined on the space $\mathfrak{X}^{n+1} = \mathfrak{X} \times \cdots \times \mathfrak{X}$ and with values in D, assigns some (nonrandomized) control at time n. More precisely, if the results of observations at time $t = 0, 1, \cdots, n$ are $x_0, \cdots, x_n$ and a chosen control is characterized by the function $d_n(\cdot)$, then the state of the system at time $n + 1$ is determined by the transition probability

$$P(\xi_{n+1} = x_{n+1} \mid \xi_n = x_n,\ d_n(x_0, \ldots, x_n)). \tag{3}$$

Definition 1. We say that the collection $\delta = \{d_n, n \geq 0\}$ of functions

$$\begin{aligned} d_0 &= d_0(x_0), \\ d_1 &= d_1(x_0, x_1), \\ &\cdots\cdots \\ d_n &= d_n(x_0, \ldots, x_n) \end{aligned} \tag{4}$$

assigns the control δ.

If a control δ is chosen, then the collection of transition probabilities (2) for each fixed initial state x_0 defines [1] a probability measure in the space of sequences $(\xi_0, \xi_1, \cdots)$. The process thus obtained, call it (ξ, δ), is said to be a *controlled* (by means of the control δ) *chain*, to emphasize its dependence on the control δ.

Since the functions $d_n = d_n(\cdot)$ depend, generally speaking, not only on the values x_n, but also on the preceding values $x_0, \cdots, x_{n-1}$, a controlled chain will not be Markov in the general case.

Remark. It is natural to call our control δ a control based on complete data, since all past values of the states of the system are assumed precisely known. Another interesting case is that of so-called controls based on incomplete data, when at each time n we do not know the values $x_0, \cdots, x_n$, but, for example, only some function of them. Some related problems can be found in [2].

Definition 2. The control δ is said to be Markov if every function $d_n = d_n(\cdot)$ depends only on the last argument: $d_n = d_n(x_n)$. More precisely, the control δ is Markov if $d_n(x_0', \cdots, x_{n-1}', x_n) = d_n(x_0'', \cdots, x_{n-1}'', x_n)$ for all values $x_0', \cdots, x_{n-1}'$ and $x_0'', \cdots, x_{n-1}''$.

Definition 3. The control δ is called homogeneous Markov if $d_{n'}(x) \equiv d_{n''}(x)$ for all n' and n''.

Moreover, a homogeneous Markov control δ is completely determined by assigning only one function, say $d = d(x)$, $x \in \mathfrak{X}$ with values in D.

From Definitions 2 and 3 it is obvious that an inhomogeneous Markov control δ defines an inhomogeneous Markov chain (ξ, δ), while a homogeneous Markov control leads to a homogeneous (stationary) Markov chain.

2. Let Δ denote the collection of all controls δ. One often encounters problems in which, in finding controls which are optimal in some sense, we need to consider not the entire class of controls Δ, but only some subclass of it Δ', whose elements, by Definition 4 below, are called admissible controls.

Let $D_n(x)$, $n \geq 0$, be fixed subsets (generally distinct for various values of $\xi_n = x$) of the space D, each of which bounds a region of values of the control d_n at time n if $\xi_n = x$.

Definition 4. The control $\delta \in \Delta$ is called admissible if it is such that for each n the values of the functions

$$d_n(x_0, \ldots, x_{n-1}, x) \in D_n(x) \subseteq D.$$

Naturally the determination of the class Δ' of admissible controls characterized by the collection $\{D_n(x), n \geq 0, x \in \mathfrak{X}\}$, depends on the nature of each concrete problem. We let Δ'' denote the class of admissible controls for which for all n

$$D_n(x) \equiv D(x) (\subseteq D).$$

3. In order to discuss the quality of a control, we introduce the function $W(x, d)$, defined on $\mathfrak{X} \times D$, characterizing the losses from accepting the solution d, when the system is in state x. Now choose a control δ. For it consider the quantity

$$R_x^\delta = \limsup_{T \to \infty} \frac{1}{T} \sum_{t=0}^{T} M_x^\delta W(x_t, d_t), \tag{5}$$

which determines mean losses for δ in a unit of time. In (5), M_x^δ is the mathematical expectation for the process (ξ, δ) given that $\xi(0) = x$.

Set

$$\rho(x) = \inf R_x^\delta, \tag{6}$$

where the inf is taken over the class of admissible controls Δ'.

Definition 5. The control $\tilde{\delta} \in \Delta'$ is called ϵ-optimal if for all $x \in \mathfrak{X}$

$$\rho(x) > R_x^{\tilde{\delta}} - \varepsilon.$$

A 0-optimal control is simply called optimal. Obviously an ϵ-optimal control always exists. The problem of *existence of an optimal control* and the larger question of when an optimal *homogeneous Markov* control exists is not so obvious. In the following section we answer this question, but only in the very special case of *finite* spaces $\mathfrak{X}$ and D.

§2. On existence of homogeneous Markov controls

Theorem 1. *If the spaces $\mathfrak{X}$ and D consist of a finite number of points, $0 \leq W(x, d) \leq C < \infty$, then an optimal homogeneous Markov control exists in the class Δ''.*

To establish the theorem for each control $\delta \in \Delta'$, alongside the "risk" R_x^δ we consider the new "risk"

$$\mathfrak{R}_x^\delta(\beta) = M_x^\delta \sum_{k=0}^{\infty} \beta^k W(x_k, d_k), \tag{7}$$

where $d_k = d_k(x_0, \cdots, x_n)$, and set

$$\mathfrak{R}_x(\beta) = \inf_{\delta \in \Delta'} \mathfrak{R}_x^\delta(\beta). \tag{8}$$

Lemma 1. *Under the assumptions of Theorem* 1, *an optimal control* δ_β^* *exists*[1] *in the class* Δ'', *for which*

$$\mathfrak{R}_x^{\delta_\beta^*}(\beta) = \mathfrak{R}_x(\beta), \quad 0 < \beta < 1, \; x \in \mathfrak{X}. \tag{9}$$

Proof is based on the following general considerations. Every control $\delta \in \Delta'$ is a collection of functions

$$\{d_0(x_0), \; d_1(x_0, x_1), \; \ldots, \; d_n(x_0, \ldots, x_n), \; \ldots\}.$$

Take the function $d_n(x_0, \cdots, x_n)$; it can be considered as a point in the space

$$\Delta_n = D_n(\mathfrak{X})^{\mathfrak{X}^{n+1}}, \quad \mathfrak{X}^{n+1} = \underbrace{\mathfrak{X} \times \ldots \times \mathfrak{X}}_{n+1},$$

where $D_n(\mathfrak{X})$ is the set of (admissible) values of the control at time n. Then the control δ is an element of the compact topological space $\Delta' = (\Delta_0 \times \Delta_1 \times \cdots)$, since each of the topological spaces (with discrete topology) Δ_n is compact, as a space consisting of a finite number of points, while by Tihonov's theorem the topological product of any family of compact spaces is also compact (see [3], p. 21).

The function $\mathfrak{R}_x^\delta(\beta)$ for every $x \in \mathfrak{X}$ and $0 < \beta < 1$, considered as a function of $\delta \in \Delta'$, is continuous (since every term in (7) is a continuous function on Δ' with a discrete topology on it, and the series (7) converges). Consequently, it attains its minimal value on Δ'. Moreover, there is a control $\delta_\beta^* \in \Delta'$,

$$\delta_\beta^* = \{d_0^*(x_0), \; d_1^*(x_0, x_1), \; \ldots\},$$

such that (9) is satisfied for it.

Lemma 2. *Consider the risks* $\mathfrak{R}_x^\delta(\beta)$ *for* $\delta \in \Delta'$. *Under the assumptions of Theorem* 1, *an optimal Markov control exists in the class* Δ'.

Proof. The problem of existence of an optimal control in the class Δ' is solved positively by Lemma 1. Now we establish that for each optimal control $\delta^* = \delta_\beta^*$ (the value of the parameter β is now fixed and therefore in the proof of Lemma 2 the index β for δ_β^* is omitted) there is a *Markov* control $\tilde{\delta}$ equivalent to it, i.e., a control for which

$$\mathfrak{R}_x^{\tilde{\delta}} = \mathfrak{R}_x^{\delta^*}, \quad x \in \mathfrak{X}. \tag{10}$$

---

1) It can be seen from the proof of the lemma that it remains valid if $\mathfrak{X}$ is a topological Hausdorff space, D is a compact Hausdorff space, and $M_{x_0}^\delta W(x_k, d_k)$ is a continuous function on Δ_k for each x_0.

(Indication of the dependence on β in (10) will also be omitted for the same reasons.) For the proof of this assertion we use considerations due to I. V. Girsanov [4].

Let δ^* be the optimal control whose existence has been established, and let δ be any control from the class Δ':

$$\delta = \{d_0(x_0),\ d_1(x_0,\ x_1)\ \ldots\}.$$

The optimal control δ^* is characterized by the fact that for it, for all $x \in \mathfrak{X}$

$$\mathfrak{R}_x^{\delta^*} \leqslant \mathfrak{R}_x^{\delta},\ \delta \in \Delta'. \tag{11}$$

By the formula for conditional mathematical expectations, for all $m \geq 0$

$$\mathfrak{R}_{x_0}^{\delta} = M_{x_0}^{\delta}\left\{\sum_{k=0}^{m-1} \beta^k W(x_k,\ d_k) + M^{\delta}\left[\sum_{k=m}^{\infty} \beta^k W(x_k,\ d_k) \mid x_0,\ \ldots,\ x_m\right]\right\}. \tag{12}$$

The following characteristic property of the optimal control δ^*, used below, is often called Bellman's principle for additive functionals: if $x_0, \cdots, x_m, \cdots$ are results of observations on the process $(\xi,\ \delta^*)$, then for any other control δ almost everywhere in the measure of the process $(\xi,\ \delta^*)$ (see formula (12))

$$M^{\delta^*}\left\{\sum_{k=m}^{\infty} \beta^k W(x_k,\ d_k^*) \mid x_0,\ \ldots,\ x_m\right\} \leqslant M^{\delta}\left\{\sum_{k=m}^{\infty} \beta^k W(x_k,\ d_k) \mid x_0,\ \ldots,\ x_m\right\}. \tag{13}$$

Indeed, suppose there exists a control δ and a set $\mathfrak{M}_m \in \mathfrak{B}_m$, where $\mathfrak{B}_m$ is the σ-algebra of sets of the space $\mathfrak{X}^{m+1}$ of positive $P_{x_0}^{\delta^*}$-measure of the process $(\xi,\ \delta^*)$, on which

$$M^{\delta^*}\left\{\sum_{k=m}^{\infty} \beta^k W(x_k,\ d_k^*) \mid x_0,\ \ldots,\ x_m\right\} > M^{\delta}\left\{\sum_{k=m}^{\infty} \beta^k W(x_k,\ d_k) \mid x_0,\ \ldots,\ x_m\right\}, \tag{14}$$

and on its complement (13) is satisfied. We form a new control $\tilde{\delta}$ such that for it

$$\tilde{d}_i(x_0,\ \ldots,\ x_i) = d_i^*(x_0,\ \ldots,\ x_i),\quad i < m,$$

$$\tilde{d}_{m+i}(x_0,\ \ldots,\ x_m,\ \ldots,\ x_{m+i}) = \begin{cases} d_{m+i}(x_0,\ \ldots,\ x_{m+i}), & (x_0,\ \ldots,\ x_m) \in \mathfrak{M}_m, \\ d_{m+i}^*(x_0,\ \ldots,\ x_{m+i}), & (x_0,\ \ldots,\ x_m) \in \overline{\mathfrak{M}}_m. \end{cases} \tag{15}$$

Then for $\tilde{\delta}$ it follows from (12), (14), (15) that

$$\mathfrak{R}_{x_0}^{\tilde\delta} = M_{x_0}^{\tilde\delta}\left\{\sum_{k=0}^{m-1} W(x_k, \tilde d_k) + M^{\tilde\delta}\left[\sum_{k=m}^{\infty} \beta^k W(x_k, \tilde d_k) \mid x_0, \ldots, x_m\right]\right\}$$

$$= M_{x_0}^{\tilde\delta}\sum_{k=0}^{m-1} \beta^k W(x_k, \tilde d_k) + \int_{\mathfrak{M}_m} M^{\tilde\delta}\left[\sum_{k=m}^{\infty} \beta^k W(x_k, \tilde d_k) \mid x_0, \ldots, x_m\right] dP_{x_0}^{\delta^*}(x_0, \ldots, x_m)$$

$$+ \int_{\overline{\mathfrak{M}}_m} M^{\tilde\delta}\left[\sum_{k=m}^{\infty} \beta^k W(x_k, \tilde d_k) \mid x_0, \ldots, x_m\right] dP_{x_0}^{\delta^*}(x_0, \ldots, x_m)$$

$$< M_{x_0}^{\delta^*}\sum_{k=0}^{m-1} \beta^k W(x_k, d_k^*)$$

$$+ \int_{\mathfrak{M}_m} M^{\delta^*}\left[\sum_{k=m}^{\infty} \beta^k W(x_k, d_k^*) \mid x_0, \ldots, x_m\right] dP_{x_0}^{\delta^*}(x_0, \ldots, x_m)$$

$$+ \int_{\overline{\mathfrak{M}}_m} M^{\delta^*}\left[\sum_{k=m}^{\infty} \beta^k W(x_k, d_k^*) \mid x_0, \ldots, x_m\right] dP_{x_0}^{\delta^*}(x_0, \ldots, x_m) = \mathfrak{R}_{x_0}^{\delta^*},$$

i.e.,

$$\mathfrak{R}_{x_0}^{\tilde\delta} < \mathfrak{R}_{x_0}^{\delta^*},$$

which contradicts the optimality of the control δ^*.

Thus for each realization $(x_0, \cdots, x_m)$ of nonzero probability $P_{x_0}^{\delta^*}$ (recall that such realizations exist due to the fact that the space is finite), property (13) is satisfied for the optimal control δ^*.

Now we establish that for the optimal control δ^* the conditional mathematical expectation

$$\Phi^{\delta^*}(x_0, \ldots, x_m) = M^{\delta^*}\left\{\sum_{k=m}^{\infty} \beta^k W(x_k, d_k^*) \mid x_0, \ldots, x_m\right\}$$

does not depend on $x_0, \cdots, x_{m-1}$, i.e., for each two realizations (of positive $P_{x_0'}^{\delta^*}$- and $P_{x_0''}^{\delta^*}$-measure respectively) $x_0', \cdots, x_{m-1}', x_m$ and $x_0'', x_1'', \cdots, x_{m-1}'', x_m$

$$\Phi^{\delta^*}(x_0', \ldots, x_{m-1}', x_m) = \Phi^{\delta^*}(x_0'', \ldots, x_{m-1}'', x_m).$$

Indeed, consider all realizations of the form $(x_0, \cdots, x_m)$, ending at a fixed point x_m, on which $\Phi^{\delta^*}(x_0, \cdots, x_m)$ depends explicitly in view of (3). In our case, since the space $\mathfrak{X}$ is finite, there is a finite number of them. Consequently, there is a realization of positive $P_{\hat{x}_0}^{\delta^*}$-measure $\hat{x}_0, \cdots, \hat{x}_{m-1}, \hat{x}_m$, where $\hat{x}_i = \hat{x}_i(x_m)$, $i = 0, \cdots, m-1$, such that

$$\Phi^{\delta^*}(\hat{x}_0, \ldots, \hat{x}_{m-1}, x_m) \leqslant \Phi^{\delta^*}(x_0, \ldots, x_{m-1}, x_m) \qquad (14')$$

for any realization $(x_0, \cdots, x_m)$ of positive $P_{x0}^{\delta^*}$-measure ending at a fixed point x_m.

But there cannot be a strict inequality in relation (14′). For if, say for the realization $x_0^0, \cdots, x_{m-1}^0, x_m$ of positive $P_{x0}^{\delta^*}$-measure, we had

$$\Phi^{\delta^*}(\hat{x}_0, \ldots, \hat{x}_{m-1}, x_m) < \Phi^{\delta^*}(x_0^0, \ldots, x_{m-1}^0, x_m),$$

then, after forming the new control δ^{**} coinciding with δ^* with the exception that for any realization with beginning $(x_0^0, \cdots, x_{m-1}^0, x_m)$

$$\begin{aligned} &d_{m+k}^{**}(x_0^0, \ldots, x_{m-1}^0, x_m, \ldots, x_{m+k}) \\ &= d_{m+k}^*(\hat{x}_0, \ldots, \hat{x}_{m-1}, x_m, \ldots, x_{m+k}), \quad k \geqslant 0, \end{aligned}$$

we would obtain for δ^{**}

$$M^{\delta^{**}}\left\{\sum_{k=m}^{\infty} \beta^k W(x_k, d_k^{**}) \,|\, x_0, \ldots, x_m\right\} < M^{\delta^*}\left\{\sum_{k=m}^{\infty} \beta^k W(x_k, d_k^*) \,|\, x_0, \ldots, x_m\right\},$$

in contradiction to the necessary condition (13) established above, which is satisfied by the optimal control δ^*.

Moreover, the conditional mathematical expectation $\Phi^{\delta^*}(x_0, \cdots, x_m)$ does not depend on $x_0, \cdots, x_{m-1}$, i.e., it does not depend on the path of arrival to the point x_m. Therefore, if for any realization $x_0, \cdots, x_m, \cdots, x_{m+k}$, instead of the value $d_{m+k}^*(x_0, \cdots, x_m, \cdots, x_{m+k})$ of the control δ^* at time $m+k$, we consider the value $d_{m+k}^*(\hat{x}_0(x_m), \cdots, \hat{x}_{m-1}(x_m), x_m, \cdots, x_{m+k})$, then such a change does not change the value of the conditional mathematical expectation $\Phi^{\delta^*}(x_0, \cdots, x_m)$. Thus we can say that the control $\delta^{(m)}$, which coincides with δ^* up to and including time m, and is such that $d_{m+k}^{(m)}(x_0, \cdots, x_{m+k})$ does not depend on $x_0, \cdots, x_{m-1}$ but is given by the formula

$$d_{m+k}^{(m)}(x_m, \ldots, x_{m+k}) = d_{m+k}^*(\hat{x}_0(x_m), \ldots, \hat{x}_{m-1}(x_m), x_m, \ldots, x_{m+k}), \quad k \geqslant 1,$$

guarantees the same mathematical expectation $\Phi^{\delta^*}(x_0, \cdots, x_m)$ as does the optimal control δ^*.

But now we do the following. Set $m = 1$. Then the control

$$\delta^* = \{d_0^*(x_0), d_1^*(x_0, x_1), \ldots\}$$

can be converted into the control

$$\delta_{(1)}^* = \{d_0^{(1)}(x_0), d_1^{(1)}(x_1), d_2^{(1)}(x_1, x_2), d_3^{(1)}(x_1, x_2, x_3), \ldots\},$$

equivalent to δ^*, where $d_0^{(1)}(x_0) = d_0^*(x_0)$. Now set $m = 2$ and apply the above arguments to the control $\delta^*_{(1)}$. Then we obtain the control

$$\delta^*_{(2)} = \{d_0^{(2)}(x_0),\ d_1^{(2)}(x_1),\ d_2^{(2)}(x_2),\ d_3^{(2)}(x_2, x_3), \ldots\},$$

where

$$d_0^{(2)}(x_0) = d_0^{(1)}(x_0),$$
$$d_1^{(2)}(x_1) = d_1^{(1)}(x_1).$$

In similar fashion we construct successive controls $\delta^*_{(n)}$. Consider this sequence of controls $\{\delta^*_{(n)}, n \geq 1\}$. Using the Weierstrass diagonal method, we construct the control $\bar{\delta} - \delta^*_{(\infty)}$,

$$\delta^*_{(\infty)} = \{d_0^{(\infty)}(x_0),\ d_1^{(\infty)}(x_1),\ \ldots,\ d_i^{(\infty)}(x_i), \ldots\},$$

for which

$$d_i^{(\infty)}(x_i) = d_i^{(i+1)}(x_i), \quad i \geqslant 0.$$

It is obvious that for each control $\delta^*_{(n)}$

$$\mathfrak{R}_{x_0}^{\overset{*}{\delta}(n)} = \mathfrak{R}_{x_0}^{\delta^*}, \quad x_0 \in \mathfrak{X}.$$

It only remains to show that

$$\lim_{n\to\infty} \mathfrak{R}_{x_0}^{\overset{*}{\delta}(n)} = \mathfrak{R}_{x_0}^{\overset{*}{\delta}(\infty)}, \tag{16}$$

since the facts that the control $\delta^*_{(\infty)}$ is Markov and

$$\mathfrak{R}_{x_0}^{\overset{*}{\delta}(\infty)} = \mathfrak{R}_{x_0}^{\delta^*},$$

imply existence of a Markov optimal control.

But the validity of relation (16) follows easily from the remarks which follow. The function $\mathfrak{R}^{\delta}_{x_0}$, considered as a function of $\delta \in \Delta'$, is continuous in a Tihonov (discrete) topology which is in Δ'. But since in this topology the point $\bar{\delta} = \delta^*_{(\infty)}$ is, by construction, a limit point of the sequence $\{\delta^*_{(n)}, n \geq 1\}$, then relation (16) is actually satisfied.

Thus Lemma 2 is proved.

Lemma 3. *Suppose the conditions of Theorem* 1 *are satisfied. Then in the class* Δ'' *there exists a homogeneous Markov optimal control* $\tilde{\delta}$ *(generally speaking, depending on* β*), for which*

$$\mathfrak{R}_{x_0}^{\tilde{\delta}}(\beta) = \min_{\delta\in\Delta''} \mathfrak{R}_{x_0}^{\delta}(\beta).$$

Proof. Existence of an optimal Markov control follows from Lemmas 1 and 2.

We shall show that an optimal Markov control indeed can be chosen in the class of *homogeneous* controls. Let $\delta^*_{(\infty)}$ be the control constructed in Lemma 2:

$$\{d_0^{(\infty)}(x_0),\ d_1^{(\infty)}(x_1),\ \ldots,\ d_i^{(\infty)}(x_i),\ \ldots\}.$$

According to the Bellman equation for $\Re_x(\beta)$ (see (8))

$$\Re_{x_0}(\beta) = \min_{d_0 \in D(x_0)} \left\{ MW(x_0, d_0(x_0)) + \beta \sum_y \Re_y(\beta) P(y \mid x_0, d_0(x_0)) \right\} \qquad (17)$$

(its derivation is well known and therefore not given here). Since for the optimal control $\delta^*_{(\infty)}$

$$\Re_{x_0}^{\delta^*_{(\infty)}}(\beta) = \Re_{x_0}(\beta),$$

the finding of the optimal (for each x_0) value $d_0^{(\infty)}(x_0)$ reduces to solving equation (17), in which $D(x_0) \subseteq D$ (see §1). Now assume that we have found optimal controls $d_0^{(\infty)}(x_0), \ldots, d_{m-1}^{(\infty)}(x_{m-1})$, and they led to the observations $x_0, \ldots, x_{m-1}, x_m$. Then by the characteristic property (13) of the optimal controls, the extension

$$\delta^*_m = \{d_m^{(\infty)}(x_m),\ d_{m+1}^{(\infty)}(x_{m+1}),\ \ldots\}$$

of the optimal control $\delta^*_{(\infty)}$ is such that

$$M^{\delta^*_m}\left\{ \sum_{k=m}^{\infty} \beta^k W(x_k, d_k^{(\infty)}) \mid x_m \right\} \leqslant M^{\delta_m}\left\{ \sum_{k=m}^{\infty} \beta^k W(x_k, d_k) \mid x_m \right\}$$

for any extension

$$\delta_m = \{d_m(x_m),\ d_{m+1}(x_{m+1}),\ \ldots\}$$

of the Markov control δ. It is easy to see that

$$M_{x_m}^{\delta^*_m}\left\{ \sum_{k=m}^{\infty} \beta^k W(x_k, d_k^{(\infty)}) \mid x_m \right\} = \beta^m \Re_{x_m}(\beta),$$

and the optimal value $d_m^{(\infty)}$ of the control δ for each fixed x_m can be found as a solution of the equation

$$\Re_{x_m}(\beta) = \min_{d_m \in D(x_m)} \left\{ MW(x_m, d_m) + \beta \sum_y \Re_y(\beta) P(y \mid x_m, d_m(x_m)) \right\},$$

coinciding with (17), since the space $D(x_m)$ does not depend on the number of the trial, as do the probabilities $P(y|x_m, d_m(x_m))$ and the functions $W(x_m, d_m)$.

Moreover, among Markov controls, there exists a homogeneous one, which is found as a solution of equation (17). Lemma 3 is proved.

Remark 1. We note that for finding an optimal control according to equation

(17), it is necessary to know the function $\Re_x(\beta)$, which we do not know. Therefore it is first necessary to solve equation (17), find its solution $\Re_x(\beta)$, and then to seek the optimal control. But then we encounter the difficulty that the solution of equation (17) is, generally speaking, not unique. Fortunately, in our case the solution of equation (17) is unique. Indeed, suppose this equation has some other solution Q_x distinct from $\Re_x$. Then

$$Q_{x_0} = \min_{d_0 \in D(x_0)} \left\{ MW(x_0, d_0) + \beta \sum_{x_1} Q_{x_1} P(x_1 \mid x_0, d_0) \right\}. \tag{18}$$

Hence

$$Q_{x_0} = \min_{d_0 \in D(x_0)} \left\{ MW(x_0, d_0) + \beta \sum_{x_1} \min_{d_1 \in D(x_1)} \left[MW(x_1, d_1) + \beta \sum_{x_2} Q_{x_2} P(x_2 \mid x_1, d_1) \right] P(x_1 \mid x_0, d_0) \right\}.$$

Or

$$Q_{x_0} = \min_{d_0 \in D(x_0)} \min_{d_1 \in D(x_1)} \left\{ M_{x_0} \sum_{k=0}^{1} \beta^k W(x_k, d_k) + \beta M_{x_0} (Q_{x_2} \mid x_1, d_1) \right\}.$$

And generally

$$Q_{x_0} = \min_{d_0 \in D(x_0), \ldots, d_m \in D(x_m)} \left\{ M_{x_0} \sum_{k=0}^{m} \beta^k W(x_k, d_k) + \beta^m M_{x_0} Q_{x_{m+1}} \right\}.$$

Hence, passing to the limit

$$Q_{x_0} = \Re_{x_0} + \lim_{m \to \infty} \min_{d_0 \in D(x_0), \ldots, d_m \in D(x_m)} \beta^m M_{x_0} Q_{x_{m+1}}. \tag{19}$$

Since $0 < \beta < 1$ and it is sufficient to seek the solution of equation (17) in the class of functions uniformly in x not greater than $c/(1-\beta)$, then the last term in the right part of relation (19) goes to zero, which proves uniqueness of the solution of equation (17).

Proof of Theorem 1. From Lemmas 1, 2 and 3 it follows that for each $0 < \beta < 1$ there exists an optimal homogeneous Markov control $\tilde{\delta} = \delta^*_\beta$. On the strength of the homogeneity, as was mentioned above, the control δ^*_β is completely determined by the assignment of a unique function $d^*_\beta = d^*_\beta(x)$. This function is defined on the *finite* set $\mathfrak{X}$ and takes on a *finite* number of values. Consequently, there is an infinite sequence of numbers $\beta_\alpha \to 1$ such that for all of them the controls $\delta^*_{\beta_\alpha}$ coincide: $\delta^*_{\beta_\alpha} = \delta^*$, $\delta^* = \{d^*(x)\}$.

Consider this control δ^*. Since it is Markov and homogeneous, the chain corresponding to this control will be Markov and homogeneous with transition probabilities

$$P(x \mid y, d^*(y)).$$

For this control δ^* consider

$$R_x^{\delta^*} = \limsup_{T\to\infty} \frac{1}{T} \sum_{t=0}^{T} M_x^{\delta^*} W(x_t, d_t^*), \quad d_t^* = d^*(x_t).$$

Let $\omega(x) = W(x, d^*(x))$. Then

$$\frac{1}{T} \sum_{k=0}^{T} M_x^{\delta^*} W(x_k, d_k^*) = \sum_{y\in\mathfrak{X}} \omega(y) \left(\frac{1}{T} \sum_{k=0}^{T} P_{xy}^{(k)} \right),$$

where $P_{xy}^{(k)}$ is the k-step transition probability from the point x to the point y for the process (ξ, δ^*).

By the well-known ergodic theorem (see, for example, [1], Theorem 2.1, Chapter V) for homogeneous Markov chains, there exists a matrix π_{xy} such that

$$\lim_{T\to\infty} \frac{1}{T} \sum_{k=0}^{T} P_{xy}^{(k)} = \pi_{xy}.$$

Moreover, for the controlled chain (ξ, δ^*) the following limit exists

$$\lim_{T\to\infty} \frac{1}{T} \sum_{k=0}^{T} M_x^{\delta^*} W(x_k, d_k^*) = R_x^{\delta^*}, \tag{20}$$

whence, by the abelian theorem (see, for example, [5], p. 180), from existence of the limit (20) follows the existence of

$$\lim_{\beta_\alpha\to 1} (1-\beta_\alpha) \mathfrak{R}_x^{\delta^*}(\beta_\alpha) \tag{21}$$

and their equality, i.e.,

$$\lim_{\beta_\alpha\to 1} (1-\beta_\alpha) \mathfrak{R}_x^{\delta^*}(\beta_\alpha) = R_x^{\delta^*}. \tag{22}$$

Here for an arbitrary control δ

$$R_x^{\delta} = \limsup_{T\to\infty} \frac{1}{T} \sum_{k=0}^{T} M_x^{\delta} W(x_k, d_k) \geqslant \limsup_{\beta_\alpha\to 1} (1-\beta_\alpha) \mathfrak{R}_x^{\delta}(\beta_\alpha) \tag{23}$$

(see [5], p. 183).

Now it is entirely easy to show that the homogeneous Markov control δ^* is optimal in the problem of minimizing the functional (5). Indeed, let δ be any control of the class Δ'', and let β_α be the sequence of numbers chosen above, tending to unity. Then from formula (23), from the fact that $\delta_{\beta_\alpha}^* = \delta^*$, and from formula (22), it follows that

$$R_x^\delta = \lim_{T\to\infty} \sup \frac{1}{T} \sum_{k=0}^{T} M_x^\delta W(x_k, d_k) \geqslant \lim_{\beta_\alpha\to 1} \sup (1-\beta_\alpha) \mathfrak{R}_x^\delta(\beta_\alpha)$$

$$\geqslant \lim_{\beta_\alpha\to 1} \sup (1-\beta_\alpha) \mathfrak{R}_x^{\delta^*} = \lim_{T\to\infty} \frac{1}{T} \sum_{k=0}^{T} M_x^{\delta^*} W(x_k, d_k^*) = R_x^{\delta^*},$$

i.e.,

$$R_x^\delta \geqslant R_x^{\delta^*}$$

for any $x \subset \mathfrak{X}$ and any control $\delta \in \Delta''$. Consequently, the control δ^* is optimal, and also homogeneous and Markov, in the problem of minimizing the functional (5). Theorem 1 is proved.

Remark 2. From the very beginning we have considered averaged functionals (5), because $S^\delta = \lim_{T\to\infty} \sup \Sigma_{k=0}^{T} M^\delta W(x_k, d_k)$ for all controls δ is, as a rule, infinite. A different formulation of the problem of optimal control in the case of infinite S^δ was studied by D. Blackwell in [6].

Remark 3. The monograph [7] treats problems of finding optimal controls; the author considers only homogeneous Markov controls, not justifying their optimal nature. Our Theorem 1 gives conditions under which, in actually finding optimal controls, it is sufficient to consider only homogeneous Markov controls.

Remark 4. Everywhere above we restricted consideration to nonrandomized controls only. Using Bellman's equation and standard techniques, it is easy to show that, since the space D is finite, it is not necessary to consider randomized controls, since the extremum that is of interest to us is attained in the class of nonrandomized controls.

BIBLIOGRAPHY

[1] J. L. Doob, *Stochastic processes*, Wiley, New York, 1953; Russian transl., IL, Moscow, 1956. MR **15**, 445; MR **19**, 71.

[2] A. N. Širjaev, *On the theory of resolving functions and control of processes of observation with incomplete data*, Transactions of the Third Prague Conference on Information Theory, Statistical Decision Functions, Random Processes, 1962, Publ. House Czechoslovak Acad. Sci., Prague, 1964. MR **28** #3910.

[3] L. Loomis, *Introduction to abstract harmonic analysis*, Van Nostrand, New York, 1953; Russian transl., IL, Moscow, 1956. MR **14**, 352; MR **18**, 53.

[4] I. V. Girsanov, *Some minimal problems in the theory of controlled Markov processes*, Proceedings of the Fourth All-Union Mathematical Congress, 1961.

[5] D. V. Widder, *The Laplace transform*, Princeton Math. Series, Vol. 6, Princeton Univ. Press, Princeton, N. J., 1941. MR **3**, 232.

[6] D. Blackwell, *Discrete dynamic programming*, Ann. Math. Statist. 33 (1962), 719–726. MR 26 #7449.

[7] R. Howard, *Dynamic programming and Markov Processes*, Technology Press of M. I. T., Cambridge, Mass. and Wiley, New York, 1960. MR 22 #9287.

Translated by:
L. Rosenblatt

ON REPRESENTATION OF STABLE LAWS BY INTEGRALS

V. M. ZOLOTAREV

The class of stable distribution laws frequently encountered in various limit theorems in probability theory is given, as is well known, by its corresponding family of characteristic functions $f(t)$. To begin with, we shall use the following form for writing this family [1]:

$$\log f(t) = \begin{cases} it\gamma - \lambda |t|^{\alpha} \exp\left\{-i \frac{\pi}{2} \beta K(\alpha) \operatorname{sgn} t\right\}, & \text{if } \alpha \neq 1, \\ it\gamma - \lambda |t| \left\{\frac{\pi}{2} + i\beta \operatorname{sgn} t \log |t|\right\}, & \text{if } \alpha = 1, \end{cases} \tag{1}$$

where $K(\alpha) = 1 - |1 - \alpha|$, $0 < \alpha \leq 2$, $-1 \leq \beta \leq 1$, $\lambda > 0$ and λ is any real number.

Having posed the problem of representation for stable laws, we can, without loss of generality, let $\gamma = 0$, $\lambda = 1$, since the general case reduces to this one by a linear transformation of the argument in the distribution function. We let $G(x, \alpha, \beta)$ and $g(x, \alpha, \beta)$ denote, respectively, the distribution function and the density of the stable law with parameters α, β, $\gamma = 0$, $\lambda = 1$. We note that in view of a well-known property [1]:

$$1 - G(-x, \alpha, \beta) = G(x, \alpha, -\beta),$$

where again without loss of generality we can assume $x \geq 0$ or $\beta \geq 0$. Actually, assuming that one of the variables x, β is nonnegative, we must vary the second one in the entire region of admissable values.

The aim of the present note is the proof of the following representation of the distribution function of a stable law.

Theorem. 1) *If* $\alpha \neq 1$ *and* $x > 0$, *then*

$$G(x, \alpha, \beta) = \begin{cases} \frac{1}{2}(1 - \beta) + \frac{1}{\pi} \int\limits_{-\frac{\pi}{2}\beta}^{\frac{\pi}{2}} \exp\{-V_{\alpha}(x, \varphi)\}\, d\varphi, & \text{if } \alpha < 1, \\ 1 - \frac{1}{\pi} \int\limits_{-\frac{\pi}{2}\beta \frac{K(\alpha)}{\alpha}}^{\frac{\pi}{2}} \exp\{-V_{\alpha}(x, \varphi)\}\, d\varphi, & \text{if } \alpha > 1, \end{cases}$$

where

$$V_{\alpha}(x, \varphi) = x^{\frac{\alpha}{\alpha - 1}} \left(\frac{\sin\left(\alpha\varphi + \frac{\pi}{2}\beta K(\alpha)\right)}{\cos\varphi} \right)^{\frac{\alpha}{1-\alpha}} \frac{\cos\left[(\alpha - 1)\varphi + \frac{\pi}{2}\beta K(\alpha)\right]}{\cos\varphi}.$$

Also, for $\alpha \neq 1$ *and* $x = 0$ *we have*

$$G(0, \alpha, \beta) = \frac{1}{2}\left(1 - \beta\frac{K(\alpha)}{\alpha}\right).$$

2) *If* $\alpha = 1$ *and* $\beta > 0$, *then*

$$G(x, 1, \beta) = \frac{1}{\pi}\int_{-\frac{\pi}{2}}^{\frac{\pi}{2}} \exp\{-V_1(x, \varphi)\}\, d\varphi,$$

where

$$V_1(x, \varphi) = \frac{\frac{\pi}{2} + \beta\varphi}{\cos\varphi}\exp\left\{-\frac{x}{\beta} + \left(\varphi + \frac{\pi}{2\beta}\right)\operatorname{tg}\varphi\right\}.$$

The case $\alpha = 1$, $\beta = 0$ is well known (Cauchy law):

$$G(x, 1, 0) = \frac{1}{2} + \frac{1}{\pi}\operatorname{arc\,tg} x.$$

Proof. From the inversion formula we have

$$g(x, \alpha, \beta) = \frac{1}{\pi}\operatorname{Re}\int_0^\infty e^{-itx} f(t)\, dt. \tag{2}$$

From the form (1) of the function $f(t)$ it is easily seen that it admits an analytic extension from the positive half-axis into the complex plane with a cut along the negative half-axis. This extension is denoted by $f^+(z, \alpha, \beta)$. It is easy to verify that

$$\log f^+(z, \alpha, \beta) = \begin{cases} -z^\alpha \exp\left\{-i\frac{\pi}{2}\beta K(\alpha)\right\}, & \text{if } \alpha \neq 1, \\ -z\left(\frac{\pi}{2} + i\beta\log z\right), & \text{if } \alpha = 1. \end{cases}$$

Consequently, every representation having as its basis the inversion formula (2) is related to an integral of the form

$$\int_L \exp(-ixz) f^+(z, \alpha, \beta)\, dz, \tag{3}$$

where the contour of integration begins at zero and ends at infinity. In order to make it possible to replace one contour by another on the basis of the Cauchy theorem, we need the following lemma.

Lemma. *Let* ϵ *be an arbitrarily small positive number. Then as* $R \to \infty$

$$Q(C_R) = \int_{C_R} e^{-ixz} f^+(z, \alpha, \beta)\, dz \to 0,$$

where C_R *denotes a sequence of contours of the following form:*

1. *For* $\alpha < 1$, $x < 0$ *and any* β, *or* $\alpha = 1$, $\beta < 0$ *and any* x

$$C_R = \{z : |z| = R,\ 0 \leqslant \arg z \leqslant \pi - \varepsilon\}.$$

2. *For* $\alpha < 1$, $x < 0$, $\beta > 2 - 1/\alpha$

$$C_R = \{z : |z| = R,\ 0 \leqslant \arg z \leqslant \pi\}.$$

3. *For* $\alpha > 1$, $x < 0$ *and arbitrary* β

$$\check{C}_R = \{z : |z| = R,\ -\theta_1 + \varepsilon \leqslant \arg z \leqslant \theta_2\},$$

where

$$\theta_1 = \frac{\pi}{2\alpha}\left(1 - \beta K(\alpha)\right),\ \theta_2 = \frac{\pi}{2\alpha}(1 + \beta K(\alpha)).$$

The assertion of the lemma remains valid if instead of the sequence of contours $\{C_R\}$ *we take any sequence of their parts.*

Proof of the lemma. The following estimate is obvious:

$$|Q(C_R)| \leqslant \pi R \sup_{z \in C_R} |e^{-ixz} f^+(z, \alpha, \beta)|. \tag{4}$$

In the case $\alpha < 1$, $x < 0$ and arbitrary β we take $0 < \epsilon \leq \theta = \pi(1 - \alpha)/2$. Then the right part of (4) can be written and estimated as follows:

$$\pi R \sup_{0 \leqslant \varphi \leqslant \pi - \varepsilon} \exp\left\{xR \sin\varphi - R^\alpha \cos\alpha\left(\varphi - \frac{\pi}{2}\beta\right)\right\}$$
$$\leqslant \pi R \sup_{0 \leqslant \varphi \leqslant \theta} \exp\{\dots\} + \pi R \sup_{\theta \leqslant \varphi \leqslant \pi - \varepsilon} \exp\{\dots\}$$
$$< \pi R \exp\left\{-R^\alpha \cos\frac{\pi}{2}\alpha(2 - \alpha)\right\} + \pi R \exp\{\varepsilon x R + R^\alpha\} \to 0,$$

as $R \to \infty$.

In the case $\alpha = 1$, $\beta < 0$ and arbitrary x we take $l = \pi(1 + |x|)^{-1}/4$ and $0 < \epsilon \leq l$.

We have

$$\pi R \sup_{0 \leqslant \varphi \leqslant \pi - \varepsilon} \exp\left\{xR \sin\varphi - \frac{\pi}{2} R \cos\varphi + \beta R \log R + \beta R \varphi \cos\varphi\right\}$$
$$\leqslant \pi R \sup_{0 \leqslant \varphi \leqslant l} \exp\{\dots\} + \pi R \sup_{l \leqslant \varphi \leqslant \pi - \varepsilon} \exp\{\dots\} < \pi R \exp\left(-\frac{\pi}{4} R\right)$$
$$+ \pi R \exp\{\varepsilon \beta R \log R + 2\pi R\},$$

as $R \to \infty$.

The remaining cases are analysed analogously.

Now we turn to the proof of the theorem. As in the lemma, we assume

$v=-x>0$ in the case $\alpha\neq 1$ and $\beta<0$ in the case $\alpha=1$. We let Γ denote a contour of the following form

$$\Gamma=\left\{z\colon \operatorname{Im}(ivz+\log f^{+}(z,\alpha,\beta))=0,\ \frac{\pi}{2}\mu<\arg z<\frac{\pi}{2}\right\},$$

where $\mu=\beta K(\alpha)/\alpha$ if $\alpha\neq 1$, and $\mu=-1$ if $\alpha=1$. In the case $\alpha<1$ and $\beta=1$ the contour Γ coincides with the positive imaginary axis. In the remaining cases Γ is a curve whose equation in polar coordinates $(z=\rho e^{i\phi})$ is

$$\rho=\rho(\varphi)=\begin{cases}\left[\dfrac{\sin\left(\alpha\varphi-\frac{\pi}{2}\beta K(\alpha)\right)}{v\cos\varphi}\right]^{\frac{1}{1-\alpha}}, & \text{if } \alpha\neq 1,\\ \exp\left\{\frac{v}{\beta}+\left(\varphi-\frac{\pi}{2\beta}\right)\operatorname{tg}\varphi\right\}, & \text{if } \alpha=1.\end{cases}\tag{5}$$

Here $v=-x$ and $\pi\mu/2<\phi<\pi/2$.

Let

$$j(\alpha)=\begin{cases}1, & \text{if } \alpha\leqslant 1\\ -1, & \text{if } \alpha>1,\end{cases}\qquad \tau=\begin{cases}\left(\frac{\alpha}{v}\right)^{\frac{1}{1-\alpha}}, & \text{if } \alpha\neq 1,\\ \exp(v-1), & \text{if } \alpha=1.\end{cases}$$

One end of the contour Γ, as is easily seen, goes to infinity, having as an asymptote the ray $\phi=\pi/2$ in the case $\alpha\leq 1$, and $\phi=\pi\mu/2$ in the case $\alpha>1$. The second end of the contour reaches zero in the case $\beta\neq -j(\alpha)$ and the point $-i\tau j(\alpha)$ in the case $\beta=-j(\alpha)$. We extend Γ so that its second end is always at zero, letting

$$\Gamma^{*}=\begin{cases}\Gamma, & \text{if } \beta\neq -j(\alpha),\\ \Gamma\cup\{z\colon \operatorname{Re} z=0,\ 0\leqslant -j(\alpha)\arg z\leqslant\tau\}, & \text{if } \beta=-j(\alpha).\end{cases}$$

The lemma shows that in (3) we can replace integration along the positive half-axis by integration along the contour Γ^{*}, here keeping the value of the integral unchanged. We further note that on the contour Γ

$$\operatorname{Re}\{ivz+\log f^{+}(z,\alpha,\beta)\}=-V_{\alpha}(v,\varphi)$$

(the function $V_{\alpha}(v,\phi)$ was defined in the statement of the theorem). Also, the value of the integral (3) on the part extending Γ to Γ^{*} is pure imaginary. Hence we obtain that

$$g(-v,\alpha,\beta)=\frac{1}{\pi}j(\alpha)\int_{\frac{\pi}{2}\mu}^{\frac{\pi}{2}}\exp\{-V_{\alpha}(v,\varphi)\}\,d(\rho\cos\varphi).\tag{6}$$

Starting with the explicit expressions $\rho(\phi)$ and $V_{\alpha}(v,\phi)$, it is easy to verify the validity of the equality

$$d(\rho \cos \varphi) = -\frac{\partial}{\partial v} V_\alpha(v, \varphi)\, d\varphi,$$

which makes it possible to put relation (6) into the form

$$g(-v, \alpha, \beta) = \frac{1}{\pi} j(\alpha) \int_{\frac{\pi}{2}\mu}^{\frac{\pi}{2}} \frac{\partial}{\partial v} \exp\{-V_\alpha(v, \varphi)\}\, d\varphi. \tag{7}$$

Since as $v \longrightarrow \infty$ the function $\exp\{-V_\alpha(v, \phi)\}$ goes to $(1 + j(\alpha))/2$, then, integrating both parts of (7) from ∞ to v, we obtain

$$G(-v, \alpha, \beta) = \frac{1}{4}(1 + j(\alpha))(1 - \mu) - \frac{1}{\pi} j(\alpha) \int_{\frac{\pi}{2}\mu}^{\frac{\pi}{2}} \exp\{-V_\alpha(v, \varphi)\}\, d\varphi.$$

The values $G(0, \alpha, \beta)$ in the case $\alpha \neq 1$ are obtained from this formula by passage to the limit as $v \longrightarrow 0$.

We make several concluding remarks.

I. The idea of such a representation of the density of a stable law was partially used by I. A. Ibragimov and K. E. Černin for proof of unimodality of these distributions [2].

II. As an obvious consequence of the representation thus obtained, we find that the positive half-axis in the case $\alpha < 1$, $\beta = 1$, the negative in the case $\alpha < 1$, $\beta = -1$ and the entire x axis in the remaining cases serve as the carrier of the measure of the stable laws $G(x, \alpha, \beta)$.

III. For the well-known normal law with parameters $(0, 1)$ we obtain a representation by the finite integral

$$1 - \Phi(x) = \frac{1}{\pi} \int_0^{\frac{\pi}{2}} \exp\left\{-\frac{x^2}{2\sin^2\varphi}\right\} d\varphi.$$

BIBLIOGRAPHY

[1] V. M. Zolotarev, *Mellin-Stieltjes transformations in probability theory,* Teor. Veroyatnost. i Primenen. 2 (1957), 444–469. MR 21 #7555.

[2] I. A. Ibragimov and K. E. Černin, *On the unimodality of stable laws,* Teor. Veroyatnost. i Primenen 4 (1959), 453–456. MR 22 #7170.

Translated by:
L. Rosenblatt

DISTRIBUTION OF QUEUE LENGTH AND NUMBER OF OPERATING LINES IN A SYSTEM OF ERLANG TYPE WITH RANDOM BREAKAGE AND RESTORATION OF LINES

V. M. ZOLOTAREV

§1. Statement of the problem and notation

A randomly incoming flow of calls is serviced by m lines. The lines may leave the structure in a random fashion and be restored. Calls which cannot be serviced at the moment of arrival because the lines are busy or broken off get into a queue and wait for the line to become restored or free.

We are interested in the joint distributions and related characteristics of the following two random variables:

$\nu(t)$: the number of lines in operation at time t,

$\kappa(t)$: the number of calls occupying the system at time t (i.e., the number being serviced plus the number waiting in queue).

We impose the following conditions on the random process described above:

1. Time t varies continuously ($t \geq 0$).

2. Calls not serviced because lines are broken queue up on a universal basis, and the calls in queue are not distinguished.

3. The process is assumed homogeneous in time.

4. In the time interval dt arrival of more than one call and change of state of more than one line (i.e., arrival of a call, end of service, breakage, replacement) occurs with probability $o(dt)$.

5. In the time interval dt a call arrives with probability $\lambda dt + o(dt)$, $\lambda > 0$.

6. Lines leave the structure and are restored independently of each other, and in the interval dt:

a) service of a call is completed with probability

$$\mu dt + o(dt), \ \mu > 0,$$

b) a line leaves the structure with probability $\theta_1 dt + o(dt)$, $\theta_1 \geq 0$,

c) a nonfunctioning line is restored with probability

$$\theta_2 dt + o(dt), \ \theta_2 \geq 0.$$

7. At the initial moment of time there are n_0 lines in operation, and k_0 calls occupy the system.

Below we shall use the following notation:

$$P(t, n, k) = \mathbf{P}\{\nu(t) = n, \varkappa(t) = k\}$$

(since the system cannot be found in the states $n < 0$, $k < 0$ and $n > m$, for these values it is natural to let $P(t, n, k) = 0$),

$$\pi(s, n, k) = \int_0^\infty e^{-st} P(t, n, k)\, dt, \quad \mathrm{Re}\, s > 0.$$

For the probabilities in question we introduce the semi-generating and the genera-ing functions

$$\Psi_n(t, v) = \sum_{k=0}^{\infty} P(t, n, k) v^k, \quad \Phi(t, u, v) = \sum_{n=0}^{m} \Psi_n(t, v) u^n$$

and their corresponding Laplace transforms:

$$\psi_n(s, v), \quad \varphi(s, u, v), \quad \mathrm{Re}\, s > 0, \quad |v| \leqslant 1.$$

For the stationary state of the system we need the same notations P, Ψ_n, Φ in which, however, the variable t is omitted.

The condition of normalization of probabilities gives us

$$\begin{gathered} \sum_{n,k} P(t, n, k) = \sum_n \Psi_n(t, 1) = \Phi(t, 1, 1) = 1, \\ \sum_{n,k} \pi(s, n, k) = \sum \psi_n(s, 1) = \varphi(s, 1, 1) = \frac{1}{s}. \end{gathered} \tag{1}$$

The most interesting numerical characteristic is the mean value $(\mathbf{M}\nu(t), \mathbf{M}\kappa(t))$ of the vector $(\nu(t), \kappa(t))$. If the computation of $\mathbf{M}\nu(t)$ is not difficult because the number of functioning lines does not depend on the flow of calls and has the simple form

$$\mathbf{M}\nu(t) = \frac{m\theta_2}{\theta_1 + \theta_2} + \frac{n_0\theta_1 + (m - n_0)\theta_2}{\theta_1 + \theta_2} e^{-(\theta_1+\theta_2)t}, \tag{2}$$

then the computation of $\mathbf{M}\kappa(t)$ presents great difficulty.

The mean value $\mathbf{M}\kappa(t)$ does not have a simple expression and can be written only in terms of roots of auxiliary equations and inverse Laplace transforms (the stationary case does not require inverse Laplace transforms, but nevertheless uses roots of complicated algebraic functions), so that seeking precise expressions for $\mathbf{M}\kappa(t)$ both in the general and in the stationary case is hardly justified. We note that the mean value $\mathbf{M}\kappa(t)$ is related to the functions Ψ_n and Φ by the relations

$$\mathbf{M}\varkappa(t) = \sum_n \left.\frac{\partial \Psi_n}{\partial v}\right|_{v=1} = \left.\frac{\partial \Phi}{\partial v}\right|_{u=v=1}. \tag{3}$$

For purposes of exposition we also pose the problem of asymptotic representation of $\mathbf{M}\kappa(t)$ under various assumptions about the behavior of the parameters.

Without loss of generality we can assume that $\mu = 1$, and for each of the parameters λ, θ_1, θ_2 make one of the assumptions "$o(1)$", "$\asymp 1$" or "$\to \infty$". In the stationary case, of the 27 possible cases, 12 are rejected at once as incompatible with the condition of stationarity of the process

$$\mathbf{M}\nu = \frac{m\theta_2}{\theta_1 + \theta_2} > \frac{\lambda}{\mu}, \tag{4}$$

denoting that the mean service time of one call by the system, equal to $(\mu \mathbf{M} \nu)^{-1}$, must be less than the mean time interval λ^{-1} between two successive calls.

Below we concentrate our attention chiefly on the stationary variant of the problem. But even in this case we restrict ourselves to only one of the 15 possible asymptotic formulations of the problem. Namely,

$$\lambda = \text{const}, \quad \mu = \text{const}, \quad \theta_1 = \text{const}, \quad \theta_2 \to \infty.$$

But it will be evident from the argument that interferes with analogous investigation of the greater part of the remaining cases.

§2. Basic equations

By the argument commonly used in mass service systems, based on use of the formula of total probability, it is easy to obtain the following equations

$$\begin{aligned} &\frac{\partial}{\partial t} P(t, n, k) + [n\theta_1 + (m - n)\theta_2 + \lambda + \mu \min(n, k)] P(t, n, k) \\ &\quad = \mu \min(n, k+1) P(t, n, k+1) + (n+1)\theta_1 P(t, n+1, k) \\ &\qquad + (m - n + 1)\theta_2 P(t, n-1, k) + \lambda P(t, n, k-1). \end{aligned} \tag{5}$$

With the normalizing condition (1), uniqueness of the solution of this system of equations is guaranteed by the following initial and boundary conditions

$$\begin{aligned} &P(0, n, k) = \delta(n - n_0, k - k_0), \quad (\delta \text{ is the Kronecker symbol}) \\ &P(t, n, k) = 0, \quad \text{if} \quad k < 0, \quad \text{or} \quad n < 0, \quad \text{or} \quad n > m. \end{aligned} \tag{6}$$

From system (5) it is quite simple to obtain a system of equations for the functions Ψ_n and an equation for Φ:

$$\begin{aligned} &\frac{\partial}{\partial t} \Psi_n(t, v) + \left[n\theta_1 + (m - n)\theta_2 + \lambda(1 - v) + n\mu\left(1 - \frac{1}{v}\right)\right] \Psi_n(t, v) \\ &= (n+1)\theta_1 \Psi_{n+1}(t, v) + (m - n + 1)\theta_2 \Psi_{n-1}(t, v) + \mu\left(1 - \frac{1}{v}\right) R_n(t, v). \end{aligned} \tag{7}$$

Here $R_n(t, v) = \sum_{k=0}^{n-1} (n - k) P(t, n, k) v^k$ and the system is to be solved subject to the conditions

$$\Psi_n(0, v) = v^{k_0} \delta(n - n_0, 0); \quad \Psi_n = 0, \quad \text{if} \quad n < 0 \quad \text{or} \quad n > m.$$

$$\begin{aligned} &\frac{\partial}{\partial t} \Phi(t, u, v) + [m\theta_2(1 - u) + \lambda(1 - v)] \Phi(t, u, v) = \Big[\theta_2 u(1 - u) \\ &+ \theta_1(1 - u) + \mu u\left(\frac{1}{v} - 1\right)\Big] \frac{\partial}{\partial u} \Phi(t, u, v) + \mu\left(1 - \frac{1}{v}\right) R(t, u, v) \end{aligned} \tag{9}$$

(of course, together with (1)). Here $R(t, u, v) = \sum_{n=0}^{m} R_n(t, v) u^n$, and the initial and boundary conditions are

$$\Phi(0, u, v) = u^{n_0} v^{k_0}, \quad \Phi(t, 1, 1) = 1. \tag{10}$$

Analogous equations are written for the functions $\pi(s, n, k)$, $\psi_n(s, v)$ and $\phi(s, u, v)$ (they are derived from the obtained transforms of equations (5), (7) and (9)):

$$\begin{aligned} &[s + n\theta_1 + (m-n)\theta_2 + \lambda + \mu \min(n, k)]\,\pi(s, n, k) \\ &= \mu \min(n, k+1)\,\pi(s, n, k+1) + (n+1)\theta_1 \pi(s, n+1, k) \\ &+ (m-n+1)\theta_2 \pi(s, n-1, k) + \lambda\pi(s, n, k-1) + \delta(n-n_0, k-k_0). \end{aligned} \tag{11}$$

The functions π satisfy, besides (1), the conditions

$$\pi(s, n, k) = 0, \text{ if } k < 0, \text{ or } n < 0, \text{ or } n > m. \tag{12}$$

For the functions ψ_n we obtain the system of equations

$$q_{n-1}^{(0)} \psi_{n-1}(s, v) + q_n^{(1)} \psi_n(s, v) + q_{n+1}^{(2)} \psi_{n+1}(s, v) = l_n(s, v), \tag{13}$$

where

$$q_n^{(0)} = (m-n)\theta_2, \quad q_n^{(1)} = -\left[s + n\theta_1 + (m-n)\theta_2 + \lambda(1-v) + n\mu\left(1 - \frac{1}{v}\right)\right],$$

$$q_n^{(2)} = n\theta_1, \quad l_n(s, v) = \mu\left(\frac{1}{v} - 1\right) r_n(s, v) - v^{k_0}\delta(n - n_0, 0),$$

$r_n(s, v)$ is the Laplace transform of the function $R_n(t, v)$. Besides (1) the functions ψ_n satisfy the conditions

$$\psi_n(s, v) = 0, \quad \text{if} \quad n < 0 \quad \text{or} \quad n > m. \tag{14}$$

The function ϕ satisfies the equation

$$\left[\theta_1(1-u) + \theta_2 u(1-u) + \mu u\left(\frac{1}{v} - 1\right)\right] \frac{\partial}{\partial u} \varphi(s, u, v)$$

$$= [s + m\theta_2(1-u) + \lambda(1-v)]\,\varphi(s, u, v) + \mu\left(\frac{1}{v} - 1\right) r(s, u, v) - u^{n_0} v^{k_0} \tag{15}$$

and the additional conditions

$$s\varphi(s, 1, 1) = 1, \quad \lim_{s \to 0} s\varphi(s, u, v) = u^{n_0} v^{k_0}. \tag{16}$$

In (15) the function $r(s, u, v)$ is the Laplace transform of the function $R(t, u, v)$.

In the equations written for Ψ_n, Φ, ψ_n and ϕ there are unknown functions $P(t, n, k)$, $0 \leq k \leq n - 1 \leq m - 1$ or their Laplace transforms. There are $m(m+1)/2$ such functions, and therefore for determining them we need the same number of independent relations between these functions. Equations (5) and (11) furnish us with $m(m-1)/2$ of the needed relations. One more is obtained from the normalizing condition (1). The remaining $m - 1$ relations are obtained from the regularity properties of the functions depending on v, in the circle $|v| < 1$ (this will be discussed in greater detail in §4).

We conclude this section with a remark on the role played by the function $R(t, u, v)$ in finding $M\kappa(t)$. Since $M\nu(t) = \partial\Phi/\partial u|_{u=v=1}$ and $M\kappa(t) = \partial\Phi/\partial v|_{u=v=1}$, equation (9) immediately makes it possible to establish the dependence between these means and the function $R(t, 1, 1)$:

$$\frac{d}{dt}\mathbf{M}\varkappa(t)+\mu\mathbf{M}\nu(t)-\lambda=\mu R(t, 1, 1). \tag{17}$$

Equality (17) shows that finding the function $R(t, 1, 1)$ is equivalent to finding the mean $M\kappa(t)$. In the stationary case (17) takes the form

$$R(1, 1)=\sum_{n=0}^{m}\sum_{k=0}^{n-1}(n-k)P(n, k)=\frac{m\theta_2}{\theta_1+\theta_2}-\frac{\lambda}{\mu}. \tag{18}$$

In the problem of finding the unknown functions P which generate the functions R_n and R, this relation turns out to be equivalent to the condition of normalization (1) but is considerably simpler, since it contains only reformable functions $P(n, k)$ with $k \leq n-1 \leq m-1$.

§3. The case of two lines working in a stationary state

System (7) (taking stationarity into account) can be written in the form of three equations

$$\begin{gathered}[\lambda v-(2\theta_2+\lambda)]\Psi_0+\theta_1\Psi_1=0,\\ 2\theta_2 v\Psi_0+[(\lambda v-\mu)(v-1)-(\theta_1+\theta_2)v]\Psi_1+2\theta_1 v\Psi_2=\mu(1-v)R_1,\\ \theta_2 v\Psi_1+[(\lambda v-2\mu)(v-1)-2\theta_1 v]\Psi_2=\mu(1-v)R_2,\end{gathered}$$

where $R_1 = P(1, 0) = p_1$ and $R_2 = 2P(2, 0) + P(2, 1)v = 2p_2 + p_3 v$. After a few transformations the determinant of this system is easily represented in the form $(v-1)D(v)$, where $D(v) = \sum_{k=0}^{4}(-1)^k a_k v^k$, and the coefficients are equal to

$$\begin{gathered}a_4=\lambda^3,\\ a_3=\lambda^2[2\lambda+3\mu+3(\theta_1+\theta_2)],\\ a_2=\lambda[2(\theta_1+\theta_2)^2+(3\lambda+4\mu)(\theta_1+\theta_2)+4\mu\theta_2+6\lambda\mu+2\mu^2+\lambda^2],\\ a_1=\mu[4\lambda\mu+3\lambda^2+4\mu\theta_2+4\lambda\theta_1+4\theta_1\theta_2+4\theta_2^2+8\lambda\theta_2],\\ a_0=2\mu^2(2\theta_2+\lambda).\end{gathered}$$

Since, as is easy to compute,

$$D(0)=a_0>0, \quad D(1)=-2\mu(\theta_1+\theta_2)^2\omega^2<0, \tag{19}$$

where $\omega^2 = 2\theta_2/(\theta_1+\theta_2) - \lambda/\mu$, the function $D(v)$ has at least one root α in the interval $(0, 1)$. (Such a root is unique, but we do not need this below.)

The functions $\Psi_n(v)$ can be written explicitly as the solution of a system of linear equations of the yet undetermined functions R_1 and R_2:

$$\Psi_n(v)=\frac{M_n(v)}{D(v)}, \quad n=0, 1, 2, \tag{20}$$

where

$$M_0(v) = \mu\theta_1 \begin{vmatrix} R_1(v) & 2\theta_1 v \\ R_2(v) & [(\lambda v - 2\mu)(v-1) - 2\theta_1 v] \end{vmatrix},$$

$$M_1(v) = \frac{1}{\theta_1}[(2\theta_2 + \lambda) - \lambda v]\, M_0(v),$$

$$M_2(v) = \mu \begin{vmatrix} (2\theta_2 + \lambda) - \lambda v & \theta_1 & 0 \\ -2\theta_2 v & [(\mu - \lambda v)(1 - v) - (\theta_1 + \theta_2) v] & R_1(v) \\ 0 & \theta_2 v & R_2(v) \end{vmatrix},$$

$$\Psi_0(1) = \frac{\theta_1^2 G}{(\theta_1 + \theta_2)^2}, \quad \Psi_1(1) = \frac{2\theta_1\theta_2 G}{(\theta_1 + \theta_2)^2}, \quad \Psi_2(1) = \frac{\theta_2^2 G}{(\theta_1 + \theta_2)^2}, \tag{21}$$

where

$$G = \frac{1}{\omega^2}[R_1(1) + R_2(1)], \quad \omega^2 = \frac{2\theta_2}{\theta_1 + \theta_2} - \frac{\lambda}{\mu}.$$

Hence it can be seen that the normalizing condition (1) is equivalent to relation (18) (i.e., $G = 1$).

For finding the unknown probabilities p_1, p_2, and p_3, we can now write the following relations. From the recursion relations (5):

$$\theta_2 p_1 - (2\theta_1 + \lambda) p_2 + \mu p_3 = 0.$$

According to (18):

$$p_1 + 2p_2 + p_3 = \omega^2.$$

As we have seen, $D(v)$ has a root α in the interval (0, 1). Also the function $\Psi_0(v)$ is regular in the circle $|v| < 1$. Consequently α is a zero of the determinant in the numerator of (20) containing the functions $R_n(v)$. This gives us

$$[(\alpha\lambda - 2\mu)(1 - \alpha) + 2\alpha\theta_1] p_1 + 4\alpha\theta_1 p_2 + 2\alpha^2\theta_1 p_3 = 0.$$

Once again we have a system of linear equations. The determinant of this system is equal to

$$K = (1 - \alpha)[(2\mu - \alpha\lambda)(2\theta_1 + 2\mu + \lambda) - 2\alpha\theta_1(2\theta_1 + 2\theta_2 + \lambda)], \tag{22}$$

and our desired quantities p_k are respectively equal to

$$p_1 = 2\alpha\theta_1[(2\theta_1 + \lambda)\alpha + 2\mu]\frac{\omega^2}{K},$$

$$\mu(2\mu - \alpha\lambda)(1 - \alpha) + 2\alpha^2\theta_1\theta_2 - 2\alpha\mu\theta_1]\frac{\omega^2}{K}, \tag{23}$$

$$p_3 = [(2\theta_1 + \lambda)(2\mu - \alpha\lambda)(1 - \alpha) - 2\alpha\theta_1(2\theta_1 + \lambda) - 4\alpha\theta_1\theta_2]\frac{\omega^2}{K}.$$

This properly completes the solution of the problem of finding the distribution of the vector (ν, κ) in the stationary case, since the expressions (20) for the functions $\Psi_n(v)$ contain only quantities known to us.

In computing the mean value $\mathbf{M}\kappa$ we start with the transformed representation

$$\mathbf{M}\kappa = \sum_n \Psi'_n(1) = \frac{1}{D(1)}\left\{\sum_n M'_n(1) - D'(1)\right\}. \tag{24}$$

In the case of two lines

$$\sum_n M'_n(1) = -\mu\{[2(\theta_1+\theta_2)^2 - 3\lambda(\theta_1+\theta_2) + 2\mu\theta_2]\omega^2 + 2\mu(\theta_1+\theta_2)p_1 + 2(\theta_1+\theta_2)^2 p_2\}.$$

Also $D(1) = -2\mu(\theta_1+\theta_2)^2\omega^2$ and

$$D'(1) = 4\lambda(\theta_1+\theta_2)^2 + \lambda(4\mu - 3\lambda)(\theta_1+\theta_2) - 4\mu\theta_2(\theta_1+\theta_2+\mu-\lambda).$$

Consequently it is possible to express the mean value $\mathbf{M}\kappa$ in terms of the parameters and the root α.

§ 4. General algorithm for solution of the problem

The case of two lines operating in a stationary state, treated in the preceding section, makes it possible to talk about the qualitative behavior of the process in the general case and by analogy construct an algorithm leading to a precise (though not necessarily explicit) solution of the general problem.

Indeed, systems (7) and the equivalent (13) yield the solution of our problem if by some means or other we are able to determine the functions $R_n(t, v)$ (or equivalent functions $r_n(s, v)$). But this is equivalent to finding the probabilities $P(t, n, k)$, $0 \leq k \leq n-1 \leq m-1$ determining them, or the equivalent functions $\pi(s, n, k)$. There are always $m(m+1)/2$ such functions; therefore to find them we need as many independent equations.

The recursion relations (5) (or (11)), if n and k vary in the indicated limits, gives us $m(m-1)/2$ such equations. One more relation is furnished by the normalizing condition (1), which in the stationary case can be replaced by the more convenient condition (18). The remaining $m-1$ equations are obtained from the following considerations. It is easy to see that equations (13) form a system of linear algebraic equations (in the stationary case equations (7) have this same property) and therefore the functions $\psi_n(s, v)$ can be written in the form

$$\psi_n(s, v) = \frac{b_n(s, v)}{d(s, v)} = \frac{1}{d(s, v)}\sum_{j=0}^{m} a_{jn}(s, v)\, l_j(s, v),$$

where $d(s, v) = \det\|\gamma_{ij}\|$, $\gamma_{ij} = q_{j-1}^{(i-j+1)}$, if $|i-j| \leq 1$, $j = 1, 2, \cdots, m+1$; $\gamma_{ij} = 0$ otherwise, and the functions $a_{jn}(s, v)$ are algebraic cofactors of the elements γ_{jn}.

It can be shown that in the circle $|v| < 1$ there are precisely $m-1$ roots of the equation

$$d(s, v)=0.$$

Since the functions $\psi_n(s, v)$ are regular in the circle $|v| < 1$, this means that the sum

$$\sum_{j=0}^{m} a_{jn}(s, v)\, l_j(s, v)=0$$

at all $m-1$ points at which the function $d(s, v)$ goes to zero, which provides us with the $m-1$ missing equations. In the stationary case the picture is analogous (we saw this in the example of two lines) except that we have to start out with equation (7) and the functions $\Psi_n(v)$.

In analogy with the expression (24) for $M\kappa(t)$, in the stationary case we have

$$\int_0^\infty e^{-st} M\varkappa(t)\, dt=\frac{1}{d(s, 1)}\left\{\sum_n \frac{\partial b_n}{\partial v}-\frac{\partial d}{\partial v}\right\}_{v=1}. \tag{25}$$

§5. Asymptotic analysis of a system working in a stationary state when $\theta_2 \to \infty$

In the stationary case equations (5) take the form

$$(m-n+1)\theta_2\Psi_{n-1}(v)-\left[n\theta_1+(m-n)\theta_2+\lambda(1-v)+n\mu\left(1-\frac{1}{v}\right)\right]\Psi_n(v)$$
$$+(n+1)\theta_1\Psi_{n+1}(v)=\mu\left(\frac{1}{v}-1\right)R_n(v), \quad 0\leqslant n\leqslant m. \tag{26}$$

It is easy to see that as $\theta_2 \to \infty$ each of the quantities $P(n, k)$, $\Psi_n(v)$, $R_n(v)$ has order $O(\theta_2^{n-m})$. Consequently, being interested in asymptotic representations of these quantities and their related characteristics to within $O(\theta_2^{-2})$, we are justified in considering only those for which $n=m,\ m-1$. We let

$$\begin{aligned}
P(m, k)&=\alpha_k+\frac{\theta_1}{\theta_2}\beta_k+O(\theta_2^{-2}),\\
P(m-1, k)&=\frac{\theta_1}{\theta_2}\gamma_k+O(\theta_2^{-2}),\\
\Psi_m(v)&=A(v)+\frac{\theta_1}{\theta_2}B(v)+O(\theta_2^{-2}),\\
\Psi_{m-1}(v)&=\frac{\theta_1}{\theta_2}C(v)+O(\theta_2^{-2}),\\
R_m(v)&=P(v)+\frac{\theta_1}{\theta_2}Q(v)+O(\theta_2^{-2}),\\
R_{m-1}(v)&=\frac{\theta_1}{\theta_2}S(v)+O(\theta_2^{-2}).
\end{aligned} \tag{27}$$

Simple computations on equations (26), considered for the values $n = m,\ m-1,$ $m-2$, lead us to relations between the functions just introduced, from which we obtain

$$A(v)=\left(m-\frac{\lambda}{\mu}v\right)^{-1}P(v),\quad C(v)=mA(v),$$
$$B(v)=-C(v)+\left(m-\frac{\lambda}{\mu}v\right)^{-1}[Q(v)+S(v)]+\left(m-\frac{\lambda}{\mu}v\right)^{-1}A(v). \tag{28}$$

Condition (18), equivalent to the normalization (1), in turn yields the two relations

$$P(1)=\sum_{k=0}^{m-1}(m-k)\,\alpha_k=m-\frac{\lambda}{\mu},$$
$$Q(1)+S(1)=\sum_{k=0}^{m-1}[(m-k)\,\beta_k+(m-k-1)\,\gamma_k]=-m. \tag{29}$$

To compute the probabilities α_k, β_k, γ_k with values $k \le m-1$, which would completely solve the problem of explicit expression of A, B, and C, we need to set up and effectively solve a system of $3m-1$ equations. We already have two of them. These are equalities (29); the remaining are obtained from the recursion relations between the probabilities (5). The expansion in them of the probabilities $P(n,k)$ in powers of θ_1/θ_2 and comparison of the coefficients leads us, after a few more transformations, to the following equalities ($k = 0, 1, \cdots, m-2$):

$$\gamma_k=m\alpha_k,$$
$$\lambda\alpha_{k-1}-(k\mu+\lambda)\,\alpha_k+(k+1)\,\mu\alpha_{k+1}=0, \tag{30}$$
$$\lambda\beta_{k-1}-(k\mu+\lambda)\,\beta_k+(k+1)\,\mu\beta_{k+1}=0.$$

These equalities form three systems of equations, each of which is easily solved with consideration of conditions (29). As a result, for $k \le m-1$ we have

$$\alpha_k=\frac{1}{k!}\left(\frac{\lambda}{\mu}\right)^k\frac{P(1)}{\sum_{s=0}^{m-1}\frac{(m-s)}{s!}\left(\frac{\lambda}{\mu}\right)^s},\quad P(1)=m-\frac{\lambda}{\mu}, \tag{31}$$
$$\beta_k=\frac{1}{k!}\left(\frac{\lambda}{\mu}\right)^k\frac{Q(1)}{\sum_{s=0}^{m-1}\frac{(m-s)}{s!}\left(\frac{\lambda}{\mu}\right)^s}.$$

Finding the value $Q(1)$ involves finding the quantities α_k for $k \ge m$. Indeed, from (29) and (31) we have

$$Q(1) = -m - S(1) = -m\left[1 + P(1) - \sum_{k=0}^{m-1} \alpha_k\right]$$

$$= -m\left[\left(m - \frac{\lambda}{\mu}\right) + \sum_{k=m}^{\infty} \alpha_k\right].$$

We introduce the functions

$$H_m(z) = \sum_{k=0}^{m-1} (m-k)\frac{z^k}{k!}.$$

Then from (28), (31) we obtain

$$A(v) = \frac{m - \frac{\lambda}{\mu}}{m - \frac{\lambda}{\mu} v} \cdot \frac{H_m\left(v\frac{\lambda}{\mu}\right)}{H_m\left(\frac{\lambda}{\mu}\right)}, \tag{32}$$

whence it is easy to find α_k for $k \geq m$. Appropriate transformations yield

$$\alpha_k = \left(1 - \frac{\lambda}{\mu m}\right)\left(\frac{\lambda}{\mu m}\right)^k \frac{H_m(m)}{H_m\left(\frac{\lambda}{\mu}\right)} \tag{33}$$

and

$$\sum_{k=m}^{\infty} \alpha_k = \left(\frac{\lambda}{\mu m}\right)^m \frac{H_m(m)}{H_m\left(\frac{\lambda}{\mu}\right)}.$$

Consequently

$$Q(1) = -m\left[\left(m - \frac{\lambda}{\mu}\right) + \left(\frac{\lambda}{\mu m}\right)^m \frac{H_m(m)}{H_m\left(\frac{\lambda}{\mu}\right)}\right]. \tag{34}$$

Further, since for $k \leq m-1$

$$\beta_k = \frac{Q(1)}{P(1)} \alpha_k, \tag{35}$$

we have

$$Q(v) + S(v) = \frac{Q(1)}{P(1)} P(v) + mP(v) - m\sum_{k=0}^{m-1} \alpha_k v^k$$

$$= \left[m + \frac{Q(1)}{P(1)}\right] P(v) - mA(v) + m\sum_{k=m}^{\infty} \alpha_k v^k.$$

Hence, by (28),

$$B(v)=\frac{Q(1)}{P(1)}A(v)+\left(1-\frac{\lambda}{\mu m}v\right)^{-1}\sum_{k=m}^{\infty}\alpha_k v^k$$

$$=\frac{Q(1)}{P(1)}A(v)+\left(1-\frac{\lambda}{\mu m}\right)\frac{H_m(m)}{H_m\left(\frac{\lambda}{\mu}\right)}\frac{\left(\frac{\lambda}{\mu m}v\right)^m}{\left(1-\frac{\lambda}{\mu m}v\right)^2}. \qquad (36)$$

Now expanding $B(v)$ in powers of v, we obtain for $k\geq m$:

$$\beta_k=\left(\frac{\lambda}{\mu m}\right)^k\frac{H_m(m)}{H_m\left(\frac{\lambda}{\mu}\right)}\left\{(k-m+1)\left(1-\frac{\lambda}{\mu m}\right)-\left(\frac{\lambda}{\mu m}\right)^m\frac{H_m(m)}{H_m\left(\frac{\lambda}{\mu}\right)}\right\}. \qquad (37)$$

From the relation (29) between the generating functions $A(v)$ and $C(v)$ it follows that for all k

$$\gamma_k=m\alpha_k. \qquad (38)$$

This essentially completes the asymptotic study of the problem we posed.

We give an additional asymptotic representation for $\mathbf{M}\kappa$. Since we already have the asymptotic representation of the generating function $\mathbf{M}v^{\kappa}=\sum_n\Psi_n(v)$, this is simple to accomplish. Indeed,

$$\mathbf{M}v^{\kappa}=A(v)+\frac{\theta_1}{\theta_2}[B(v)+C(v)]+O(\theta_2^{-2}).$$

If we denote

$$\mathbf{M}\kappa=U_0+\frac{\theta_1}{\theta_2}U_1+O(\theta_2^{-2}),$$

then obviously

$$U_0=A'(1),\quad U_1=B'(1)+C'(1).$$

Using (28), (32) and (36), we obtain

$$U_0=\frac{\lambda}{\mu m}\left[\left(1-\frac{\lambda}{\mu m}\right)^{-1}+m\frac{H'_m\left(\frac{\lambda}{\mu}\right)}{H_m\left(\frac{\lambda}{\mu}\right)}\right], \qquad (39)$$

$$U_1=\left(1-\frac{\lambda}{\mu m}\right)^{-2}\left(\frac{\lambda}{\mu m}\right)^m\frac{H_m(m)}{H_m\left(\frac{\lambda}{\mu}\right)}\left\{\left(1-\frac{\lambda}{\mu m}\right)\left(m-\frac{\lambda}{\mu}\frac{H'_m\left(\frac{\lambda}{\mu}\right)}{H_m\left(\frac{\lambda}{\mu}\right)}\right)+\frac{\lambda}{\mu m}\right\}. \qquad (40)$$

In conclusion we note that formulas (39) and (40) can be reduced to a form in which tabulated Erlang functions are involved, which considerably simplifies the computational procedure.

Translated by:
L. Rosenblatt

SOLUTION OF A TYPE OF DIFFUSION EQUATIONS

G. I. KALMYKOV

1. In this article we consider a stationary Markov process $\{\xi_t\}$. We let $p(x)$ denote the density of the stationary distribution, and $f(t, x, y)$ the density of the transition probability. The density of the joint distribution of the random variables ξ_s and ξ_{s+t} is denoted by $p(t, x, y) = p(x)f(t, x, y)$. We assume that the process $\{\xi_t\}$ satisfies the forward and backward equations of A. N. Kolmogorov

$$\frac{\partial f}{\partial t} = \frac{\partial^2}{\partial y^2}[a(y)f(t, x, y)] - \frac{\partial}{\partial y}[b(y)f(t, x, y)]; \tag{1}$$

$$\frac{\partial f}{\partial t} = a(x)\frac{\partial^2}{\partial x^2}f(t, x, y) + b(x)\frac{\partial}{\partial x}f(t, x, y). \tag{2}$$

In this paper we consider stationary Markov processes satisfying equations (1) and (2) with coefficients

$$a(x) = 1 + x^2, \quad b(x) = -\gamma x, \tag{3}$$

where γ is a constant.

We shall consider the process for the maximal boundaries for which it can be defined on the interval $(-\infty, \infty)$.

Below it will be shown that a stationary Markov process satisfying the equations of A. N. Kolmogorov with coefficients (3) and defined on the interval $(-\infty, \infty)$ exists only in the case where

$$\gamma > -1. \tag{4}$$

Below we assume that γ satisfies condition (4).

The aim of the present article is to find, in explicit form, both the density of the stationary distribution and the density $f(t, x, y)$ of the transition probability of the given process, and also to investigate certain probabilistic properties of this process.

2. As is well known, the density $p(x)$ of the stationary distribution satisfies the equation (see [1]):

$$\frac{d}{dy}[(1 + y^2)p(y)] + \gamma y p(y) = 0. \tag{5}$$

Integrating (5), we obtain the following expression for the density of the stationary distribution:

$$p(y) = C(1 + y^2)^{-\left(1 + \frac{\gamma}{2}\right)}, \tag{6}$$

where C is the normalizing constant:

$$C=\left[\int_{-\infty}^{\infty}(1+x^2)^{-\left(1+\frac{\gamma}{2}\right)}dx\right]^{-1}.$$

Since $p(y)$ is integrable on the interval $(-\infty, \infty)$, it should be assumed that $\gamma>-1$.

We compute the constant C.

$$\int_{-\infty}^{\infty}(1+x^2)^{-\left(1+\frac{\gamma}{2}\right)}dx=B\left(\frac{1}{2}, \frac{\gamma+1}{2}\right)=\frac{\sqrt{\pi}\,\Gamma\left(\frac{\gamma+1}{2}\right)}{\Gamma\left(\frac{\gamma+2}{2}\right)}$$

(see [2], 3.251). Hence

$$C=\frac{\Gamma\left(\frac{\gamma+2}{2}\right)}{\sqrt{\pi}\,\Gamma\left(\frac{\gamma+1}{2}\right)}. \qquad (7)$$

3. We determine the type of boundaries according to the classification of W (Feller [3]). With this aim we form the function:

$$W(x)=\exp\left[-\int_0^x\frac{b(s)}{a(s)}\,ds\right].$$

In the given case:

$$W(x)=(1+x^2)^{\frac{\gamma}{2}}.$$

It is obvious that $W(x)$ is not integrable on the intervals $(-\infty, 0)$ and $(0, \infty)$. On the other hand, the function $a^{-1}(x)\,W^{-1}(x)=(1+x^2)^{-(1+\gamma/2)}\in L(-\infty, \infty)$. It is easy to see that the function

$$a^{-1}(x)\,W^{-1}(x)\times\int_0^x W(s)\,ds=(1+x^2)^{-\left(1+\frac{\gamma}{2}\right)}\int_0^x(1+s^2)^{\frac{\gamma}{2}}\,ds$$

is not integrable on the intervals $(-\infty, 0)$ and $(0, \infty)$. According to the classification of W. Feller, the boundaries are inaccessible (natural). Therefore the density $f(t, x, y)$ of the transition probability is determined by each of the equations (1) and (2) in a unique fashion.

4. It turns out that in the given case it is impossible to find the density $f(t, x, y)$ of the transition probability by the method of expansion into series in eigenfunctions used by O. V. Sarmanov [4]. This method is inapplicable here because, as will be shown below, the integral

$$\int_{-\infty}^{\infty}\int_{-\infty}^{\infty}\frac{p^2(t, x, y)}{p(x)\,p(y)}\,dxdy$$

does not exist.

W. Feller in [3] and [5] developed a theory for solution of diffusion equations by the method of Laplace transformation. By this method it is possible to solve the given equations (1) and (2) only in the case when γ is an odd number. Therefore we propose using another method to solve the given equation, namely the method of conditional characteristic functions.

5. To the density $f(t, x, y)$ of the transition probability corresponds the conditional characteristic function

$$F(t, x, u) = \int_{-\infty}^{\infty} e^{iuy} f(t, x, y)\, dy. \tag{8}$$

Obviously $F(0, x, u) = e^{iux}$. Since $F(t, x, -u) = \overline{F(t, x, u)}$, it suffices to consider $F(t, x, u)$ only for positive u.

Below we assume that the second moment

$$\int_{-\infty}^{\infty} y^2 f(t, x, y)\, dy \tag{9}$$

of the conditional distribution exists. We assume further that the functions

$$\frac{\partial}{\partial y}[yf(t, x, y)], \quad \frac{\partial^2}{\partial y^2}[y^2 f(t, x, y)]$$

are Lebesgue-integrable on the interval $(-\infty, \infty)$. In this case the Fourier transforms of these functions exist, and

$$\int_{-\infty}^{\infty} e^{iuy} \frac{\partial}{\partial y}[yf(t, x, y)]\, dy = -u\, \frac{\partial F(t, x, u)}{\partial u};$$

$$\int_{-\infty}^{\infty} e^{iuy} \frac{\partial^2}{\partial y^2}[(1 + y^2) f(t, x, y)]\, dy = u^2 \frac{\partial^2 F(t, x, u)}{\partial u^2} - u^2 F(t, x, u).$$

Applying the Fourier transformation to equation (2), we obtain the following equation for the conditional characteristic function:

$$\frac{\partial F}{\partial t} = u^2 \frac{\partial^2 F}{\partial u^2} - \gamma u \frac{\partial F}{\partial u} - u^2 F. \tag{10}$$

Applying the Laplace transformation in t to both parts of (10), we obtain the equation

$$\lambda G - e^{iux} = u^2 \frac{\partial^2 G}{\partial u^2} - \gamma u \frac{\partial G}{\partial u} - u^2 G, \tag{11}$$

where

$$G = G(\lambda, x, u) = \int_0^{\infty} e^{-\lambda t} F(t, x, u)\, dt, \ \lambda > 0. \tag{12}$$

For the function $G(\lambda, x, u)$ we have the following estimate:

$$|G(\lambda, x, u)| \leqslant \frac{1}{\lambda}. \tag{12'}$$

Indeed

$$|G(\lambda, x, u)| \leqslant \int_0^{\infty} e^{-\lambda t} |F(t, x, u)|\, dt \leqslant \int_0^{\infty} e^{-\lambda t}\, dt = \frac{1}{\lambda}.$$

Equation (11) can be rewritten thus:

$$u^2\frac{\partial^2 G}{\partial u^2} - \gamma u\frac{\partial G}{\partial u} - (u^2+\lambda)G = -e^{iux}. \tag{11'}$$

6. We find solutions of the homogeneous equation

$$u^2\frac{\partial^2 G}{\partial u^2} - \gamma u\frac{\partial G}{\partial u} - (u^2+\lambda)G = 0 \tag{11''}$$

corresponding to (11′).

This equation has the solution ([6], Chapter III, §II, 2.162, 1a):

$$G_1(u) = u^{\frac{\gamma+1}{2}} I_p(u), \tag{13}$$

$$G_2(u) = u^{\frac{\gamma+1}{2}} K_p(u), \tag{13'}$$

where

$$p = \sqrt{\left(\frac{\gamma+1}{2}\right)^2 + \lambda}. \tag{14}$$

Now we find the solution of the inhomogeneous equation (11′) with the help of the Green's function. The Wronski determinant of the functions $G_1(u)$ and $G_2(u)$ has the form ([2], 8.474)

$$W(G_1, G_2) = u^{\gamma+1} W(I_p, K_p) = -u^{\gamma}.$$

The solution of the inhomogeneous equation has the form

$$G(\lambda, x, u) = C_1 \int_0^\infty K(u, s)(-e^{isx})\, ds, \tag{15}$$

where the kernel $K(u, s)$ has the form

$$K(u, s) = \begin{cases} -K_p(u)\, u^{\frac{\gamma+1}{2}} I_p(s)\, s^{-\frac{\gamma+3}{2}}, & 0 \leqslant s \leqslant u < \infty, \\ -I_p(u)\, u^{\frac{\gamma+1}{2}} K_p(s)\, s^{-\frac{\gamma+3}{2}}, & 0 \leqslant u \leqslant s < \infty. \end{cases} \tag{16}$$

From the fact that $K_p(x)$ and $I_p(x)$ are positive for real p and positive x, it follows that $K(u, s)$ is real and, moreover, negative.

Below we shall prove that the integrals

$$\int_0^u I_p(s)\, s^{-\frac{\gamma+3}{2}}\, ds, \qquad \int_u^\infty K_p(s)\, s^{-\frac{\gamma+3}{2}}\, ds \tag{17}$$

involved in the definition (15) of the function $G(\lambda, x, u)$ exist. With the help of this result it will be proved that $K(u, s)$ is actually the Green's function.

7. We need to estimate the growth of the functions $I_p(x)$ and $K_p(x)$ as $x \to 0$ and as $x \to \infty$.

Using known expansions of the functions $I_p(x)$ and $K_p(x)$ in powers of x

(see [2], 8.445 and 8.446; and also [7], 7.8), as $x \to 0$ we obtain the following estimates

$$I_p(x) = \frac{1}{\Gamma(p+1)}\left(\frac{x}{2}\right)^p [1 + o(1)]; \tag{18}$$

$$K_p(x) = k(p)\, x^{-p} [1 + o(1)], \tag{19}$$

where $k(p)$ is a positive constant depending on p.

Using estimates found in [7] (§§7.3 and 7.8), as $x \to \infty$ we obtain

$$K_p(x) = \sqrt{\frac{\pi}{2}}\, e^{-x} x^{-\frac{1}{2}} (1 + o(1)); \tag{20}$$

$$I_p(x) = \sqrt{\frac{1}{2\pi}}\, e^{x} x^{-\frac{1}{2}} (1 + o(1)). \tag{21}$$

According to the definition (14), $p > (\gamma + 1)/2$. Therefore it follows from the estimates (18) and (20) that the integrals (17) exist.

8. We now prove that $G(\lambda, x, u)$, considered as a function of u, is uniformly bounded and absolutely integrable on the interval $(0, \infty)$. For this it is sufficient to prove that $G(\lambda, x, u)$ is bounded in a neighborhood of zero and $|G(\lambda, x, u)| \leq O(u^{-2})$ as $u \to \infty$.

We have

$$|G(\lambda, x, u) \leqslant |C_1| \int_0^\infty |K(u, s)|\, ds$$

$$= |G_1| \left[u^{\frac{\gamma+1}{2}} K_p(u) \int_0^u I_p(s)\, s^{-\frac{\gamma+3}{2}}\, ds + u^{\frac{\gamma+1}{2}} I_p(u) \int_u^\infty K_p(s)\, s^{-\frac{\gamma+3}{2}}\, ds \right]. \tag{22}$$

Let a be a fixed number, greater than u.

Using the estimates for $K_p(x)$ and $I_p(x)$ ((18) and (19) respectively), we obtain from inequalities (22):

$$|G(\lambda, x, u)| \leqslant |C_1| \left[u^{\frac{\gamma+1}{2} - p} k(p) (1 + o(1)) \int_0^u s^{p - \frac{\gamma+3}{2}} \cdot \frac{1 + o(1)}{\Gamma(p+1)\, 2^p}\, ds \right.$$

$$+ \frac{u^{\frac{\gamma+1}{2}+p} (1 + o(1))}{\Gamma(p+1)\, 2^p} \int_a^\infty K_p(s)\, s^{-\frac{\gamma+3}{2}}\, ds + \frac{u^{\frac{\gamma+1}{2}+p}}{\Gamma(p+1)\, 2^p} (1 + o(1))$$

$$\left. \times \int_u^a k(p)\, s^{-\left(p + \frac{\gamma+3}{2}\right)} (1 + o(1))\, ds \right]. \tag{22'}$$

We denote $\int_a^\infty K_p(s)\, s^{-(\gamma+3)/2} ds$ by $k_1(p)$. Then, integrating in the right part of (22′), we have:

$$|G(\lambda, x, u)| \leqslant |C_1| \left[k(p) \frac{1+o(1)}{2^p \left(p - \frac{\gamma+1}{2}\right) \Gamma(p+1)} \right.$$

$$\left. + k_1(p) \frac{[1+o(1)]\, u^{\frac{\gamma+1}{2}+p}}{2^p \Gamma(p+1)} + k(p) \frac{1+o(1)}{2^p \Gamma(p+1)\left(p+\frac{\gamma+1}{2}\right)} \right]. \qquad (22'')$$

Since a is fixed, the second term in the square brackets goes to zero as $u \to 0$. We obtain the following estimate as $u \to 0$:

$$|G(\lambda, x, u)| \leqslant |C_1| [k_2(p) + k_3(p) + o(1)], \qquad (23)$$

where

$$k_2(p) = \frac{k(p)}{\Gamma(p+1)\, 2^p \left(p - \frac{\gamma+1}{2}\right)}$$

and

$$k_3(p) = \frac{k(p)}{\Gamma(p+1)\, 2^p \left(p + \frac{\gamma+1}{2}\right)}.$$

Now we estimate $G(\lambda, x, u)$ as $u \to \infty$.

$$|G(\lambda, x, u)| < |C_1| \int_0^\infty |K(u, s)|\, ds = |C_1| \left[K_p(u)\, u^{\frac{\gamma+1}{2}} \int_0^a I_p(s)\, s^{-\frac{\gamma+3}{2}}\, ds \right.$$

$$+ K_p(u)\, u^{\frac{\gamma+1}{2}} \int_a^{\frac{u}{2}} I_p(s)\, s^{-\frac{\gamma+3}{2}}\, ds + K_p(u)\, u^{\frac{\gamma+1}{2}} \int_{\frac{u}{2}}^{u} I_p(s)\, s^{-\frac{\gamma+3}{2}}\, ds$$

$$\left. + I_p(u)\, u^{\frac{\gamma+1}{2}} \int_u^\infty K_p(s)\, s^{-\frac{\gamma+3}{2}}\, ds \right], \qquad (24)$$

where a is fixed, $0 < a < u$.

We consider each of the summands in the square brackets in the right part of (24). The integral $\int_0^a I_p(s)\, s^{-(\gamma+3)/2} ds$ is denoted by $k_4(p)$.

Starting from (20) for the first summand we obtain the estimate

$$K_p(u)\, u^{\frac{\gamma+1}{2}} \int_0^a I_p(s)\, s^{-\frac{\gamma+3}{2}}\, ds = k_4(p) \sqrt{\frac{\pi}{2}}\, e^{-u} u^{\frac{\gamma}{2}} (1 + o(1)). \qquad (25)$$

Consider the second summand in the square brackets. Using estimates (20) and (21), we have

$$K_p(u)\,u^{\frac{\gamma+1}{2}}\int\limits_a^{\frac{u}{2}} I_p(s)\,s^{-\frac{\gamma+3}{2}}\,ds=\frac{1}{2}u^{\frac{\gamma}{2}}e^{-u}(1+o(1))\int\limits_a^{\frac{u}{2}} e^s s^{-\frac{\gamma+4}{2}}(1+o(1))\,ds$$

$$<\frac{1}{2}e^{-\frac{u}{2}}u^{\frac{\gamma}{2}}\int\limits_a^{\frac{u}{2}} s^{-\frac{\gamma+4}{2}}[1+o(1)]\,ds$$

$$=\frac{1}{\gamma+2}2^{\frac{\gamma+2}{2}}u^{-1}e^{-\frac{u}{2}}(1+o(1))=o(u^{-2}). \tag{26}$$

We estimate the third summand in the square brackets, starting from the estimates (20) and (21):

$$K_p(u)\,u^{\frac{\gamma+1}{2}}\int\limits_{\frac{u}{2}}^{u} I_p(s)\,s^{-\frac{\gamma+3}{2}}\,ds=\frac{1}{2}u^{\frac{\gamma}{2}}e^{-u}(1+o(1))\int\limits_{\frac{u}{2}}^{u} e^s s^{-\frac{\gamma+4}{2}}(1+o(1))\,ds$$

$$<\frac{1}{2}u^{\frac{\gamma}{2}}e^{-u}(1+o(1))\int\limits_{\frac{u}{2}}^{u} e^s\left(\frac{u}{2}\right)^{-\frac{\gamma+4}{2}}(1+o(1))\,ds$$

$$=2^{\frac{\gamma+2}{2}}u^{-2}(1+o(1))=O(u^{-2}). \tag{27}$$

We estimate the fourth summand:

$$I_p(u)\,u^{\frac{\gamma+1}{2}}\int\limits_u^{\infty} K_p(s)\,s^{-\frac{\gamma+3}{2}}\,ds=\frac{1}{2}e^u u^{\frac{\gamma}{2}}(1+o(1))\int\limits_u^{\infty} e^{-s}s^{-\frac{\gamma+4}{2}}(1+o(1))\,ds$$

$$<\frac{e^u}{2}u^{-2}(1+o(1))\int\limits_u^{\infty} e^{-s}(1+o(1))\,ds=\frac{1}{2}u^{-2}(1+o(1))=O(u^{-2}). \tag{28}$$

From the estimates (25)–(28) it follows that as $u\to\infty$

$$|G(\lambda,\ x,\ u)|\leqslant O(u^{-2}). \tag{29}$$

It follows from (23) and (29) that $G(\lambda, x, u)$, as a function of u, is uniformly bounded and absolutely integrable on the interval $(0, \infty)$.

9. The function $G(\lambda, x, u)$ is a bounded solution of equation (11′). We shall prove that solutions of the corresponding homogeneous equation (11″) are unbounded.

Starting from (21), we obtain the following estimate for the growth of the function $G_1(u)$ as $u\to\infty$:

$$G_1(u)=u^{\frac{\gamma+1}{2}}I_p(u)=\sqrt{\frac{1}{2\pi}}\,e^u u^{\frac{\gamma}{2}}(1+o(1)).$$

It is obvious that $G_1(u)$ grows unboundedly as $u\to\infty$.

We shall prove that $G_2(u)$ is not bounded in a neighborhood of zero. Indeed, starting with estimate (19), we have the following estimate for $G_2(u)$:

$$G_2(u)=u^{\frac{\gamma+1}{2}}K_p(u)=k(p)\,u^{\frac{\gamma+1}{2}-p}(1+o(1)),$$

where $k(p)$ is the constant defined above. Since λ is positive, from formula (14) defining p it follows that

$$\frac{\gamma+1}{2}-p<0.$$

Therefore the function $G_2(u)$ is not bounded in a neighborhood of zero.

In view of the fact that $G_1(u)$ and $G_2(u)$ are unbounded solutions, the function $G(\lambda, x, u)$ is, to within a constant factor, the unique bounded solution of equation (11′). By (12′), the desired solution is bounded. Consequently $G(\lambda, x, u)$ is the desired solution.

10. We use the conventional notation for the Laplace transform operator:

$$Lf(\lambda)=\int_0^\infty e^{-\lambda t}f(t)\,dt.$$

We let L^{-1} denote the inverse Laplace transform. We prove the formula:

$$G(\lambda, x, u)=LF(\lambda, x, u)=\int_{-\infty}^{\infty} e^{iyu}Lf(\lambda, x, y)\,dy. \tag{30}$$

Indeed,

$$G(\lambda, x, u)=LF(\lambda, x, u)=\int_0^\infty e^{-\lambda t}\int_{-\infty}^{\infty} e^{iyu}f(t, x, y)\,dydt.$$

By Fubini's theorem we can interchange the order of integration here:

$$\int_0^\infty e^{-\lambda t}\int_{-\infty}^{\infty} e^{iyu}f(t, x, y)\,dydt=\int_{-\infty}^{\infty} e^{iyu}\int_0^\infty e^{-\lambda t}f(t, x, y)\,dtdy$$

$$=\int_{-\infty}^{\infty} e^{iyu}Lf(\lambda, x, y)\,dy.$$

Consequently (30) holds.

We introduce the notation: $\Phi(\lambda, x, y)=Lf(\lambda, x, y)$. In view of the absolute integrability of the function $G(\lambda, x, u)$, it has the inverse Fourier transform

$$\Phi(\lambda, x, y)=\frac{1}{2\pi}\int_{-\infty}^{\infty} e^{-iyu}G(\lambda, x, u)\,du. \tag{31}$$

Further, we note that

$$G(\lambda, x, -u)=\overline{G(\lambda, x, u)}. \tag{32}$$

Thus

$$\Phi(\lambda, x, y)=\frac{1}{2\pi}\left[\int_0^\infty e^{-iyu}G(\lambda, x, u)\,du+\int_0^\infty e^{iyu}\,\overline{G(\lambda, x, u)}\,du\right]$$

$$=\frac{1}{\pi}\operatorname{Re}\int_0^\infty e^{-iyu}G(\lambda, x, u)\,du.$$

Replacing $G(\lambda, x, u)$ by its expression according to formula (15) here, we obtain the following expression for $\Phi(\lambda, x, y)$:

$$\Phi(\lambda, x, y)=C_2\operatorname{Re}\int_0^\infty\int_0^\infty e^{-iyu}(-e^{ixs})\,K(u, s)\,ds\,du, \tag{33}$$

where

$$C_2=\frac{C_1}{\pi}.$$

Using expression (16) for the kernel $K(u, s)$, we express $\Phi(\lambda, x, y)$ as follows

$$\Phi(\lambda, x, y)=C_2\left[\int_0^\infty\int_0^u \frac{\cos(sx-uy)}{s^{\frac{\gamma+3}{2}}}u^{\frac{\gamma+1}{2}}K_p(u)\,I_p(s)\,ds\,du\right.$$

$$\left.+\int_0^\infty\int_u^\infty \frac{\cos(sx-uy)}{s^{\frac{\gamma+3}{2}}}u^{\frac{\gamma+1}{2}}K_p(s)\,I_p(u)\,ds\,du\right]. \tag{34}$$

We consider the first integral in the square brackets and prove that this integral converges absolutely:

$$\int_0^\infty\int_0^u\left|\frac{\cos(sx-uy)}{s^{\frac{\gamma+3}{2}}}u^{\frac{\gamma+1}{2}}K_p(u)\,I_p(s)\right|ds\,du$$

$$<\int_0^\infty\int_0^u K_p(u)\,u^{\frac{\gamma+1}{2}}I_p(s)\,s^{-\frac{\gamma+3}{2}}\,ds\,du<\int_0^\infty\frac{|G(\lambda, 0, u)|}{|C_2|}\,du.$$

Since $G(\lambda, x, u)$ is absolutely integrable in u on the interval $(0, \infty)$, the above integral converges absolutely. By Fubini's theorem we can change the order of integration in the given integral. In the right part of (34) we make the following change: in the first integral we change the order of integration, and in the second integral we change the notation of the variables of integration. As a result we obtain

$$\Phi(\lambda, x, y)=C_2\int_0^\infty\int_s^\infty\left[\cos(sx-uy)\cdot s^{-\frac{\gamma+3}{2}}u^{\frac{\gamma+1}{2}}\right.$$

$$\left.+\cos(ux-sy)\cdot s^{\frac{\gamma+1}{2}}u^{-\frac{\gamma+3}{2}}\right]K_p(u)\,I_p(s)\,du\,ds.$$

We make the change of variable $s/u = v$. As a result we have ([2], 6.672.4):

$$\Phi(\lambda, x, y) = C_2 \int_0^1 \int_0^\infty \Big\{ \cos[u(vx - y)] \cdot v^{-\frac{\gamma+3}{2}}$$

$$+ \cos[u(x - vy)] \cdot v^{\frac{\gamma+1}{2}} \Big\} K_p(u) I_p(uv)\, du dv$$

$$= \frac{C_2}{2} \left[\int_0^1 v^{-\frac{\gamma+4}{2}} Q_{p-\frac{1}{2}} \left(\frac{1 + v^2 + (vx - y)^2}{2v} \right) dv \right.$$

$$\left. + \int_0^1 v^{\frac{\gamma}{2}} Q_{p-\frac{1}{2}} \left(\frac{1 + v^2 + (x - vy)^2}{2v} \right) dv \right]. \tag{35}$$

11. We find the inverse Laplace transform of the function $Q_{p-1/2}(\beta)$, where

$$p = \sqrt{\left(\frac{\gamma+1}{2}\right)^2 + \lambda}.$$

We shall write

$$f(t) \doteqdot F(\lambda),$$

if $F(\lambda)$ is the Laplace transform of the function $f(t)$.

We have ([8], t. II C. 10.6 or [9], 5.13.15):

$$L^{-1}\left[Q_{\lambda-\frac{1}{2}}(\beta)\right] = \begin{cases} [2(\operatorname{ch} t - \beta)]^{-\frac{1}{2}} & \text{for } t > \operatorname{arch} \beta, \\ 0 & \text{for } t \leqslant \operatorname{arch} \beta. \end{cases} \tag{36}$$

Further ([10], 9.33)

$$L^{-1}\left[Q_{\sqrt{\lambda}-\frac{1}{2}}(\beta)\right] = \frac{1}{2\sqrt{2\pi t^3}} \int_{\operatorname{arch}\beta}^{\infty} \tau e^{-\frac{\tau^2}{4t}} (\operatorname{ch} \tau - \beta)^{-\frac{1}{2}} d\tau. \tag{37}$$

Further, we need the well-known formula for transform shift: $e^{\beta t}\phi(t) \doteqdot F(\lambda - \beta)$, if $\phi(t) \doteqdot F(\lambda)$. According to this formula

$$L^{-1}\left[Q_{p-\frac{1}{2}}(\beta)\right] = \frac{e^{-\left(\frac{1+\gamma}{2}\right)^2 t}}{2\sqrt{2\pi t^3}} \int_{\operatorname{arch}\beta}^{\infty} \tau e^{-\frac{\tau^2}{4t}} (\operatorname{ch} \tau - \beta)^{-\frac{1}{2}} d\tau, \tag{38}$$

where p is defined by formula (14).

12. We need the following

Lemma. *Let $F(x, \lambda)$ be the Laplace transform in t of the function $f(x, t)$, positive in the region*

$$D = \begin{bmatrix} a \leqslant x \leqslant b \\ 0 < t < \infty \end{bmatrix}.$$

If the integral $\int_a^b F(x, \lambda)\,dx$ *exists, then*

$$L^{-1}\left[\int_a^b F(x, \lambda)\,dx\right]=\int_a^b f(x, t)\,dx. \tag{39}$$

Proof. Since $F(x, \lambda)$ is the Laplace transform of the function $f(x, t)$, the following equality is obvious:

$$\int_a^b F(x, \lambda)\,dx=\int_a^b\int_0^\infty e^{-\lambda t} f(x, t)\,dt\,dx. \tag{40}$$

Since there is a positive function under the integral in the right part of (40), by Fubini's theorem we can change the order of integration. After changing the order of integration we obtain

$$\int_a^b F(x, \lambda)\,dx=\int_0^\infty e^{-\lambda t}\int_a^b f(x, t)\,dx\,dt. \tag{41}$$

Applying the inverse Laplace transform to both parts of (41), we obtain (39). The lemma is proved.

13. Applying the lemma and making use of formula (38), we obtain, starting from (35):

$$\begin{aligned} f(t, x, y)&=L^{-1}[\Phi(\lambda, x, y)]\\ &=\frac{C_2}{2}\left\{\int_0^1 v^{-\frac{\gamma+4}{2}} L^{-1}\left[Q_{p-\frac{1}{2}}\left(\frac{1+v^2+(vx-y)^2}{2v}\right)\right] dv\right.\\ &\quad\left.+\int_0^1 v^{\frac{\gamma}{2}} L^{-1}\left[Q_{p-\frac{1}{2}}\left(\frac{1+v^2+(x-vy)^2}{2v}\right)\right] dv\right\}\\ &=C_3\frac{e^{-\left(\frac{\gamma+1}{2}\right)^2 t}}{\sqrt{2\pi t^3}}\left\{\int_0^1 v^{-\frac{\gamma+4}{2}}\int_{\operatorname{arch}\beta(v, y)}^\infty \tau e^{-\frac{\tau^2}{4t}}[\operatorname{ch}\tau-\beta(v, y)]^{-\frac{1}{2}}\,d\tau\,dv\right.\\ &\quad\left.+\int_0^1 v^{\frac{\gamma}{2}}\int_{\operatorname{arch}\beta_1(v, y)}^\infty \tau e^{-\frac{\tau^2}{4t}}[\operatorname{ch}\tau-\beta_1(v, y)]^{-\frac{1}{2}}\,d\tau\,dv\right\}, \end{aligned} \tag{42}$$

where

$$\beta(v, y)=\frac{1+v^2+(vx-y)^2}{2v}, \tag{43}$$

$$\beta_1(v, y)=\frac{1+v^2+(x-vy)^2}{2v}, \tag{44}$$

$$C_3=\frac{C_2}{4}=\frac{C_1}{4\pi}. \tag{45}$$

For further study we need to represent the density $f(t, x, y)$ of the transition probability in the following form:

$$f(t,\ x,\ y)=C_3\frac{e^{-\left(\frac{\gamma+1}{2}\right)^2 t}}{\sqrt{2\pi t^3}}[f_1(t,\ x,\ y)+f_2(t,\ x,\ y)], \tag{46}$$

where

$$f_1(t,\ x,\ y)=\int_0^1 v^{\frac{\gamma}{2}}\int_{\operatorname{arch}\beta_1(v,\ y)}^{\infty}\tau e^{-\frac{\tau^2}{4t}}[\operatorname{ch}\tau-\beta_1(v,\ y)]^{-\frac{1}{2}}\,d\tau dv, \tag{47}$$

$$f_2(t,\ x,\ y)=\int_0^1 v^{-\frac{\gamma+4}{2}}\int_{\operatorname{arch}\beta(v,\ y)}^{\infty}\tau e^{-\frac{\tau^2}{4t}}[\operatorname{ch}\tau-\beta(v,\ y)]^{-\frac{1}{2}}\,d\tau dv. \tag{48}$$

The functions $\beta(v,\ y)$ and $\beta_1(v,\ y)$ involved in (47) and (48) are defined by formulas (43) and (44).

14. We determine the constant C_3 involved in (46) and also the first moment

$$M_{x(s)}\xi_{s+t}=\int_{-\infty}^{\infty} yf(t,\ x,\ y)\,dy$$

of the conditional distribution.

To this end, we need to compute the two integrals

$$I_1=\int_{-\infty}^{\infty}[f_1(t,\ x,\ y)+f_2(t,\ x,\ y)]\,dy \tag{49}$$

and

$$I_2=\int_{-\infty}^{\infty} y[f_1(t,\ x,\ y)+f_2(t,\ x,\ y)]\,dy. \tag{50}$$

Consider the integral

$$I_3=\int_{-\infty}^{\infty} f_1(t,\ x,\ y)\,dy=\int_{-\infty}^{\infty}\int_0^1 v^{\frac{\gamma}{2}}\int_{\operatorname{arch}\beta_1(v,\ y)}^{\infty}\tau e^{-\frac{\tau^2}{4t}}[\operatorname{ch}\tau-\beta_1(v,\ y)]^{-\frac{1}{2}}\,d\tau dv dy. \tag{51}$$

Here we make the change of variable:

$$z=-\frac{x-vy}{\sqrt{2v}},\quad \frac{\partial z}{\partial y}=\frac{v}{\sqrt{2v}}. \tag{52}$$

The integral I_3 takes the form

$$I_3=\int_0^1\sqrt{2}\,v^{\frac{\gamma-1}{2}}\int_{\varphi(v)}^{\infty}\tau e^{-\frac{\tau^2}{4t}}\int_{-\alpha(\tau,\ v)}^{\alpha(\tau,\ v)}\left(\operatorname{ch}\tau-\frac{1+v^2}{2v}-z^2\right)^{-\frac{1}{2}}dz\,d\tau dv, \tag{53}$$

where

$$\varphi(v)=\operatorname{arch}\frac{1+v^2}{2v}, \tag{54}$$

$$\alpha(\tau, v) = \sqrt{\operatorname{ch}\tau - \frac{1+v^2}{2v}}. \tag{54'}$$

Integrating in z in the right part of (53), we obtain

$$I_3 = \pi\sqrt{2}\int_0^1 v^{\frac{\gamma-1}{2}} \int_{\varphi(v)}^{\infty} \tau e^{-\frac{\tau^2}{4t}} d\tau dv. \tag{55}$$

Consider the integral

$$I_4 = \int_{-\infty}^{\infty} y f_1(t, x, y)\, dy. \tag{56}$$

Here we make the change of variable (52). Since $y = (\sqrt{2v}\, z + x)/v$, we have

$$I_4 = \int_0^1 v^{\frac{\gamma-2}{2}} \int_{\varphi(v)}^{\infty} \tau e^{-\frac{\tau^2}{4t}} \int_{-\alpha(\tau, v)}^{\alpha(\tau, v)} 2z\left(\operatorname{ch}\tau - \frac{1+v^2}{2v} - z^2\right)^{-\frac{1}{2}} dz d\tau dv +$$

$$+ x\sqrt{2}\int_0^1 v^{\frac{\gamma-3}{2}} \int_{\varphi(v)}^{\infty} \tau e^{-\frac{\tau^2}{4t}} \int_{-\alpha(\tau, v)}^{\alpha(\tau, v)} \left(\operatorname{ch}\tau - \frac{1+v^2}{2v} - z^2\right)^{-\frac{1}{2}} dz d\tau dv. \tag{57}$$

Since $z(\operatorname{ch}\tau - (1+v^2)/2v - z^2)^{-1/2}$ is an odd function of z, the first integral in the right part of (57) vanishes. Integrating in the right part of (57) in z, we obtain:

$$I_4 = x\pi\sqrt{2}\int_0^1 v^{\frac{\gamma-3}{2}} \int_{\varphi(v)}^{\infty} \tau e^{-\frac{\tau^2}{4t}} d\tau dv. \tag{58}$$

Finally, consider the integrals

$$I_5 = \int_{-\infty}^{\infty} f_2(t, x, y)\, dy = \int_{-\infty}^{\infty}\int_0^1 v^{-\frac{\gamma+4}{2}}$$

$$\times \int_{\operatorname{arch}\beta(v, y)}^{\infty} \tau e^{-\frac{\tau^2}{4t}} [\operatorname{ch}\tau - \beta(v, y)]^{-\frac{1}{2}} d\tau dv dy; \tag{59}$$

$$I_6 = \int_{-\infty}^{\infty} y f_2(t, x, y)\, dy = \int_{-\infty}^{\infty} y \int_0^1 v^{-\frac{\gamma+4}{2}}$$

$$\times \int_{\operatorname{arch}\beta(v, y)}^{\infty} \tau e^{-\frac{\tau^2}{4t}} [\operatorname{ch}\tau - \beta(v, y)]^{-\frac{1}{2}} d\tau dv dy. \tag{60}$$

We make the change of variable

$$z = -\frac{vx - y}{\sqrt{2v}}, \quad \frac{\partial z}{\partial y} = \frac{1}{\sqrt{2v}}. \tag{61}$$

Since $y = \sqrt{2v}\, z + vx$, the integrals I_5 and I_6 take the form

$$I_5=\int_0^1 \sqrt{2}\, v^{-\frac{\gamma+5}{2}} \int_{\varphi(v)}^{\infty} \tau e^{-\frac{\tau^2}{4t}} \int_{-\alpha(\tau, v)}^{\alpha(\tau, v)} \left(\operatorname{ch}\tau-\frac{1+v^2}{2v}-z^2\right)^{-\frac{1}{2}} dz d\tau dv, \tag{62}$$

$$I_6=2\int_0^1 v^{-\frac{\gamma+2}{2}} \int_{\varphi(v)}^{\infty} \tau e^{-\frac{\tau^2}{4t}} \int_{-\alpha(\tau, v)}^{\alpha(\tau, v)} \left(\operatorname{ch}\tau-\frac{1+v^2}{2v}-z^2\right)^{-\frac{1}{2}} dz d\tau dv$$
$$+x\sqrt{2}\int_0^1 v^{-\frac{\gamma+1}{2}} \int_{\varphi(v)}^{\infty} \tau e^{-\frac{\tau^2}{4t}} \int_{-\alpha(\tau, v)}^{\alpha(\tau, v)} \left(\operatorname{ch}\tau-\frac{1+v^2}{2v}-z^2\right)^{-\frac{1}{2}} dz d\tau dv. \tag{63}$$

We consider the first of the integrals in the right part of (63). Since the function $z(\operatorname{ch} \tau - (1+v^2)/2v - z^2)^{-1/2}$, considered as a function of z, is odd, the integral is equal to zero. Integrating in the right parts of (62) and (63) in z, we obtain

$$I_5=\pi\sqrt{2}\int_0^1 v^{-\frac{\gamma+3}{2}} \int_{\varphi(v)}^{\infty} \tau e^{-\frac{\tau^2}{4t}} d\tau dv, \tag{64}$$

$$I_6=x\pi\sqrt{2}\int_0^1 v^{-\frac{\gamma+1}{2}} \int_{\varphi(v)}^{\infty} \tau e^{-\frac{\tau^2}{4t}} dt dv. \tag{65}$$

Consider the integral

$$I=\int_0^1 v^{\alpha} \int_{\varphi(v)}^{\infty} \tau e^{-\frac{\tau^2}{4t}} d\tau dv.$$

Here, changing the order of integration and integrating in v, we obtain

$$I=\begin{cases} \dfrac{1}{\alpha+1}\displaystyle\int_0^{\infty} \tau e^{-\frac{\tau^2}{4t}} \left(1-e^{-(\alpha+1)\tau}\right) d\tau, & \text{if } \alpha\neq-1, \\ \displaystyle\int_0^{\infty} \tau^2 e^{-\frac{\tau^2}{4t}} d\tau, & \text{if } \alpha=-1. \end{cases} \tag{66}$$

We now compute I_1. We have

$$I_1=I_3+I_5=\pi\sqrt{2}\left[\int_0^1 v^{\frac{\gamma-1}{2}} \int_{\varphi(v)}^{\infty} \tau e^{-\frac{\tau^2}{4t}} d\tau+\int_0^1 v^{-\frac{\gamma+3}{2}} \int_{\varphi(v)}^{\infty} \tau e^{-\frac{\tau^2}{4t}} d\tau\right]. \tag{67}$$

Since $\gamma > -1$, neither $(\gamma-1)/2$ nor $-(\gamma+3)/2$ is equal to (-1). Applying formula (66), we obtain:

$$I_1=\frac{4\pi\sqrt{2}}{\gamma+1}\int_0^{\infty} \tau e^{-\frac{\tau^2}{4t}} \operatorname{sh}\left(\frac{\gamma+1}{2}\tau\right) d\tau. \tag{68}$$

Finally, integrating in (68) in τ ([2], 3.562.3), we obtain

$$I_1 = 4\pi\sqrt{2\pi t^3}\, e^{\left(\frac{\gamma+1}{2}\right)^2 t}. \tag{69}$$

We compute the integral I_2.

According to formulas (58) and (65) we have:

$$I_2 = I_4 + I_6 = x\pi\sqrt{2}\left[\int_0^1 v^{\frac{\gamma-3}{2}} \int_{\varphi(v)}^{\infty} \tau e^{-\frac{\tau^2}{4t}}\, d\tau dv + \int_0^1 v^{-\frac{\gamma+1}{2}} \int_{\varphi(v)}^{\infty} \tau e^{-\frac{\tau^2}{4t}}\, d\tau dv\right]. \tag{70}$$

Assume that $\gamma \neq 1$. Then neither of the numbers $(\gamma - 3)/2$ and $(\gamma + 1)/2$ is equal to -1. Applying formula (66), we obtain

$$I_2 = \frac{4x\pi\sqrt{2}}{\gamma - 1}\int_0^{\infty} \tau e^{-\frac{\tau^2}{4t}}\, \mathrm{sh}\left(\frac{\gamma-1}{2}\tau\right) d\tau. \tag{71}$$

Finally, integrating in (71) in τ, we obtain ([2], 3.562.3):

$$I_2 = x4\pi\sqrt{2\pi t^3}\, e^{\left(\frac{\gamma-1}{2}\right)^2 t}. \tag{72}$$

We consider the case $\gamma = 1$. Equality (70) in this case takes the form

$$I_2 = I_4 + I_6 = 2x\pi\sqrt{2}\int_0^1 v^{-1} \int_{\varphi(v)}^{\infty} \tau e^{-\frac{\tau^2}{4t}}\, d\tau dv. \tag{70$'$}$$

Applying formula (66), we obtain

$$I_2 = x2\pi\sqrt{2}\int_0^{\infty} \tau^2 e^{-\frac{\tau^2}{4t}}\, d\tau. \tag{71$'$}$$

As a result of integrating in the right part of (71′) we obtain ([2], 3.461.2)

$$I_2 = x4\pi\sqrt{2\pi t^3}. \tag{72$'$}$$

It is obvious that formula (72′) is a special case of formula (72) for $\gamma = 1$.

Thus, using the expression (69) for I_1, we obtain

$$\int_{-\infty}^{\infty} f(t, x, y)\, ay = C_3 \frac{e^{-\left(\frac{\gamma+1}{2}\right)^2 t}}{\sqrt{2\pi t^3}} \int_{-\infty}^{\infty} [f_1(t, x, y) + f_2(t, x, y)]\, dy = C_3 4\pi = 1. \tag{73}$$

Consequently,

$$C_3 = \frac{1}{4\pi}, \quad C_1 = 1. \tag{74}$$

Using the expression (72) for I_2, we obtain

$$\int_{-\infty}^{\infty} yf(t, x, y)\,dy = \frac{e^{-\left(\frac{\gamma+1}{2}\right)^2 t}}{4\pi\sqrt{2\pi t^3}} \int_{-\infty}^{\infty} y\,[f_1(t, x, y) + f_2(t, x, y)]\,dy = e^{-\gamma t}x. \quad (75)$$

15. Thus the function $\phi(y) = y$ is an eigenfunction of the integral operator with kernel $f(t, x, y)$, and $e^{-\gamma t}$ is the corresponding eigenvalue.

We introduce the operator whose kernel is

$$K(t, x, y) = \frac{p(t, x, y)}{\sqrt{p(x)\,p(y)}} = \frac{\sqrt{p(x)}\,f(t, x, y)}{\sqrt{p(y)}}.$$

The function $y\sqrt{p(y)}$ is an eigenfunction belonging to the operator whose kernel is $K(t, x, y)$, and $e^{-\gamma t}$ is the corresponding eigenvalue. Indeed,

$$\int_{-\infty}^{\infty} K(t, x, y)\,y\sqrt{p(y)}\,dy = \int_{-\infty}^{\infty} \sqrt{p(x)}\,f(t, x, y)\,y\,dy = e^{-\gamma t}x\sqrt{p(x)}.$$

From formula (6) for the density of the stationary distribution it is obvious that the random variable ξ_t has a second moment if $\gamma > 1$. If $\gamma \leq 1$, then ξ_t does not have a second moment. Indeed,

$$\mathbf{M}\xi_t^2 = C\int_{-\infty}^{\infty} \frac{y^2}{(1+y^2)^{\frac{1+\gamma}{2}}}\,dy.$$

The last integral converges for $\gamma > 1$ and diverges for $\gamma \leq 1$. Thus for $\gamma > 1$ the correlation function for the process $\{\xi_t\}$ exists. For $\gamma \leq 1$ the correlation function does not exist.

Since for $\gamma \leq 1$ the square of the eigenfunction $y\sqrt{p(y)}$ is not integrable, the square of the kernel $K(t, x, y)$ is also not integrable. In other words, the integral

$$\int_{-\infty}^{\infty}\int_{-\infty}^{\infty} \frac{p^2(t, x, y)}{p(x)\,p(y)}\,dx\,dy$$

does not exist.

We compute the correlation function of the process

$$\mathbf{M}\xi_s\xi_{s+t} = \mathbf{M}\,[\xi_s\mathbf{M}\,(\xi_{s+t} \mid \xi_s)].$$

According to (75) we have the following expression:

$$\mathbf{M}\,(\xi_{s+t} \mid \xi_s = x) = \int_{-\infty}^{\infty} yf(t, x, y)\,dy = e^{-\gamma t}x$$

for the conditional mathematical expectation.

Thus

$$\mathbf{M}\xi_s\xi_{s+t} = \int_{-\infty}^{\infty} e^{-\gamma t}x^2 p(x)\,dx.$$

The density of the stationary distribution is determined by formulas (6) and (7). Using these formulas, we obtain

$$\int_{-\infty}^{\infty} x^2 p(x)\,dx = 2\frac{\Gamma\left(\frac{\gamma+2}{2}\right)}{\sqrt{\pi}\,\Gamma\left(\frac{\gamma+1}{2}\right)}\int_{0}^{\infty}\frac{x^2dx}{(1+x^2)^{\frac{1+\gamma}{2}}}. \tag{76}$$

After integration we obtain ([2], 3.251.2):

$$\int_{-\infty}^{\infty} x^2 p(x)\,dx = \frac{\Gamma\left(\frac{\gamma+2}{2}\right)}{\sqrt{\pi}\,\Gamma\left(\frac{\gamma+1}{2}\right)}\,\mathrm{B}\left(\frac{3}{2},\ \frac{\gamma-1}{2}\right) = \frac{1}{\gamma-1}. \tag{77}$$

Consequently, for $\gamma > 1$ the correlation function $r(t)$ (not normalized) exists and has the form

$$r(t) = \mathrm{M}\xi_s\xi_{s+t} = e^{-\gamma t}\cdot\frac{1}{\gamma-1}. \tag{78}$$

We also note that the random variables ξ_s and ξ_{s+t} are in linear correlation, that is, the process $\{\xi_t\}$ is linear ([4]).

In conclusion, the author considers it his pleasant duty to express his gratitude to O. V. Sarmanov for his guidance with this work. The author also thanks A. P. Prudnikov for helpful discussions.

BIBLIOGRAPHY

[1] S. N. Bernšteĭn, *Principes de la théorie des équations différentielles stochastiques*, Trudy Mat. Inst. Steklov. 5 (1934), 95–124.

[2] I. M. Ryžik and I. I. Gradšteĭn, *Tables of integrals, sums, series and products*, GITTL, Moscow, 1951; 4th ed., Fizmatgiz, Moscow, 1963; English and German transl., 2nd rev. ed., VEB Deutscher Verlag der Wissenschaften, Berlin and Plenum Press, New York, 1963. MR 14, 643; MR 22 #3120; MR 28 #1326.

[3] W. Feller, *Parabolic differential equations and their corresponding semigroups of transformations*, Mat. Sb. 1 (1957), 105–153.

[4] O. V. Sarmanov, *Investigation of stationary Markov processes by the method of eigenfunction expansion*, Trudy Mat. Inst. Steklov. 60 (1961), 238–261. MR 26 #1935.

[5] W. Feller, *One-dimensional diffusion processes*, Mat. Sb. 2 (1958), 120–146.

[6] E. Kamke, *Handbook of ordinary differential equations*, Fizmatgiz, Moscow, 1961.

[7] A. Kratzer and V. Franz, *Transzendente Funktionen*, Geest & Portig K.-G.,

Leipzig, 1960; Russian transl., IL, Moscow, 1963. MR 23 #A1843.

[8] G. Doetsch, *Tabellen zur Laplace-Transformation und Anleitung zum Gebrauch*, Springer, Berlin, 1947. MR 9, 237.

[9] A. Erdélyi, W. Magnus, F. Oberhettinger and F. G. Tricomi, *Tables of integral transforms*, I, II, McGraw-Hill, New York, 1954, MR 16, 468.

[10] V. A. Ditkin and A. P. Prudnikov, *Integral transforms and operational calculus*, Fizmatgiz, Moscow, 1961.

Translated by:
L. Rosenblatt

TIME OF STAY OF A SYSTEM IN A FAULTY STATE

È. L. PRESMAN

1. In practice we often encounter situations where we are interested in the time a system of m devices is in a faulty state; the devices fail and are restored over random time intervals. If at time t, k devices are operating, then the probability that at time $t + dt$ one device is restored and $k + 1$ devices operate is proportional to the number of faulty devices and equal to $Q_2(m - k)dt$; the probability that a device fails and $k - 1$ devices operate is proportional to the number of operating devices and equals $Q_1 kdt$; while the probability that no changes take place equals $1 - [Q_1 k + Q_2(m - k)]dt$. By a faulty state we mean one where less than a given number n of devices are operating.

Consider a Markov process with states $E_0, E_1, E_2, \cdots, E_n$ beginning with state E_{n-1} and broken off upon entering into state E_n. The probability of passing from state E_k to E_{k+1} in time dt equals $\lambda_k dt$ $(0 \le k \le n - 1)$; the probability of passing from E_k to E_{k-1} equals $M_k dt$ $(1 \le k \le n - 1)$; the probability of remaining in state E_k equals $1 - (\lambda_k + M_k)\,dt$ $(1 \le k \le n - 1)$. The time up to the "breaking off" of such a process τ_n (with substitution of the values $Q_2(m - k)$ and $Q_1 k$ for λ_k and M_k respectively) is the random variable of interest.

Suppose that the probability that the system is in state E_k at time t is equal to $p_k(t)$. Then according to [1] for our process we have the following system of differential equations

$$\frac{dp_n(t)}{dt} = \lambda_{n-1} p_{n-1}(t),$$

$$\frac{dp_{n-1}(t)}{dt} = \lambda_{n-2} p_{n-2}(t) - (\lambda_{n-1} + M_{n-1})\, p_{n-1}(t),$$

$$\frac{dp_k(t)}{dt} = \lambda_{k-1} p_{k-1}(t) - (\lambda_k + M_k)\, p_k(t) + M_{k+1} p_{k+1}(t) \quad (1 \leqslant k \leqslant n-2), \tag{1.1}$$

$$\frac{dp_0(t)}{dt} = -\lambda_0 p_0(t) + M_1 p_1(t),$$

with initial conditions

$$p_{n-1}(0) = 1; \quad p_k(0) = 0 \quad (k \neq n-1), \tag{1.2}$$

where

$$p_n(t) = \mathbf{P}\{\tau_n < t\} = F_{\tau_n}(t). \tag{1.3}$$

2. Basic formulas.

$$\frac{dF_{\tau_n}(t)}{dt} = \sum_{i=1}^{n} a_i e^{-\lambda_i^{(n)} t}, \tag{2.1}$$

where $0 < \lambda_1^{(n)} < \lambda_2^{(m)} < \cdots < \lambda_n^{(n)}$.

Further,

$$\mathbf{M}(\tau_n) = \sum_{k=1}^{n} \frac{M_n M_{n-1} \ldots M_{n-k+1}}{M_n \lambda_{n-1} \lambda_{n-2} \ldots \lambda_{n-k}}. \tag{2.2}$$

For a process with r states, we define

$$\mathbf{M}(\tau_r) = \sum_{k=1}^{r} \frac{M_r M_{r-1} \ldots M_{r-k+1}}{M_r \lambda_{r-1} \lambda_{r-2} \ldots \lambda_{r-k}} \quad (1 \leqslant r \leqslant n-1)$$

in the obvious way. Then

$$\mathbf{M}(\tau_n^2) = 2 \sum_{r=1}^{n} [\mathbf{M}(\tau_r)] \cdot \sum_{k=n-r+1}^{n} \frac{M_n M_{n-1} \ldots M_{n-k+1}}{M_n \lambda_{n-1} \lambda_{n-2} \ldots \lambda_{n-k}}. \tag{2.3}$$

From (2.1) it is evident that as $t \to \infty$ the probability distribution for τ_n is determined only by the value $\lambda_1^{(n)}$. Therefore to obtain the asymptotics as $t \to \infty$ it is desirable to have an estimate for $\lambda_1^{(n)}$ from below. Such an estimate is given by

$$\lambda_1^{(n)} > \frac{1}{\sum_{k=1}^{n} \mathbf{M}\tau_k}. \tag{2.4}$$

For the case $\lambda_k = Q_2(m-k)$ and $M_k = Q_1 k$ we introduce the random variable ψ, taking on integral values and having a binomial distribution, with probability of success $\alpha = Q_2/(Q_1 + Q_2)$. (This distribution can be considered as the stationary distribution of the number of operating lines.) Letting $p_k = C_m^k \alpha^k (1-\alpha)^{m-k}$, we obtain

$$\mathbf{M}(\tau_n) = \frac{1}{(m-n+1)Q_2} \frac{\mathbf{P}\{\psi \leqslant n-1\}}{\mathbf{P}\{\psi = n-1\}} = \frac{1}{(m-n+1)Q_2} \sum_{k=0}^{n-1} \frac{p_k}{p_{n-1}}, \tag{2.5}$$

$$\mathbf{M}(\tau_n^2) - 2 \sum_{k=1}^{n} \frac{m-k+1}{m-n+1} \frac{p_{k-1}}{p_{n-1}} \mathbf{M}^2(\tau_k). \tag{2.6}$$

3. **Proofs of the basic formulas.** We apply the Laplace transform to the system (1.1), (1.2). Let $p_k^*(s)$ denote the Laplace transform of $p_k(t)$ for $0 \leq k \leq n$. Then for $p_k^*(s)$ we have the following recursion relations:

$$p_r^*(s)\,\varphi_{r+1}(s) = M_{r+1} p_{r+1}^*(s)\,\varphi_r(s) \quad (0 \leqslant r \leqslant n-2), \tag{3.1}$$

$$p_{n-1}^*(s)\,\varphi_n(s) = \varphi_{n-1}(s), \tag{3.2}$$

$$\left(\frac{dp_n(t)}{dt}\right)^* = \lambda_{n-1} p_{n-1}^*(s) = \frac{\lambda_{n-1}\varphi_{n-1}(s)}{\varphi_n(s)}, \tag{3.3}$$

where $\phi_r(s)$ are rth degree polynomials defined according to [2] by the following recursion relations:

$$\varphi_0(s)=1,\ \varphi_1(s)=s+\lambda_0,$$

$$\varphi_r(s)=\varphi_{r-1}(s)[s+\lambda_{r-1}+M_{r-1}]-\lambda_{r-2}M_{r-1}\varphi_{r-2}(s)\ (r>1). \tag{3.4}$$

According to [2] the polynomials $\phi_n(s)=s^n+d_1^{(n)}s^{n-1}+\dots+d_n^{(n)}$ have the following properties:

a) all of their roots lie on the negative part of the real axis and are distinct;

b) the roots of the $(n-1)$st polynomial alternate strictly with the roots of the nth; i.e., if $\phi_n(s)=\Pi_{k=1}^{n}(s+\lambda_k^{(n)})$, then

$$0<\lambda_1^{(n)}<\lambda_1^{(n-1)}<\lambda_2^{(n)}<\lambda_2^{(n-1)}<\dots<\lambda_{n-1}^{(n)}<\lambda_{n-1}^{(n-1)}<\lambda_n^{(n)}. \tag{3.5}$$

It is also easy to prove the following properties by distinction:

c)

$$\frac{d_r^{(n)}}{d_{r-1}^{(n)}}<\frac{d_{r-1}^{(n-1)}}{d_{r-2}^{(n-1)}}<\frac{d_{r-1}^{(n)}}{d_{r-2}^{(n)}}, \tag{3.6}$$

d)

$$d_r^{(r)}=\lambda_0\lambda_1\dots\lambda_{r-1}, \tag{3.7}$$

e)

$$d_{r-1}^{(r)}-\lambda_{r-1}d_{r-2}^{(r-1)}=\sum_{k=1}^{r}d_{r-k}^{(r-k)}\frac{M_rM_{r-1}\dots M_{r-k+1}}{M_r}, \tag{3.8}$$

f)

$$d_{r-2}^{(r)}-\lambda_{r-1}d_{r-2}^{(r-1)}=\sum_{k=1}^{r}d_{r-k-1}^{(r-k)}\frac{M_rM_{r-1}\dots M_{r-k+1}}{M_r}. \tag{3.9}$$

From (3.3) and (3.5) it follows that

$$\left(\frac{dp_n(t)}{dt}\right)^*=\sum_{i=1}^{n}\frac{a_i}{s+\lambda_i^{(n)}}.$$

Inversion of the Laplace transform proves formula (2.1).

For proof of (2.2) and (2.3) we note that according to (3.3),

$$\left(\frac{dp_n(t)}{dt}\right)^*=\frac{\lambda_{n-1}\varphi_{n-1}(s)}{\varphi_n(s)}=\lambda_{n-1}\frac{s^{n-1}+d_1^{(n-1)}s^{n-2}+\dots+d_{n-1}^{(n-1)}}{s^n+d_1^{(n)}s^{n-1}+\dots+d_n^{(n)}}$$

$$=\frac{\lambda_{n-1}}{d_n^{(n)}}[d_{n-1}^{(n-1)}+d_{n-2}^{(n-1)}s+d_{n-3}^{(n-1)}s^2+o(s^2)]$$

$$\times\left[1-\frac{d_{n-1}^{(n)}}{d_n^{(n)}}s-\frac{d_{n-2}^{(n)}}{d_n^{(n)}}s^2+\left(\frac{d_{n-1}^{(n)}}{d_n^{(n)}}\right)^2s^2+o(s^2)\right]$$

$$=\frac{\lambda_{n-1}}{d_n^{(n)}}\left\{d_{n-1}^{(n-1)}-\left(\frac{d_{n-1}^{(n)}}{d_n^{(n)}}d_{n-1}^{(n-1)}-d_{n-2}^{(n-1)}\right)s\right.$$

$$\left.+\left[\left(\frac{d_{n-1}^{(n)}}{d_n^{(n)}}\right)^2d_{n-1}^{(n-1)}-\frac{d_{n-2}^{(n)}}{d_n^{(n)}}d_{n-1}^{(n-1)}-\frac{d_{n-1}^{(n)}}{d_n^{(n)}}d_{n-2}^{(n-1)}+d_{n-3}^{(n-1)}\right]s^2+o(s^2)\right\}. \tag{3.10}$$

Hence, using properties d) and e), we obtain

$$\mathbf{M}(\tau_n)=\frac{\lambda_{n-1}}{d_n^{(n)}}\left(\frac{d_{n-1}^{(n)}}{d_n^{(n)}}\,d_{n-1}^{(n-1)}-d_{n-2}^{(n-1)}\right)=\frac{d_{n-1}^{(n)}-\lambda_{n-1}d_{n-2}^{(n-1)}}{d_n^{(n)}}$$

$$=\sum_{k=1}^{n}\frac{d_{n-k}^{(n-k)}}{d_n^{(n)}}\,\frac{M_nM_{n-1}\dots M_{n-k+1}}{M_n}=\sum_{k=1}^{n}\frac{M_nM_{n-1}\dots M_{n-k+1}}{M_n\lambda_{n-1}\lambda_{n-2}\dots\lambda_{n-k}},$$

which also proves (2.2). Hence it is easy to obtain

$$d_{n-1}^{(n)}=d_n^{(n)}\sum_{k=1}^{n}\mathbf{M}\tau_k. \tag{3.11}$$

From (3.10), using (3.11) and property f), we obtain

$$\frac{1}{2}\mathbf{M}(\tau_n^2)=\frac{\lambda_{n-1}}{d_n^{(n)}}\left[\frac{d_{n-1}^{(n)}}{d_n^{(n)}}\left(\frac{d_{n-1}^{(n)}}{d_n^{(n)}}\,d_{n-1}^{(n-1)}-d_{n-1}^{(n-1)}\right)\right]-\frac{\lambda_{n-1}}{d_n^{(n)}}\left[\frac{d_{n-2}^{(n)}}{d_n^{(n)}}\,d_{n-1}^{(n-1)}-d_{n-3}^{(n-1)}\right]$$

$$=\frac{\lambda_{n-1}}{d_n^{(n)}}\,\frac{d_{n-1}^{(n)}}{d_n^{(n)}}\left(\frac{d_{n-1}^{(n)}-\lambda_{n-1}d_{n-2}^{(n-1)}}{\lambda_{n-1}}\right)-\frac{\lambda_{n-1}}{d_n^{(n)}}\left(\frac{d_{n-2}^{(n)}-\lambda_{n-1}d_{n-3}^{(n-1)}}{\lambda_{n-1}}\right)$$

$$=\frac{d_{n-1}^{(n)}}{d_n^{(n)}}\mathbf{M}(\tau_n)-\frac{d_{n-2}^{(n)}-\lambda_{n-1}d_{n-3}^{(n-1)}}{d_n^{(n)}}=\mathbf{M}(\tau_n)\sum_{k=1}^{n}\mathbf{M}(\tau_k)$$

$$-\sum_{k=1}^{n}\frac{d_{n-k-1}^{(n-k)}}{d_{n-k}^{(n-k)}}\,\frac{M_nM_{n-1}\dots M_{n-k+1}}{M_n\lambda_{n-1}\lambda_{n-2}\dots\lambda_{n-k}}=\mathbf{M}(\tau_n)\sum_{r=1}^{n}\mathbf{M}(\tau_r)-\sum_{k=1}^{n}\frac{M_nM_{n-1}\dots M_{n-k+1}}{M_n\lambda_{n-1}\lambda_{n-2}\dots\lambda_{n-k}}$$

$$\times\sum_{r=1}^{n-k}\mathbf{M}(\tau_2)=\mathbf{M}(\tau_n)\sum_{r=1}^{n}\mathbf{M}(\tau_r)-\sum_{r=1}^{n}\mathbf{M}(\tau_r)\sum_{k=1}^{n-r}\frac{M_nM_{n-1}\dots M_{n-k+1}}{M_n\lambda_{n-1}\lambda_{n-2}\dots\lambda_{n-k}}$$

$$=\sum_{r=1}^{n}\mathbf{M}(\tau_r)\sum_{k=n-r+1}^{n}\frac{M_nM_{n-1}\dots M_{n-k+1}}{M_n\lambda_{n-1}\lambda_{n-2}\dots\lambda_{n-k}},$$

which proves formula (2.3). Formulas (2.5) and (2.6) are obtained from (2.2) and (2.3) via the substitution $\lambda_k=\theta_2(m-k)$, $M_k=\theta_1\cdot k$ and simple transformations.

For proof of (2.4) we note that, according to ([3], Part III, No. 23):

$$\lambda_1^{(n)}>\min\left(d_1^{(n)};\ \frac{d_2^{(n)}}{d_1^{(n)}};\ \frac{d_3^{(n)}}{d_2^{(n)}};\ \dots;\ \frac{d_{n-1}^{(n)}}{d_{n-2}^{(n)}};\ \frac{d_n^{(n)}}{d_{n-1}^{(n)}}\right). \tag{3.12}$$

But by property c) the numbers in (3.12) are written in decreasing order, which proves (2.4), taking (3.11) into account.

In conclusion I wish to thank my advisor, Professor Ju. V. Prohorov, for the statement of this problem.

BIBLIOGRAPHY

[1] N. T. J. Bailey, *Some further results in the non-equilibruim theory of a simple queue*, J. Roy. Statist. Soc. Ser. 19 (1957).

[2] W. Ledermann and G. E. H. Reuter, *Spectral theory for the differential equations of simple birth and death processes*, Philos. Trans. Roy. Soc. London. Ser. A. 246 (1954), 321–369. MR 15, 625.

[3] G. Pólya and G. Szegö, *Problems and theorems of analysis*. Chapter 1, Fizmatgiz, Moscow, 1956.

Translated by:
L. Rosenblatt

CONTROL OF A WIENER PROCESS WITH BOUNDED NUMBER OF SWITCHINGS

Ju. V. PROHOROV

Introduction

In this note we consider a simple class of controls of a Wiener process. Each of these controls requires only a finite number of switchings per unit of time. The aim of the control is either to keep the process within given bounds as long as possible, or to minimize the [stationary] probability of its being beyond the given bounds. With the help of standard procedures from the theory of random walk, we find relations determining the optimal control. From my point of view, Theorems 1 and 2 are of basic interest; they show that the precise formulas turn into simple asymptotic expressions under the assumption of "high quality" of the control (the exact meaning of the latter will be explained below).

We may note that in recent years many publications have appeared on controlled Markov processes. The aim of the control is to maximize (minimize) the mean value of some additive function of the realization of the process. Here the control most often reduces to a change in the bounded limits of the "drift coefficients." Questions of existence of optimal controls and their construction reduce to a study of parabolic equations with discontinuous coefficients. In the case of one-dimensional processes, optimal behavior often consists of "switching" of the drift coefficient from one extreme position to another, which leads to the necessity of unboundedly frequent switching in a neighborhood of the equilibrium position.

The restrictions here imposed on the mean number of switchings per unit time necessarily lead to non-Markov controls. If desired, this restriction can be interpreted as a reflection of the "inertia" property of the controlling structure.

§1. **Statement of the problem.** Let $\xi(t)$ be a Wiener process, i.e., a Gaussian process with homogeneous and independent increments

$$\xi(0)=0, \quad \mathrm{M}\Delta\xi=0, \quad \mathrm{D}\Delta\xi=\sigma^2\Delta t.$$

We are given *basic* boundaries, namely at the points $-H$ and H.

A *control* is given by two parameters, denoted by x and a $(0 < x < H,\ a > 0)$. *Warning* boundaries are located at the points $-x$ and x. When the process first reaches the warning boundaries, a controlling influence is applied to it, namely a systematic downwards bias when x is reached and upwards when $-x$ is reached. The rate of shift is assumed constant and in absolute value equal to a. With subsequent attainment of the opposite boundary, the direction of the shift is reversed etc. (It is clear that between two switchings the process may leave the interval with bounds $-x$ and x.)

Aim of the control. First formulation. For given H choose x and a so as to maximize the mean time ϑ to the first exit from the basic bounds.

Second formulation. For given H choose x and a so as to minimize the stationary probability p of being beyond the boundaries $(-H, H)$.

Choosing the parameter a sufficiently large for given x, $0 < x < H$, it is easy to see that we can make the time ϑ arbitrarily large and the probability p arbitrarily small. In order to make the problem meaningful, we impose the

Restriction. The mean time δ between two switchings must be $\geq \delta_0$, where δ_0 is a given positive number. This is clearly equivalent to a restriction from above on the mean number N of switchings per unit time.

We note that it we replace the process $\xi(t)$ by the process

$$\xi'(t) = \frac{1}{H}\xi(tH^2\sigma^{-2}),$$

we can reduce the general case to the special one: $H' = 1$, $\sigma' = 1$. Here

$$x' = xH^{-1}, \quad a' = aH\sigma^{-2}, \quad \delta_0' = \delta_0\sigma^2H^{-2},$$

$$\vartheta' = \vartheta\sigma^2H^{-2}, \quad p' = p.$$

It can be shown that ϑ' has a simple meaning; it is the ratio between the time up to exit from the basic bounds under regulation and the analogous quantity in the absence of regulation. Therefore it is natural to call ϑ' the "improvement coefficient." Obviously

$$\vartheta' = \vartheta'(x', a') = \vartheta'(xH^{-1}, aH\sigma^{-2}).$$

The "high quality" of control noted above corresponds to the assumption

$$\vartheta' \to \infty \quad \text{or} \quad p' = p \to 0.$$

§2. Precise formulas. It can be shown that in the first formulation (maximization of mean time up to exit from the basic bounds)

$$\vartheta'(x', a') = x'^2 - \frac{1}{a'^2}V + \frac{2x'}{a'}\frac{e^{2V}-1}{e^{4a'x'}-1}$$

and

$$\delta_1'(x', a') = \frac{1}{a'^2}[-U \operatorname{cth} U + V \operatorname{cth} V],$$

where $U = a'(1-x')$, $V = a'(1+x')$ (the subscript on δ' indicates which formulation of the problem is being considered).

Further, in the second formulation (maximization of the stationary probability)

$$p'(x', a') = e^{-2a'}\frac{\operatorname{sh}(2a'x')}{2a'x'}$$

and

$$\delta_2'(x', a') = \frac{2x'}{a'}.$$

These formulas can be obtained either by methods developed especially for diffusion processes [1] or by a limit passage from known relations for discrete random walk [2], Chapter XIV, §§2–6. We briefly review the latter.

a. **Simulating random walk.** Suppose $t \geq 0$, $t_{n,k} = k/n$, $\Delta t = 1/n$ and the random sequence $\xi^{(n)}(t_{n,k})$ is defined by the condition: the increments

$$\Delta\xi_k^{(n)} = \xi^{(n)}(t_{n,k+1}) - \xi^{(n)}(t_{n,k})$$

are independent for distinct k and take on the values $1/\sqrt{n}$ and $-1/\sqrt{n}$ with probabilities $p_n = 1/2 - a/2\sqrt{n}$ and $q_n = 1/2 + a/2\sqrt{n}$ respectively. Then $\mathbf{M}\Delta\xi_k^{(n)} = -a\Delta t$ and $\mathbf{D}\Delta\xi_k^{(n)} \sim \Delta t$. Thus this discrete walk "simulates" Brownian motion with

$$\mathbf{M}\Delta\xi = -a\Delta t, \quad \mathbf{D}\Delta\xi = \Delta t. \tag{2.1}$$

Together with this we consider the random sequence

$$\eta^{(n)}(k) = \sqrt{n}\,\xi^{(n)}(t_{n,k}). \tag{2.2}$$

Its increments $\Delta\eta_k^{(n)} = \sqrt{n}\,\Delta\xi_k^{(n)}$ take on the values 1 and -1 with probabilities p_n and q_n respectively. Thus $\eta^{(n)}(k)$ is random walk of the type described by Feller (op. cit.). We can now solve this or other problems involving a random Wiener process with (2.1) by reducing them to the analogous problems for random walk (2.2). Recall that the formula for $a \neq 0$ and for $a = 0$ have different forms. The case $a = 0$ is obtained from the first by the usual rules for indeterminacy.

b. Suppose we are given the boundaries $(-h', h)$ and the points x, $-h' < x < h$. Consider the process (2.1) beginning at time $t = 0$ at the point x, and assume that the boundaries $-h'$ and h are absorbing. Let $B^{\pm}(x, -h', h)$ denote the event: absorption occurs at the upper (lower) barrier and $P^{\pm}(x, -h', h, a) = P(B^{\pm})$. Further, let $\theta^{\pm}(x, -h', h)$ denote the time up to absorption on the upper (lower) barrier (if it occurred), and let θ be the time up to absorption. Obviously

$$\vartheta = \mathbf{M}\theta = \mathbf{M}(\theta^+ \mid B^+)P^+ + \mathbf{M}(\theta^- \mid B^-)P^-. \tag{2.3}$$

For computing the enumerated probabilities and mathematical expectations we consider the random walk described by the sequence (2.2) with beginning $\eta^{(n)}(0)$ at the point $z = (x + h')\sqrt{n}$ and absorbing barrier at the points 0 and $\alpha = (h + h')\sqrt{n}$. Using the formulas from the cited book of Feller on pp. 290, 294, 297, and passing to the limit as $n \to \infty$, we obtain

$$P^+ = \frac{e^{2a(x+h')} - 1}{e^{2a(h+h')} - 1}, \tag{2.4}$$

$$P^- = \frac{e^{2a(h+h')} - e^{2a(x+h')}}{e^{2a(h+h')} - 1}, \tag{2.5}$$

$$\vartheta = \mathbf{M}\theta = \begin{cases} \dfrac{x+h'}{a} - \dfrac{h+h'}{a} P^+, & a \neq 0, \quad (2.6) \\ (x+h')(h-x), & a = 0, \quad (2.7) \end{cases}$$

$$\mathbf{M}(\theta^- \mid B^-) P^- = \frac{1}{a^2} e^{a(x+h')} \frac{-U \operatorname{ch} U \operatorname{sh} V + V \operatorname{sh} U \operatorname{ch} V}{\operatorname{sh}^2 V}, \quad (2.8)$$

$$U = a(h-x), \quad V = a(h+h').$$

We note one more formula related to the process (2.1). We keep only one absorbing boundary, namely $-h'$. For $a > 0$ absorption occurs at $-h'$ with probability one. Here a certain (random) time $\tau(x, h, h')$ will be spent above the boundary h. In the same way as above, it is easy to compute that

$$\mathbf{M}\tau(x; -h', h) = \frac{1}{a^2} e^{-2a\left(\frac{h'-x}{2} + h\right)} \operatorname{sh} a(x+h'). \quad (2.9)$$

c. Proof of the formulas at the beginning of the section is adequately illustrated in the case of the first formulation. The mean time ϑ' up to exit from the boundaries $(-1, 1)$ for the chosen method of control is equal to $\mathbf{M}\tau_1 + R$. Here τ_1 is the moment of first exit from the "warning boundaries," and R is defined by the equation

$$R = \mathbf{M}(\theta^+ \mid B^+) P^+ + [\mathbf{M}(\theta^- \mid B^-) + R] P^-, \quad (2.10)$$

where $\theta^\pm, B^\pm, P^\pm$ correspond to the values of the parameters $x = x'$, $-h' = -x'$, $h = 1$, $a = a'$. From (2.10) we obtain that

$$R = \frac{\mathbf{M}\theta(x', -x', 1)}{P(B^+(x', -x', 1))}.$$

Making use of (2.6) and (2.7), we find

$$\vartheta' = x'^2 - \frac{1+x'}{a'} + \frac{2x'}{a'} \frac{e^{2a'(1+x')} - 1}{e^{4ax'} - 1}.$$

The mean time between two switchings in this model obviously equals

$$\mathbf{M}(\theta^-(x', -x', 1) \mid B^-(x', -x', 1)),$$

which, by (2.8) and (2.5), reduces to

$$\frac{1}{a'^2} \cdot \frac{-U \operatorname{ch} U \operatorname{sh} V + V \operatorname{sh} U \operatorname{ch} V}{\operatorname{sh} V \operatorname{sh} U}$$

with $U = a'(1 - x')$, $V = a'(1 + x')$.

§3. Asymptotic analysis. It is suitable to preface the formulation of the basic theorems with the following argument: wishing to consider the largest values of ϑ' (or smallest values of p'), we must necessarily assume δ_0' small. But for small δ_0' the choice $x' = 1/2$, $a' = 1/\delta_0'$ is "almost optimal"; this leads to the values

$$\vartheta' \sim \delta_0' e^{\frac{1}{\delta_0'}} = \frac{\delta_0 \sigma^2}{H^2} \exp\left[\frac{H^2}{\delta_0 \sigma^2}\right]$$

and

$$p' \sim \frac{\delta_0'}{2} e^{-1/\delta'_0} = \frac{\delta_0 \sigma^2}{2H^2} \exp\left[-\frac{H^2}{\delta_0 \sigma^2}\right].$$

Theorem 1. *Let* $X' = 1/2$, $A' = 1/\delta_0'$. *Then*

$$\delta_1'(X', A') \geqslant \delta_0'$$

and for any $\epsilon > 0$, *the inequality*

$$\vartheta'(x', a') \geqslant (1+\varepsilon)\,\vartheta'(X', A')$$

is impossible for all x' *and* a' *with* $\delta_1'(x.\ a) \geq \delta_0'$, *if only* δ_0' *is sufficiently small* $(\delta_0' \leq \delta_1(\epsilon))$.

Theorem 2. *For all* X' *and* A'

$$\delta_2'(X', A') = \delta_0'$$

and for any $\epsilon > 0$ *the inequality*

$$p'(x', a') \leqslant (1-\varepsilon)\, p'(X', A')$$

is impossible for all x' *and* a' *with* $\delta_2'(x', a') \geq \delta_0'$, *if* δ_0' *is sufficiently small* $(\delta_0' \leq \delta_2(\epsilon))$.

These theorems are complemented in a natural fashion by the following two assertions.

Theorem 3. *If* $\vartheta'(X', A') \geq k$, *then for any* x', a' *with* $\delta_1'(x', a') \geq \delta_1'(X', A')$

$$\sup \vartheta'(x', a') \leqslant \left(1 + e^{-C_1\sqrt{\ln k}}\right) \vartheta'(X', A'),$$

where C_1 *is an absolute constant.*

Theorem 4. *If* $p(X', A') \leq \epsilon$, *then for any* x', a' *with* $\delta_2'(x', a') \geq \delta_2'(X', A')$

$$\inf p(x', a') \geqslant \left(1 - e^{-C_2\sqrt{\ln\frac{1}{\varepsilon}}}\right) p(X', A'),$$

where C_2 *is an absolute constant.*

The proofs of Theorems 1–4 present no special difficulties. We omit them.

The preceding illustrates, in the simplest formulation, the means of solution, precise and approximate, of problems such as those formulated in the introduction. The author plans to treat more complicated variants in a later article. In conclusion, I consider it my pleasant duty to thank L. Bol'šev, V. Zolotarev and E. Presman for their discussions of the questions dealt with here.

BIBLIOGRAPHY

[1] D. A. Darling and A. I. Siegert, *The first passage problem for a continuous Markov process*, Ann. Math. Statist. 24(1953), 624.

[2] W. Feller, *Introduction to probability theory and its applications*, IL, Moscow, 1952.

Translated by:
L. Rosenblatt

A PROBLEM IN OPTIMAL CONTROL OF A COMPLEX OF INSTRUMENTS

Ju. A. ROZANOV

Suppose there are n devices designed to measure some quantity $a = \langle a(t) \rangle$ which is a rather arbitrary function of time t. Each of the devices has its own measurement error; namely, the reading of the kth device $X_k(t)$ at the moment of time t differs from the corresponding value $a(t)$ by $x_k(t)$:

$$X_k(t) = a(t) + x_k(t), \quad k = 1, 2, \ldots, n, \tag{1}$$

where the measurement errors $\langle x_k(t) \rangle$, which are of a random nature, do not depend on the measured quantity $a = \langle a(t) \rangle$.

There arises the problem of "combining" these n separate devices (generally speaking, different in structure) into a *complex* in such a way that its output variable $X(t)$, formulated on the basis of the separate readings $X_k(t)$ in the interval $T_0 \le t \le T$ (where T_0 may equal $-\infty$), satisfies the following requirements: *first, the measurement error*

$$x(T) = X(T) - a(T) \tag{2}$$

must not depend on the values of the measured quantity $\langle a(t) \rangle$, *and second, the magnitude of this error must be minimal,* in the sense of

$$\delta^2 = \mathbf{M}[x(T)]^2 = \min. \tag{3}$$

A function $X(T)$ satisfying these two conditions is said to be *optimal.*

In investigating this problem it will be convenient for us to use the new "variables":

$$\begin{gathered} Y_{k^*}(t) = X_{k^*}(t), \\ Y_k(t) = X_{k^*}(t) - X_k(t), \quad k \neq k^* \ (k = 1, 2, \ldots n). \end{gathered} \tag{4}$$

The collection of quantities $Y_k(t)$, $k \neq k^*$, is denoted by $Y(t)$. Here the index k^* can be chosen arbitrarily: $1 \le k^* \le n$. Obviously the collection of variables $Y(t)$ is the same as the previous one if we replace k^* by another index l^*, since the differences $X_{k^*}(t) - X_k(t)$ and $X_{l^*}(t) - X_l(t)$, $k, l = \overline{1, n}$, are expressed linearly in terms of each other. For convenience we shall often assume that $k^* = n$.

It is natural to turn our attention to the fact that the quantities $Y_k(t)$, $k \neq k^*$, do not depend on the measured function $\langle a(t) \rangle$:

$$Y_k(t) = x_{k^*}(t) - x_k(t), \quad k^* \neq k, \tag{5}$$

while

$$Y_{k^*}(t) = a(t) + x_{k^*}(t). \tag{6}$$

Assertion 1. *Every variable $X(T)$ satisfying the first of the given conditions can be represented in the form*

$$X(T) = X_{k^*}(T) - \hat{x}_{k^*}(T), \tag{7}$$

where $\hat{x}_{k^*}(T)$ is a function only of $Y(t)$, $T_0 \le t \le T$.

We prove the assertion.

The minimal requirements (of measurability or continuity type) usually imposed on random processes make it possible to restrict consideration to the situation where the number of values of $Y_k(t)$, $k = 1, 2, \cdots, n$; $T_0 \le t \le T$, is finite. The assumption of sufficient arbitrariness of the measured quantities $a(t)$ allows us to consider them as independent variables for distinct t. This immediately compels us to draw the conclusion that the quantity $X(T)$, which is a function of $Y_k(t)$, $k = 1, 2, \cdots, n$, $T_0 \le t \le T$, does not actually depend on $Y_{k^*}(t)$ for $t \ne T$, since neither the remaining $Y_k(t)$, $k \ne k^*$, nor the difference $x(T) = X(T) - a(T)$ by hypothesis depends on $\langle a(t) \rangle$.

Further, for fixed values of all the arguments, besides $Y_{k^*}(T) = a(T) + x_{k^*}(T)$, the function $f(Y)$, $f[Y_{k^*}(T)] = X(T)$ is such that the difference $f[Y_{k^*}(T)] - a(T)$ need not depend on $a(T)$. But if the difference $\hat{x}_{k^*} = (a + x) - f(a + x)$ does not depend on a, then it also does not depend on x, since $\hat{x}_{k^*} = -f(0)$ for any fixed x if we take $a = -x$. Thus for fixed $Y_k(t)$, $k \ne k^*$, $T_0 \le t \le T$, the function $X(T)$ has the form (7), where $\hat{x}_{k^*}(T)$ is the value of $X(T)$ in the case when $Y_{k^*}(T) = 0$.

Assertion 2. *The variable $X(T)$, presented in the form (7), satisfies the second of the given conditions if and only if* *

$$\hat{x}_{k^*}(T) = \mathrm{M}[x_{k^*}(T) \mid Y(t),\ T_0 \leqslant t \leqslant T]. \tag{8}$$

This assertion is an obvious consequence of relation (3), based on well-known properties of conditional mathematical expectation.

The quantity $\hat{x}_{k^*}(T)$, as is well known, has a simple geometric meaning. Namely, if we identify all random variables which are unequal only with probability 0, and consider the Hilbert space $\mathbf{H}$ of all the variables h, $\mathrm{M}|h|^2 < \infty$, with scalar product $(h_1, h_2) = \mathrm{M}h_1 \cdot \bar{h}_2$, then $\hat{x}_{k^*}(T)$ is the projection of the element $x_{k^*}(T)$ of this space $\mathbf{H}$ onto the subspace $\mathbf{H}_y(T)$ formed by all possible functions of $Y_k(t)$, $k \ne k^*$, $T_0 \le t \le T$.

The measurement error $x(T)$ is obviously

$$x(T) = x_{k^*}(T) - \hat{x}_{k^*}(T) \tag{9}$$

and is the perpendicular dropped from the element $x_{k^*}(T)$ onto the subspace $\mathbf{H}_y(T)$.

---

* Here the symbol $\mathrm{M}(x|y)$ denotes the conditional mathematical expectation of the variable x, given y.

Thus the optimal function $X(T)$ has the form (7), while the measurement error $x(T)$ for the optimal choice of the function $X(T)$ is determined by formulas (2), (8), (9).

We see that apparently both the function $X(T)$ itself and its corresponding error $x(T)$ depend on the chosen index k^*. We shall show that this is indeed not so. Namely, for any l^*, $1 \le l^* \le n$,

$$x_{k^*}(T) - \hat{x}_{k^*}(T) = x_{l^*}(T) - \hat{x}_{l^*}(T). \tag{10}$$

Actually, the subspace $\mathbf{H}_y(T)$ introduced by us remains the same if the index k^* is replaced by l^*, since the collection $Y(t)$, $T_0 \le t \le T$, in both parts of equality (10) is an element of the space $\mathbf{H}$ orthogonal to the subspace $\mathbf{H}_y(T)$, and the difference of these elements together with $\hat{x}_{k^*}(T)$, $\hat{x}_{l^*}(T)$, and $x_{k^*}(T) - x_{l^*}(T)$ belongs to $\mathbf{H}_y(T)$. But this can be true only in the case where the elements themselves coincide.

We consider several simple examples.

Example 1. Let $x_1, \cdots, x_n$ be Gaussian independent and identically distributed variables with parameters $(0, \sigma)$. We shall find the quantity

$$\hat{x}_n = \mathbf{M}[x_n \mid Y_1, \ldots, Y_{n-1}].$$

As is well known, for Gaussian variables the projection $\hat{x}_n$ onto the subspace $\mathbf{H}_y$ lies in the linear span $\mathbf{H}_y$ of the $Y_1, \cdots, Y_{n-1}$ themselves, that is,

$$\hat{x}_n = \sum_{k=1}^{n-1} c_k Y_k.$$

The coefficients c_k $(k = \overline{1, n-1})$ can be determined from the orthogonality condition of the difference $x_n - \hat{x}_n$ to all Y_j $(j = \overline{1, n-1})$:

$$\sum_{k=1}^{n-1} c_k (Y_k, Y_j) = (x_n, Y_j),$$

where

$$(Y_k, Y_j) = \begin{cases} 2\sigma^2 & \text{for} \quad k = j, \\ \sigma^2 & \text{for} \quad k \ne j, \end{cases}$$

$$(x_n, Y_j) = \sigma^2.$$

We immediately obtain that all the coefficients c_k are equal:

$$c_k = \frac{1}{n} \quad (k = \overline{1, n-1}).$$

Thus

$$\hat{x}_n = \frac{1}{n} \sum_{k=1}^{n-1} Y_k,$$

and the optimal function

$$X = X_n - \hat{x}_n$$

has the form

$$X = X_n - \frac{1}{n}\sum_{k=1}^{n-1}(X_n - X_k) = \frac{1}{n}\sum_{k=1}^{n} X_k. \tag{11}$$

Thus the optimal function X is the arithmetic mean of the separate values X_k, $k = \overline{1, n}$.

Example 2. Let $x_1, \cdots, x_n$ be mutually independent and identically distributed on the interval $(-d, d)$. The joint distribution of the variables $Y_1, \cdots, Y_{n-1}, x_n$, related to the original ones $x_1, \cdots, x_n$ $(k^* = n)$ by (5), has density of the following form

$$p(Y_1, \ldots, Y_{n-1}, x_n) = \begin{cases} \frac{1}{(2d)^n} & \text{for} \quad \alpha \leqslant x_n \leqslant \beta \\ 0 & \text{otherwise,} \end{cases}$$

where

$$\begin{aligned} \alpha &= \max(Y_1 - d, \ldots, Y_{n-1} - d, -d), \\ \beta &= \min(Y_1 + d, \ldots, Y_{n-1} + d, d). \end{aligned} \tag{12}$$

Further,

$$p(Y_1, \ldots, Y_{n-1}) = \begin{cases} \frac{\beta - \alpha}{(2d)^n} & \text{for} \quad \alpha \leqslant \beta \\ 0 & \text{for} \quad \alpha > \beta, \end{cases}$$

and consequently the density of the conditional distribution of x_n given $Y_1, \cdots, Y_{n-1}$ is

$$p(x_n \mid Y_1, \ldots, Y_{n-1}) = \begin{cases} \frac{1}{\beta - \alpha} & \text{for} \quad \alpha \leqslant \beta \\ 0 & \text{for} \quad \alpha > \beta. \end{cases}$$

From this we find $\hat{x}_n$:

$$\hat{x}_n = \mathbf{M}[x_n \mid Y_1, \ldots, Y_{n-1}] = \int_\alpha^\beta x_n p(x_n \mid Y_1, \ldots, Y_{n-1})\, dx_n = \frac{1}{2}(\alpha + \beta),$$

where α and β are defined by relations (12).

Now it is easy to see that the optimal function X,

$$X = X_n - \hat{x}_n$$

is the average of the extreme values from $X_1, \cdots, X_n$:

$$X = \frac{1}{2}(X_* + X^*), \tag{13}$$

where

$$X_* = \min_k X_k; \quad X^* = \max_k X_k.$$

Indeed, if

$$X_* < X_n < X^*,$$

then, taking into account relations (5) and (12), we have

$$\alpha = X_n - X_* - d,$$
$$\beta = X_n - X^* + d,$$
$$X = X_n - \frac{1}{2}(\alpha + \beta) = \frac{1}{2}(X_* + X^*).$$

If X_n coincides with one of the extreme terms X_* or X^* (say with X_*), then all the Y_k $(k = \overline{1, n-1})$ are negative, and consequently

$$\alpha = -d,$$

$$\beta = X_* - \frac{1}{2}(\alpha + \beta) = \frac{1}{2}(X_* + X^*).$$

It is of interest to compare the errors occurring for the choice of the function X of form (11) (it is optimal among all linear functions) and the function X of form (13). Without loss of generality we can assume that the variables $x_1, \cdots, x_n$ are distributed on the interval $(-1/2, 1/2)$. The magnitude of the first of these errors is

$$\delta_1^2 = \mathbf{M}\left[\frac{1}{n}\sum_{k=1}^{n} X_k - a\right]^2 = \mathbf{M}\left[\frac{1}{n}\sum_{k=1}^{n} x_k\right]^2 = \frac{1}{n}\int_{-1/2}^{1/2} x^2 dx = \frac{1}{12n}. \quad (14)$$

The magnitude of the second is

$$\delta_0^2 = \mathbf{M}\left[\frac{1}{2}(X_* + X^*) - a\right]^2 = \mathbf{M}\left[\frac{1}{2}(x_* + x^*)\right]^2,$$

where

$$x_* = \min_k x_k, \quad x^* = \max_k x_k.$$

For $n \geq 2$ we have:

$$\delta_0^2 = \frac{1}{4}\sum_{i,j}\int_{\{x_*=x_i,\ x^*=x_j\}} (x_i + x_j)^2\, dx_1 \ldots dx_n$$

$$= \frac{1}{4} n(n-1) \int_{\{x_*=x_1;\ x^*=x_2\}} (x_1 + x_2)^2\, dx_1 \ldots dx_n$$

$$= \frac{n(n-1)}{4}\int_{-1/2}^{1/2}\int_{x_1}^{1/2}\int_{x_1}^{x_2}\cdots\int_{x_1}^{x_2} (x_1 + x_2)^2\, dx_1 \ldots dx_n = \frac{1}{2(n+1)(n+2)}. \quad (15)$$

Thus

$$\frac{\delta_0^2}{\delta_1^2} = \frac{6n}{(n+1)(n+2)}, \quad n \geqslant 2, \quad (16)$$

which for large n compels us to give clear preference to the function X of the form (13).

Example 3. Let $x_1, \cdots, x_n$ be mutually independent and distributed according to a symmetric exponential law, i.e., the density of each of the x_k is

$$p(x) = \frac{\alpha}{2} e^{-\alpha|x|}, \quad \alpha > 0.$$

The conditional density of the probability distribution of x_k given $Y_1, \cdots, Y_{n-1}$ can be represented in the form

$$p(x_n \mid Y_1, \ldots, Y_{n-1}) = \frac{q(x_n)}{\int_{-\infty}^{\infty} q(x)\, dx},$$

where

$$q(x) = \begin{cases} \exp\left\{-\alpha \sum_{j=1}^{n} z_j\right\} e^{n\alpha x}, & -\infty < x \leqslant z, \\ \exp\left\{\alpha \sum_{i=1}^{k} z_i - \alpha \sum_{j=k+1}^{n} z_j\right\} e^{(n-2k)\alpha x}, & z_k < x \leqslant z_{k+1}, \ (k = \overline{1, n-1}), \\ \exp\left\{\alpha \sum_{i=1}^{n} z_i\right\} e^{-n\alpha x}, & z_n < x < \infty, \end{cases}$$

$z_1 \leq z_2 \leq \cdots \leq z_n$ is the variational series composed of the variables $Y_1, \cdots, Y_{n-1}, 0$.

We have:

$$\hat{x}_n = \mathbf{M}[x_n \mid Y_1, \ldots, Y_{n-1}] = \int_{-\infty}^{\infty} x_n p(x_n \mid Y_1, \ldots, Y_{n-1})\, dx_n$$

$$= \frac{\int_{-\infty}^{\infty} x q(x)\, dx}{\int_{-\infty}^{\infty} q(x)\, dx}.$$

Simple computations lead us to the following expressions:

$$\int_{-\infty}^{\infty} q(x)\, dx = \frac{1}{n\alpha} e^{-\lambda_1 \alpha} + \sum_{k=1}^{n-1} \frac{1}{(n-2k)\alpha} (e^{-\lambda_{k_2}\alpha} - e^{-\lambda_{k_1}\alpha}) + \frac{1}{n\alpha} e^{-\lambda_n \alpha}, \quad (17)$$

$$\int_{-\infty}^{\infty} xq(x)\, dx = \frac{1}{n\alpha}\left(z_1 - \frac{1}{n\alpha}\right) e^{-\lambda_1\alpha} + \sum_{k=1}^{n} \frac{1}{(n-2k)\alpha}\left[\left(z_{k+1} - \frac{1}{(n-2k)\alpha}\right) e^{-\lambda_{k_2}\alpha}\right.$$

$$\left. - \left(z_k - \frac{1}{(n-2k)\alpha}\right) e^{-\lambda_{k_1}\alpha}\right] + \frac{1}{n\alpha}\left(z_n + \frac{1}{n\alpha}\right) e^{-\lambda_n \alpha},$$

where

$$\lambda_1=\left[-(n-1)z_1+\sum_{j=2}^{n} z_j\right],$$

$$\lambda_{k1}=\left[-\sum_{i=1}^{k-1} z_i-(n-2k+1)z_k+\sum_{j=k+1}^{n} z_j\right], \tag{18}$$

$$\lambda_{k2}=\left[-\sum_{i=1}^{k} z_i-(n-2k-1)z_{k+1}+\sum_{j=k+2}^{n} z_j\right], \quad (k=\overline{1,\ n-1})$$

$$\lambda_n=\left[-\sum_{i=1}^{n-1} z_i+(n-1)z_n\right],$$

for $n = 2m$ we must replace the term corresponding to $k = m$ by $(z_{m+1} - z_m)e^{-\lambda_m \alpha}$, $\lambda_m = -\sum_{i=1}^{m} z_i + \sum_{j=m+1}^{n} z_j$.

We see that the optimal function X

$$X=X_n-\hat{x}_n$$

in this case depends in a rather complicated fashion on the differences $Y_k = X_n - X_k$ $(k = \overline{1, n-1})$.

It should be noted that this dependence in practice is often simplified for large α. Indeed, from expressions (18) it is evident that for even n $(n = 2m)$ the exponent $\lambda = \lambda_m$ is smallest, while for n odd $(n = 2m - 1)$, the smallest exponent is λ_{m1}. Hence it is easy to deduce that

$$\lim_{\alpha\to\infty}\hat{x}_n=\begin{cases} z_m, & \text{if } n=2m-1,\\ \frac{1}{2}(z_m+z_{m+1}), & \text{if } n=2m.\end{cases} \tag{19}$$

This relation makes it possible to conclude that the function X^* of the form

$$X^*=\begin{cases} Z_m, & \text{if } n=2m-1,\\ \frac{1}{2}(Z_m+Z_{m+1}), & \text{if } n=2m,\end{cases} \tag{20}$$

is quite a good approximation to the optimal function X; here Z_m and Z_{m+1} are the middle terms of the variational series

$$Z_1\leqslant Z_2\leqslant\cdots\leqslant Z_n, \tag{21}$$

composed of $X_1, X_2, \cdots, X_n$.

Now we consider the case where the measurement errors $\langle x_k(t)\rangle$ are random processes of the form

$$x_k(t)=\sum_{j=1}^{m}\theta_{kj}A_{kj}(t)+\xi_k(t), \tag{22}$$

where $\theta = \langle\theta_{kj}\rangle$ is a collection of numerical "parameters" with arbitrary probability distribution; $\langle A_{kj}(t)\rangle$ are known functions (not random) which are mutually linearly independent; $\langle \xi_k(t) \rangle$ are Gaussian stationary* processes with zero mean, statistically independent of the collection of parameters $\langle \theta_{kj} \rangle$ (all quantities are assumed real). Here for definiteness we restrict consideration to the more difficult case of continuous t.

It will be convenient for us to use the variables Y_k, $k = \overline{1, n-1}$, related to the errors $\langle x_k(t) \rangle$ by relations (5), in which $k^* = n$. Accordingly, we assume that the multidimensional random process (column vector) $Y(t) = \{Y_k(t)\}_{k=\overline{1,n-1}}$ has the form

$$Y(t) = \sum_{k=1}^{n-1} \sum_{j=1}^{m} \Delta_{kj} A_{kj}(t) + \eta(t), \tag{23}$$

where

$$\Delta_{kj} = \theta_{nj} - \theta_{kj}, \quad k = \overline{1, n-1}, \tag{24}$$

A_{kj} are vector functions of the form

$$A_{kj} = \begin{pmatrix} 0 \\ A_{kj}(t) \\ 0 \end{pmatrix}, \quad k = \overline{1, n-1}; \; j = \overline{1, m}, \tag{25}$$

all of whose elements, except for the kth, equal zero, while the kth element is a numerical function $A_k(t)$, and finally, $\eta(t) = \{\eta_k(t)\}_{k=\overline{1,n-1}}$ is a multidimensional stationary Gaussian process with components

$$\eta_k(t) = \xi_n(t) - \xi_k(t), \quad k = \overline{1, n-1}. \tag{26}$$

The problem of finding the optimal function $X(T)$, as was shown above, reduces to finding a variable $\hat{x}_n(T)$ of the form

$$\hat{x}_n(T) = \mathbf{M}[x_n(T) \mid Y(t),\; T_0 \leqslant t \leqslant T]. \tag{27}$$

Elementary properties of conditional mathematical expectations allow us to break this problem into two parts. The first consists of finding the variable

$$\hat{x}_n(T, \theta) = \mathbf{M}[x_n(T) \mid Y(t),\; T_0 \leqslant t \leqslant T,\; \langle \theta_{kj} \rangle], \tag{28}$$

and the second, of finding the variable

$$\begin{aligned} \hat{x}_n(T) &= \mathbf{M}[\hat{x}_n(T, \theta) \mid Y(t),\; T_0 \leqslant t \leqslant T] \\ &= \int_\theta \hat{x}_n(T, \theta)\, P(d\theta \mid Y(t),\; T_0 \leqslant t \leqslant T), \end{aligned} \tag{29}$$

where $P(d\theta | Y(t),\; T_0 \le t \le T)$ is the "a posteriori" probability distribution of the parameters $\langle \theta_{kj} \rangle$ (conditional distribution given $Y(t)$, $T_0 \le t \le T$).

In order to find $\hat{x}_n(T, \theta)$, we consider the variable

---

* The method of solution set forth in the present article is easily extended to the case of processes $\xi(t)$ with Nth stationary increments.

$$\xi_n(T) = x_n(T) - \sum_{j=1}^{m} \theta_{nj} A_{nj}(T) \tag{30}$$

and its projection $\hat{\xi}_n(T)$ onto the subspace $H_\eta(T)$ (the closed linear span of values $\eta_k(t)$, $T_0 \le t \le T$, $l \le k \le n-1$).

If

$$\eta(t) = \int_{-\infty}^{\infty} e^{i\lambda t} \Phi^\eta(d\lambda), \quad \Phi^\eta = \{\Phi_k^\eta\}_{k=\overline{1, n-1}} \tag{31}$$

is the spectral representation of the multidimensional stationary process $\eta(t)$, then the variable $\hat{\xi}_n(T)$ allows a spectral representation in the form

$$\hat{\xi}_n(T) = \int_{-\infty}^{\infty} \varphi(\lambda, T) \Phi^\eta(d\lambda), \tag{32}$$

where $\phi(\lambda, T) = \{\phi_k(\lambda, T)\}^{k=\overline{1,n-1}}$ is a vector function (row vector) of the space $L_T^2(F)$. This space is the closed linear span of the functions $\phi(\lambda) = e^{i\lambda t}\delta_k$ ($T_0 \le t \le T$, $1 \le k \le n-1$) (where δ_k is the vector whose kth component is 1, and the remaining components equal 0) such that *

$$\int_{-\infty}^{\infty} \varphi(\lambda) F(d\lambda) \overset{*}{\varphi}(\lambda) < \infty. \tag{33}$$

Here the measure $F(d\lambda) = \{F_{ki}(d\lambda)\}_{k=\overline{1,n-1}}^{l=\overline{1,n-1}}$ is the spectral measure of the stationary process $\pi(t)$.

The function $\phi(\lambda, T)$ in formula (32) is determined from the conditions of orthogonality of the difference $\xi_n(T) - \hat{\xi}_n(T)$ to all the variables $\eta_k(t)$, $T_0 \le t \le T$, $1 \le k \le n-1$:

$$\int e^{i\lambda t} F(d\lambda) \overset{*}{\varphi}(\lambda, T) = B^{\eta\xi}(t - T), \tag{34}$$

where $B^{\eta\xi}(t) = \{B_k^{\eta\xi}(t)\}^{k=\overline{1,n-1}}$, $T_0 \le t \le T$ is the mutual correlation function of the stationary processes $\xi_n(t)$ and $\eta(t) - \{\eta_k(t)\}^{k=\overline{1,n-1}}$ Relation (34) is an integral equation of Wiener-Hopf type relative to $\phi(\lambda, T)$; methods for solving it have been well worked out in the case of a rational spectrum [2], [3].

Symbolically it is convenient to represent the variable $\hat{\xi}_n(t)$ in the form

$$\hat{\xi}_n(T) = \int_{T_0}^{T} c(t) \eta(t)\, dt, \tag{35}$$

where the vector function $c(t) = \{c_k(t)\}^{k=\overline{1,n-1}}$ is the Fourier transform (generally speaking, in the generalized sense) of the function $\phi(\lambda, T)$. In the case of a

* Here and below $\overset{*}{\alpha}$ will denote the vector conjugate to α, i.e., transposed and complex-conjugate.

rational spectrum its components $c_k(t)$, $k = \overline{1, n-1}$, are linear combinations of δ-functions and their derivatives at the end points $t = T_0$, T and of certain ordinary functions which are square integrable on the interval $T_0 < t < T$ (see [2]).

Taking into account the Gaussian property of the variables $\xi_n(T)$ and $\langle \eta_k(t) \rangle$ and assuming that

$$\eta(t) = Y(t) - \sum_{k=1}^{n-1} \sum_{j=1}^{m} \Delta_{kj} A_{kj}(t), \tag{36}$$

it is easy to find an expression for $\hat{x}_n(T, \theta)$.

Namely,

$$\hat{x}_n(T, \theta) = \sum_{j=1}^{m} \theta_{nj} A_j(T) + \hat{\xi}_n(T). \tag{37}$$

Further, it is necessary for us fo pass from $\hat{x}_n(T, \theta)$ to the variable $\hat{x}_n(T)$, which can be done, for example, according to formula (29).

We note that sometimes some of the "parameters" $\Delta_{kj} = \theta_{nj} - \theta_{kj}$ can be uniquely determined from the present realization of the process $Y(t)$, $T_0 \le t \le T$, i.e., the a posteriori distribution $P(d\theta | Y(t), T_0 \le t \le T)$, involved in formula (29), turns out to be degenerate. Then these $\Delta_{kj} = \Delta_{kj}[Y(t), T_0 \le t \le T]$ can be excluded precisely as was done in finding the quantity $\hat{x}_n(T, \theta)$, when all the values of the parameters $\langle \theta_{kj} \rangle$ were assumed known (see relations (36), (37)). For example, if all the values Δ_{kj} $(k = \overline{1, n-1}, j = \overline{1, m})$ were determined from the realization $Y(t)$, $T_0 \le t \le T$, then formula (29) would take the form

$$\hat{x}_n(T) = \sum_{j=1}^{m} \hat{\theta}_{nj} A_j(T) + \hat{\xi}_n(T),$$

where $\hat{\theta}_{nj}$ are the mean values of the parameters θ_{nj} with respect to the a posteriori distribution, and $\hat{\xi}_n(tT)$ has the form (35) — in expression (36) for the process $\eta(t)$ it is only necessary to take those values of the parameters $\Delta_{kj} = \Delta_{kj}[Y(t), T_0 \le t \le T]$, which occur for the given realization $Y(t)$, $T_0 \le t \le T$.

The above remark allows us to assume, without loss of generality, that none of the parameters Δ_{kj} can be determined directly from the present model $Y(t)$, $T_0 \le t \le T$. The analytic essence of this assumption consists of the fact that every vector function $Y(t)$ allows a spectral representation of the form

$$A_{kj}(t) = \int_{-\infty}^{\infty} e^{i\lambda t} F(d\lambda) \overset{*}{\alpha}{}^{(kj)}(\lambda), \quad T_0 \leqslant t \leqslant T \;\; (k = \overline{1, n-1}; \; j = \overline{1, m}), \tag{38}$$

where $F(d\lambda) = \{F_{kl}(d\lambda)\}_{k=\overline{1,n-1}}^{l=\overline{1,n-1}}$ is the spectral measure of the stationary process $\eta(t)$, $\alpha^{(kj)}(\lambda) = \{\alpha_l^{(kj)}(\lambda)\}^{l=1,n-1}$ is a vector function from the space $L_T^2(F)$ (see [2]).

We note that if the representation (38) holds for each of the functions $A_{kj}(t)$, then the process $Y(t)$ itself is harmonizable:

$$Y(t)=\int_{-\infty}^{\infty} e^{i\lambda t}\Phi^{Y}(d\lambda),\quad T_0\leqslant t\leqslant T,\quad \Phi^{Y}=\{\Phi_{k}^{Y}\}_{k=\overline{1,\,n-1}}. \tag{39}$$

Here the spectral random measure $\Phi^Y(d\lambda)$ is defined by the formula

$$\Phi^{Y}(d\lambda)=F(d\lambda)\overset{*}{\alpha}(\lambda)+\Phi^{\eta}(d\lambda), \tag{40}$$

where $\alpha(\lambda)=\{\alpha_l(\lambda)\}^{l=\overline{1,n-1}}$ is a vector function from L_T^2 of the form

$$\alpha(\lambda)=\sum_{k=1}^{n-1}\sum_{j=1}^{m}\Delta_{kj}\alpha^{(kj)}(\lambda).$$

The spectral representation (39) of the process $Y(t)$ allows us to express the quantity $\hat{x}_n(T,\theta)$ as follows:

$$\hat{x}_n(T,\theta)=\int\varphi(\lambda,T)\,\Phi^{Y}(d\lambda)+\sum_{j=1}^{m}\theta_{nj}A_j(T)-\int\varphi(\lambda,T)\,F(d\lambda)\overset{*}{\alpha}(\lambda). \tag{41}$$

The first integral in this expression depends only on $Y(t)$, $T_0\le t\le T$, and therefore

$$\hat{x}_n(T)=\int\varphi(\lambda,T)\,\Phi^{Y}(d\lambda)+\sum_{j=1}^{m}A_j(T)\hat{\theta}_{nj}-\int\varphi(\lambda,T)\,F(d\lambda)\overset{*}{\hat{\alpha}}(\lambda), \tag{42}$$

where

$$\hat{\alpha}(\lambda)=\sum_{k=1}^{n-1}\sum_{j=1}^{m}\hat{\Delta}_{kj}\alpha^{(kj)}(\lambda)$$

and $\hat{\theta}_{nj}$, $\hat{\Delta}_{kj}=\hat{\theta}_{nj}-\hat{\theta}_{kj}$ $(k=\overline{1,n-1};\ j=\overline{1,m})$ are the mean values of the parameters with respect to the "a posteriori" probability distribution $P(d\theta|Y(t),\ T_0\le t\le T)$:

$$\hat{\theta}_{kj}=\mathbf{M}[\theta_{kj}\,|\,Y(t),\ T_0\leqslant t\leqslant T]. \tag{43}$$

We consider the very important case where the probability distribution of the errors $\langle x_k(t)\rangle$ is entirely Gaussian, i.e., not only the stationary components of the errors $\langle\xi_k(t)\rangle$ but also the quantities $\langle\theta_{kj}\rangle$ are Gaussian. In this case $\hat{\theta}_{kj}$, defined by formula (43), are variables from the space $H_y(T)$, namely the closed linear span of values $Y_k(t)$, $T_0\le t\le T$, $1\le k\le n-1$, and can be represented in the spectral form

$$\hat{\theta}_{kj}=\int_{-\infty}^{\infty}\varphi^{(kj)}(\lambda)\,\Phi^{Y}(d\lambda),\quad k,\ j=\overline{1,\ n-1}, \tag{44}$$

where $\phi^{(kj)}(\lambda)=\{\phi_l^{(kj)}(\lambda)\}^{l=\overline{1,n-1}}$ are the corresponding vector functions from the space $L_T^2(F)$. The unknown $\phi^{(kj)}(\lambda)$ can be determined from the conditions of orthogonality of the differences $\theta_{kj}-\hat{\theta}_{kj}$ to all possible $Y_k(t)$, $T_0\le t\le T$,

$1 \leq k \leq n-1$:

$$\mathbf{M}Y_k(t)[\theta - \hat{\theta}] = 0. \tag{45}$$

Let us see what these conditions give us.

In order not to encumber the formulas with many indices, we here denote the chosen quantities θ_{kj}, $\hat{\theta}_{kj}$ and $\phi^{(kj)}(\lambda)$ simply by θ, $\hat{\theta}$ and ϕ. According to the relations (39), (40) and (44) we have:

$$\hat{\theta} = \sum_{k=1}^{n-1} \sum_{j=1}^{m} \Delta_{kj} \int_{-\infty}^{\infty} \varphi(\lambda) F(d\lambda) \overset{*}{\alpha}{}^{(kj)}(\lambda) + \int_{-\infty}^{\infty} \varphi(\lambda) \Phi^{\eta}(d\lambda),$$

$$Y(t) = \sum_{k=1}^{n-1} \sum_{j=1}^{m} \Delta_{kj} A_{kj}(t) + \int_{-\infty}^{\infty} e^{i\lambda t} \Phi^{\eta}(d\lambda), \tag{46}$$

$$\mathbf{M}Y(t)\hat{\theta} = \int_{-\infty}^{\infty} e^{i\lambda t} F(d\lambda) \overset{*}{\varphi}(\lambda) + \sum_{k=1}^{n-1} \sum_{j=1}^{m} \Gamma_{kj} A_{kj}(t),$$

where

$$\Gamma_{ki} = \sum_{l=1}^{n-1} \sum_{j=1}^{m} \mathbf{M}(\Delta_{ki}\Delta_{lj}) \int_{-\infty}^{\infty} \alpha^{(lj)} F(d\lambda) \overset{*}{\varphi}(\lambda) \tag{47}$$

and finally

$$\mathbf{M}Y(t)\theta = \sum_{k=1}^{n-1} \sum_{j=1}^{m} \mathbf{M}(\Delta_{kj}\theta) A_{kj}(t). \tag{48}$$

Thus equalities (45) are equivalent to the integral equation involving $\phi(\lambda)$ encountered earlier:

$$\int_{-\infty}^{\infty} e^{i\lambda t} F(d\lambda) \overset{*}{\varphi}(\lambda) = \sum_{k=1}^{n-1} \sum_{j=1}^{m} C_{kj} A_{kj}(t), \quad T_0 \leqslant t \leqslant T, \tag{49}$$

where

$$C_{kj} = \mathbf{M}(\Delta_{kj}\theta) - \Gamma_{kj} \quad (k = \overline{1, n-1};\ j = \overline{1, m}). \tag{50}$$

It is true that the constants C_{kj} depend on the function $\phi(\lambda)$ itself (see the expression (47) for Γ_{kj}). In solving equation (49) it is convenient to assume C_{kj} to be indeterminate coefficients which should be determined from the additional conditions (50), which are linear equations relative to these C_{kj}.

Obviously the solution $\phi(\lambda)$ has the form:

$$\varphi(\lambda) = \sum_{k=1}^{n-1} \sum_{j=1}^{m} C_{kj} \alpha^{(kj)}(\lambda), \tag{51}$$

where $\alpha^{(kj)}(\lambda)$ are the vector functions involved in the spectral representation (38).

Example 4. Suppose the errors $x_k(t)$ are mutually independent for distinct k and have the same Gaussian probability distribution, and all the parameters θ_{kj}

are also independent and have the same Gaussian distribution law, and $A_{kj}(t) = A_j(t)$, $k = \overline{1, n-1}$.

If $F^{\xi}(d\lambda)$ is the spectral measure (the same for all) of the stationary components $\xi_k(t)$ of the errors $x_k(t)$, then the integral equation (34) for the spectral characteristic $\phi(\lambda, T)$ in formula (32) is the following:

$$\int_{-\infty}^{\infty} e^{i\lambda t} F^{\xi}(d\lambda) G \overset{*}{\varphi}(\lambda, T) = \int_{-\infty}^{\infty} e^{i\lambda t} F^{\xi}(d\lambda) \overset{*}{E} e^{-i\lambda T},$$

where $G = \{G_{kl}\}_{k=\overline{1,n-1}}^{l=\overline{1,n-1}}$ is a matrix of the form

$$G = \begin{pmatrix} 2 & 1 \dots 1 \\ 1 & 2 \dots 1 \\ \cdot & \cdot \;\; \cdot \;\; \cdot \\ 1 & 1 \dots 2 \end{pmatrix},$$

and $E = \{1\}^{k=\overline{1,n-1}}$ is a vector with unit components. From the uniqueness of the solution (as elements of the space $L_T^2(F)$) we conclude that

$$\varphi(\lambda, T) = \frac{1}{n} e^{i\lambda T} E,$$

that is

$$\varphi_k(\lambda, T) = \frac{1}{n} e^{i\lambda T}, \quad k = \overline{1, n-1}.$$

Thus

$$\hat{\xi}_n(T) = \frac{1}{n} \sum_{k=1}^{n-1} \eta_k(T)$$

and according to formula (41)

$$\hat{x}_n(T, \theta) = \frac{1}{n} \sum_{k=1}^{n-1} Y_k(T) + \sum_{j=1}^{m} \left(\frac{1}{n} \sum_{k=1}^{n} \theta_{kj} \right) A_j(T).$$

In order to arrive at the desired quantity $\hat{x}_n(T)$, we find $\hat{\theta} = (1/n) \Sigma_{k=1}^n \hat{\theta}_{kj}$. Letting $\theta = (1/n) \Sigma_{k=1}^n \theta_{kj}$. We have

$$\mathbf{M} Y_k(t) \theta = \sum_{k=1}^{n-1} \sum_{j=1}^{m} \mathbf{M}(\Delta_{kj}\theta) A_j(t) = 0,$$

since, as is easily seen,

$$\mathbf{M} \Delta_{kj} \theta = 0.$$

We see that the quantity θ is orthogonal to the subspace $H_y(T)$, and therefore its projection $\hat{\theta}$ onto this subspace is equal to 0,

$$\frac{1}{n} \sum_{k=1}^{n} \hat{\theta}_{kj} = 0 \quad (j = \overline{1, m}).$$

By formula (42),

$$\hat{x}_n(T)=\frac{1}{n}\sum_{k=1}^{n-1}Y_k(T).$$

Thus the optimal function $X(T)$ (see relations (4) and (7)) has the form

$$X(T)=X_n(T)-\frac{1}{n}\sum_{k=1}^{n-1}[X_n(T)-X_k(T)]=\frac{1}{n}\sum_{k=1}^{n}X_k(T), \tag{52}$$

that is, the optimal function $X(T)$ (as is in the simplest case, considered in Example 1) is the arithmetic mean of the separate readings $X_k(T)$, $k=\overline{1,n}$.

Now we consider the general case where the parameters $\langle\theta_{kj}\rangle$ have arbitrary probability distribution $P(d\theta)$. In order for it to be possible to find the desired variables $\hat{\theta}_{kj}$ involved in formula (42) for $\hat{x}_n(T)$, it is necessary, generally speaking, to know the "a posteriori" probability distribution $P(d\theta|Y(t),\ T_0\leq t\leq T)$, which we now proceed to compute.

Let $Q_\theta(dY)$ denote the probability measure, in the space of vector functions $Y=\langle Y(t)\rangle$, $Y(t)=\{Y_k(t)\}_{k=\overline{1,n-1}}$, $T_0\leq t\leq T$, corresponding to a random process $Y(t)$ of the form (23) with fixed values of the parameters θ_{kj} (defined on the σ-algebra of sets generated by the cylinder sets).

Because of the fact that each of the vector functions $A_{kj}(t)$ on the interval $T_0\leq t\leq T$ allows a representation of the form (38), the measure $Q_\theta(dY)$ is absolutely continuous with respect to the measure $Q(dY)$ corresponding to zero values of the parameters θ_{kj} (it corresponds to the stationary process $\eta(t)$, [2]).

Let

$$p(Y,\ \theta)=\frac{Q_\theta(dY)}{Q(dY)}. \tag{53}$$

The a posteriori distribution $P(d\theta|Y(t),\ T_0\leq t\leq T)$ is expressed in terms of the density $p(Y,\ \theta)$ as follows:

$$p(d\theta\,|\,Y)=\frac{p(Y,\ \theta)\,P(d\theta)}{\int_\theta p(Y,\ \theta)\,P(d\theta)}. \tag{54}$$

We see that the problem of finding the a posteriori distribution of the parameters reduces to finding the density $p(Y,\ \theta)$.

If we consider $\log p(Y,\ \theta)$ for fixed θ as a random variable (a function of $\langle Y(t)\rangle$), then it can be represented in the following form (see [2]):

$$\log p(Y,\ \theta)=\int_{-\infty}^{\infty}\beta(\lambda)\,\Phi^T(d\lambda)-\frac{\varepsilon^2}{2}, \tag{55}$$

where $\beta(\lambda)=\{\beta_k(\lambda)\}^{k=\overline{1,n-1}}$ is a function from the space $L_T^2(F)$, and

$$\varepsilon^2 = \int_{-\infty}^{\infty} \beta(\lambda) F(d\lambda) \overset{*}{\beta}(\lambda). \tag{56}$$

This function $\beta(\lambda)$ can be determined from the Wiener-Hopf integral equation

$$\int_{-\infty}^{\infty} e^{i\lambda t} F(d\lambda) \overset{*}{\beta}(\lambda) = \sum_{n=1}^{n-1} \sum_{j=1}^{m} \Delta_{kj} A_{kj}(t). \tag{57}$$

Taking into account the spectral representation (38) of the functions $A_{kj}(t)$, we thus have

$$\beta(\lambda) = \sum_{k=1}^{n-1} \sum_{j=1}^{n} \Delta_{kj} \alpha^{(kj)}(\lambda). \tag{58}$$

Let

$$a_{kj}[Y] = \int_{-\infty}^{\infty} \alpha^{(kj)}(\lambda) \Phi^{Y}(d\lambda)$$

$$= \int_{T_0}^{T} C^{(kj)}(t) Y(t)\, dt \quad (k = \overline{1, n-1};\ j = \overline{1, m}) \tag{59}$$

($C^{(kj)}(t)$ is the (generally speaking, generalized) Fourier transform of the vector function $\alpha^{(kj)}(\lambda)$),

$$b_{(ki),(lj)} = \int_{-\infty}^{\infty} \alpha^{(ki)}(\lambda) F(d\lambda) \overset{*}{\alpha}^{(lj)}(\lambda) \quad (k, l = \overline{1, n-1};\ i, j = \overline{1, m}). \tag{60}$$

Then the density $p(Y, \theta)$ takes the following expression:

$$p(Y, \theta) = \exp\left\{ \sum_{k=1}^{n-1} \sum_{j=1}^{m} a_{kj}[Y] \Delta_{kj} - \frac{1}{2} \sum_{k,l=1}^{n-1} \sum_{i,j=1}^{m} b_{(ki),(lj)} \Delta_{ki} \Delta_{lj} \right\}. \tag{61}$$

In conclusion we give a general formula for $\hat{\theta}_{kj}$. Namely,

$$\hat{\theta}_{kj} = \frac{\int_{\theta} \theta_{kj} p(Y, \theta) P(d\theta)}{\int_{\theta} p(Y, \theta) P(d\theta)}. \tag{62}$$

BIBLIOGRAPHY

[1] Ju. A. Rozanov, *Stationary stochastic processes*, Fizmatgiz, Moscow, 1963. MR 28 #2580.

[2] V. F. Pisarenko and Ju. A. Rozanov, "Some problems for stationary processes leading to integral equations of Wiener-Hopf type," *Problems in information transmission*, Moscow, 1963.

[3] A. M. Jaglom, *Effective solutions of linear approximation problems for multivariate stationary processes with a rational spectrum*, Teor. Verojatnost. i. Primenen. 5 (1960), 265–292. MR 26 #1934.

Translated by L. Rosenblatt.

ON DISTRIBUTIONS IN SPACES DUAL TO ORDERED LINEAR SPACES

V. V. SAZONOV

In this note we generalize results of Ju. V. Prohorov on random measures on compacta (see [1]). The aim of this generalization is to find that unique situation in which the indicated results are true. The theorems given here can be obtained by a method similar to that of Prohorov.

1. Let X, Y be a pair of dual linear spaces. We assume that Y is an ordered linear space and that the pointed convex cone Y^+ of its nonnegative elements generates Y, that is, $Y = Y^+ - Y^+$ (see [2], Chapter II, §1). Suppose further that in Y^+ there exists at least one interior point with respect to the Mackey topology $\tau(Y, X)$. Here we choose a Mackey topology simply because it is the strongest of the topologies compatible with duality; it is essential for us only that a point be interior in some topology compatible with duality. The set of nonnegative $x \in X$ (that is, such that $\langle x, y\rangle \geq 0$ [1] for all $y \in Y^+$) is denoted by X^+.

Let $\mathcal{P}$ be the set of all distributions (probability measures) on $\mathfrak{L}_{X, Y}$, namely the smallest σ-algebra of subsets of X with respect to which all linear functionals $\langle \cdot, y\rangle$, $y \in Y$, are measurable. We say that the distribution $P \in \mathcal{P}$ is nonnegative if

$$P^*(X^+) = 1^*. \text{ [2]}$$

The collection of all nonnegative distributions is denoted by $\mathcal{P}^+$.

We recall that the functional

$$\chi(y) = \int e^{i\langle x, y\rangle} P(dx), \; y \in Y,$$

is called the characteristic functional (c.f.) of the distribution P on $\mathfrak{L}_{X, Y}$ (or on a wider σ-algebra).

Theorem 1. *In order for the functional χ on Y to be the c.f. of some distribution $P \in \mathcal{P}^+$, it is necessary and sufficient that the following conditions be satisfied:*

1) *χ is nonnegative definite, $\chi(0) = 1$,*

2) *for any $y \in Y^+$ the function $\chi_y(t) = \chi(ty)$, $t \in R$ is bounding for the function of a complex variable $z = t + is$, definite, continuous, does not exceed unity in modulus for $s \geq 0$, and is analytic for $s > 0$.*

2. In our given situation we have an analog of the theorem on continuity of

---

1) $\langle \cdot, \cdot \rangle$ is a bilinear form with respect to which X and Y are dual.

2) P^* is the external distribution generated by the distribution P.

the correspondence of distributions and their characteristic functions in its classical form. It is appropriate to note that for distributions in infinite-dimensional linear spaces and their c.f.'s there is, generally speaking, no such analog.

Every distribution $P \in \mathfrak{P}^+$ can be uniquely extended to a dense distribution (here and below, X is provided with the weak topology $\sigma(X, Y)$; the distribution P is said to be dense if there exist compacta in X on which P^* is arbitrarily close to unity) on the σ-algebra $\mathfrak{B}_{\sigma(X, Y)}$, namely the smallest σ-algebra relative to which all $\sigma(X, Y)$-continuous functions are measurable. The collection of all such extensions is denoted by $\mathfrak{R}^+$. Here also, $P^*(X^+) = 1$ for any $P \in \mathfrak{R}^+$.

Let w be a topology on $\mathfrak{R}^+$ induced by the weak topology of the space of all distributions on $\mathfrak{B}_{\sigma(X, Y)}$.

Theorem 2. *If the sequence of c.f.'s $\{\chi_n(y)\}$ of distributions $\{P_n\}$ from $\mathfrak{R}^+$ converges for each $y \in Y$ to the functional χ such that for each $y \in Y$ the function $\chi_y(t) = \chi(ty)$, $t \in R$, is continuous at zero, then χ is the c.f. of some distribution $P \in \mathfrak{R}^+$ and the sequence $\{P_n\}$ converges to P in the topology w. The converse, of course, is also true.*

Remark. Since in every linear topological space neighborhoods of zero are absorbing sets, every functional χ, if it is continuous in some topology on Y relative to which Y with the operations defined in it is a linear topological space, satisfies the requirement in the statement of the theorem.

3. The condition of existence of an interior point in Y^+ relative to the topology $\tau(Y, X)$ cannot be entirely discarded.

Let $X = l$ be the space of absolutely convergent sequences and $Y = m$ be the space of uniformly bounded sequences. With the usual definition of norm these are Banach spaces and $l^* = m$. (l^* is the space conjugate to l.) The topology $\tau(Y, X)$ in Y here coincides with the topology generated by the norm. For $y' = (y_1', y_2', \cdots)$, $y'' = (y_1'', y_2'', \cdots) \in Y$ we let

$$y' \leq y''$$

if

$$y_n' \leq y_n''$$

for all $n = 1, 2, \cdots$. It is obvious that with this ordering, $Y = Y^+ - Y^+$. It is easy to see that in the given case there are no points in Y^+ which are interior to $\tau(Y, X)$.

The functional

$$\chi(y) = \prod_{n=1}^{\infty} \frac{1}{1 - iy_n}, \quad y = (y_1, y_2, \ldots) \in Y$$

is not the c.f. of a distribution even though it satisfies the conditions of Theorem 1, so that Theorem 1 does not hold here. It is easy to show that Theorem 2 is also false in this situation.

BIBLIOGRAPHY

[1] Ju. V. Prohorov, *On random measures on a compactum*, Dokl. Akad. Nauk SSSR 138 (1961), 53–55 = Soviet Math. Dokl. 2 (1961), 539–544. MR 25 #4327.

[2] N. Bourbaki, *Espaces Vectoriels topologiques*, Livre V, *Éléments de Mathématiques*, Actualités Sci. Ind. Nos. 1189, 1220, 1230, Hermann, Paris, 1953, 1955; Russian transl., IL, Moscow, 1959. MR 14, 880; MR 17, 1109.

[3] L. LeCam, *Convergence in distribution of random processes*, Matematika 4 (1960), 107–142.

Translated by:
L. Rosenblatt

SOME THEOREMS ON THE LIMITING BEHAVIOR OF THE EMPIRICAL DISTRIBUTION FUNCTION

D. M. ČIBISOV

1. Let $G_n^*(u)$, $0 \leq u \leq 1$ be the empirical distribution function constructed from a sample of n independent random variables distributed uniformly on $[0, 1]$; let $\beta_n(u) = \sqrt{n}\,(G_n^*(u) - u)$. Let $\beta(u)$, $0 \leq u \leq 1$, be a separable Gaussian process with $\mathbf{M}\beta(u) = 0$, $\mathbf{M}[\beta(u)\beta(v)] = \min(u, v) - uv$ (conditional Wiener process under the condition $\beta(0) = \beta(1) = 0$). There are many well-known assertions that $\beta(u)$ is in some sense or other limiting for $\beta_n(u)$. We investigate the local behavior of the processes $\beta_n(u)$ and $\beta(t)$ close to the points $u = 0$ and $u = 1$.

Let $\overline{\lim}_{u\to 0}\psi(u)\beta(u) = l$, where $\psi(u)$ $(u \geq 0)$ is some function. It is known* that the random variable l is degenerate, i.e., takes on a constant value (possibly ∞) with probability 1. Roughly speaking, the basic result consists of the fact that $\beta_n(u)$ in the limit as $n \to \infty$ has an analogous property. A precise statement of this result is contained in Theorems 1 and 2. With the help of these theorems we obtain a generalization of known results on weak convergence of distributions corresponding to $\beta_n(u)$ (Theorem 3).

In §2 some results are given relating to the process $\beta(u)$; §§3 and 4 contain the statements of the results; the proofs are given in §5.

The functions $\psi(u)$, and also $\phi(u)$ and $h(u)$ encountered below, will be assumed growing regularly in a neighborhood of $u = 0$. The function $g(u)$ is said to be regularly growing if for each $k > 0$

$$\lim_{u\to 0}\frac{g(ku)}{g(u)} = k^{\gamma},$$

where γ is a constant (see [5]).

2. For the results given here see P. Levy [2].

2.1. **Definition.** 1°. The function $\phi(u)$ $(u > 0)$ belongs to the upper class if

$$\lim_{\tau\to 0}\mathbf{P}\left\{\sup_{0<u\leqslant\tau}\frac{\beta(u)}{\varphi(u)} > 1\right\} = 0.$$

2°. The function $\phi(u)$ belongs to the lower class if for any $\tau > 0$

$$\mathbf{P}\left\{\sup_{0<u\leqslant\tau}\frac{\beta(u)}{\varphi(u)} > 1\right\} = 1.$$

*The results which we present here and below as known for $\beta(u)$ are indeed known for the unconditional Wiener process $x(u)$. However, it is easy to verify that they are also true for $\beta(u)$, for example, by making use of the relation $\beta(u) = (1-u)x(u/(1-u))$.

The upper and lower classes are denoted by S and L respectively.

2.2. *Let* $\phi(u) = \sqrt{u}\, h(u)$. *If*

$$\int_0^{u_0} e^{-\frac{h^2(u)}{2}} \frac{du}{u} \tag{1}$$

converges (*diverges*), *then* $(1+\delta)\phi(u) \in S$ $((1-\delta)\phi(u) \in L)$ *for any* $\delta > 0$.

The proof is analogous to the proof of A. Ja. Hinčin [1] for a local law of the iterated logarithm, which is a special case of this assertion.

2.3. Corollary. *Let* $\psi(u) = 1/\sqrt{u}\, h(u)$. *Let the number* l^* *be such that*

$$\int_0^{u_0} e^{-\frac{l^2 h^2(u)}{2}} \frac{du}{u} \tag{2}$$

converges for each $l > l^*$ *and diverges for each* $0 < l < l^*$ (*if the integral* (2) *converges* (*diverges*) *for each* $l > 0$, *then we let* $l^* = 0$ $(l^* = \infty)$). *Then with probability* 1

$$\overline{\lim_{u\to 0}}\, \psi(u)\,|\beta(u)| = l^*.$$

Since the distribution $\beta(u)$ does not change when u is replaced by $1-u$, we can replace $\beta(u)$ by $\beta(1-u)$ in the above statements.

Remark. In [2] the following criterion is given: $\phi(u) \in S$ $(\phi(u) \in L)$ if the following integral converges (diverges):

$$\int_0^{u_0} h(u)\, e^{-\frac{h^2(u)}{2}} \frac{du}{u}.$$

However, this criterion is given without proof or reference to published papers. For our purposes an assertion weaker than 2.2 suffices.

3. Theorem 1. *Let* $\phi(u) = \sqrt{u}\, h(u)$. *If the integral* (1) *converges, then for each* $\delta > 0$

$$\lim_{\substack{\tau\to 0\\ n\to\infty}} \mathbf{P}\left\{ \sup_{0<u\leq\tau} \frac{|\beta_n(u)|}{(1+\delta)\varphi(u)} > 1 \right\} = \lim_{\substack{\tau\to 0\\ n\to\infty}} \mathbf{P}\left\{ \sup_{0<u\leq\tau} \frac{|\beta_n(1-u)|}{(1+\delta)\varphi(u)} > 1 \right\} = 0.$$

Theorem 2. *If* $\phi(u) \in L$, *then for each* $\tau > 0$

$$\lim_{n\to\infty} \mathbf{P}\left\{ \sup_{0<u\leq\tau} \frac{|\beta_n(u)|}{\varphi(u)} > 1 \right\} = \lim_{n\to\infty} \mathbf{P}\left\{ \sup_{0<u\leq\tau} \frac{|\beta_n(1-u)|}{\varphi(u)} > 1 \right\} = 1.$$

Both theorems remain valid if $|\beta_n(u)|$ is replaced by $\beta_n(u)$ or $-\beta_n(u)$.

4. Using Theorem 1, we can obtain a generalization of known results on convergence of the sequence $\beta_n(u)$ to $\beta(u)$. We let $L[\xi]$ denote the distribution of the random variable ξ. Donsker [3] proved that if the functional f is continuous in a uniform metric, then $L[f(\beta_n)] \to L[f(\beta)]$ (convergence is taken in the usual

sense of weak convergence of distributions on the line). In many articles (see, for example, [4]) a study is made of functionals (for example, $\sup_{0<u<1} \psi(u)\beta(u)$) which are continuous in a stronger metric ρ_ψ (see (3) below), and of the convergence of their distributions to $L[f(\beta)]$. We shall establish conditions on the function $\psi(u)$ under which this convergence takes place.

Suppose the function $\psi(u) \geq 0$ is continuous on the interval $(0, 1)$; for the function $x(u)$, $y(u)$ defined on $[0, 1]$, we let

$$\rho_\psi(x(u), y(u)) = \sup_{0<u<1} \psi(u)\,|x(u) - y(u)|. \tag{3}$$

Let R_ψ be the metric space of functions defined on $[0, 1]$ which do not have discontinuities of second kind, with metric ρ_ψ. Let $\psi(u) = 1/\sqrt{u}h_1(u)$, $\psi(1-u) = 1/\sqrt{u}h_2(u)$.

Theorem 3. *A necessary and sufficient condition so that for any functional f continuous on R_ψ,*

$$L[f(\beta_n(u))] \to L[f(\beta(u))],$$

is convergence of the integral

$$\int_0^{u_0} e^{-\frac{l^2 h_i^2(u)}{2}} \frac{du}{u} \quad (i = 1, 2) \tag{4}$$

for any $l > 0$.

By Corollary 2.3 this condition is equivalent to

$$\overline{\lim_{u\to 0}}\, \psi(u)\,|\beta(u)| = \overline{\lim_{u\to 1}}\, \psi(u)\,|\beta(u)| = 0.$$

5. Proof of Theorem 1. First we investigate certain properties of the function $h(u)$ involved in the statement of the theorem.

Lemma 1. *If the integral* (1) *converges, then $h(u) \to \infty$ as $u \to 0$.*

Proof. We assume the contrary, and prove that then $c_1\sqrt{u}h(u) \in L$ for any $c_1 < \infty$, which on the strength of 2.2 contradicts the hypothesis. We shall prove that $c_1\sqrt{u}h(u)$ is a lower function for the unconditional Wiener process $x(u)$ (see footnote on p. 1).

By the hypothesis $h(u) \not\to \infty$ there exists c_2 and a sequence $u_1, u_2, \cdots, u_k, \cdots \cdots \to 0$ such that $h(u_k) < c_2$, where $u_{k+1} < u_k/2$ (which can be achieved by choosing a sufficiently sparse subsequence). Let $\Phi(u)$ denote a standard normal distribution function. From

$$\mathbf{P}\{|x(u_k) - x(u_{k+1})| > 2c\sqrt{u_k}\} = 2\left[1 - \Phi\left(\frac{2c\sqrt{u_k}}{\sqrt{u_k - u_{k+1}}}\right)\right] \geqslant 2[1 - \Phi(2c\sqrt{2})]$$

and the independence of the events $A = \{|x(u_k) - x(u_{k+1})| > 2c\sqrt{u_k}\}$ it follows that the events A_k occur for an infinitely large number of values k. Since from

A_k for $c = c_1c_2$ follows at least one of the inequalities $|x(u_j)| > c_1\sqrt{u_j}h(u_j)$, $j = k, k+1$, the lemma is proved.

It is known (see [5]) that a regularly growing function can be represented in the form

$$g(u) = c(u)\, u^{\gamma} \exp\left\{\int_u^{u_1} \frac{\varepsilon(v)}{v}\, dv\right\},$$

where γ is a constant, $\epsilon(u) \to 0$, $c(u) \to c \neq 0$ as $u \to 0$. From this representation we obtain directly

Lemma 2. *For any* $\alpha > 0$ *there exist* $\tau > 0$ *and* $0 < q < 1$ *such that* $0 < qu'' < u' < u'' \leq \tau$,

$$\left|\frac{h(u')}{h(u'')} - 1\right| < \alpha.$$

Remark. For the proof of the theorem we use no property of regularly growing functions except that given in Lemma 2. Accordingly, the statement of the theorem can be somewhat broadened.

Let $\xi_1, \xi_2, \cdots$ be independent random variables having exponential distribution with density equal to ne^{-nu} for $u \geq 0$ and 0 for $u < 0$. Let $\eta_k = \xi_1 + \cdots + \xi_k$, and let $\nu_n(u)$ be the number of variables η_k smaller than u; $\nu_n(u)$ is a Poisson process with parameter n. Let

$$\gamma_n(u) = \sqrt{n}\left(\frac{\nu_n(u)}{n} - u\right).$$

Lemma 3. *The theorem is valid if it is valid for* ν_n *(in place of* $\beta_n(u)$*).*

Proof. The variables $V_i^{(n)} = \eta_i/\eta_{n+1}$; $i = 1, 2, \cdots, n$ are distributed jointly as an ordered sample of n independent uniformly distributed random variables. Therefore as $G_n^*(u)$ we can take the empirical distribution function of the variables $V_i^{(n)}$. Then $G_n^*(u) = \nu_n(u\eta_{n+1})/n$ and

$$\beta_n(u) = \gamma_n(u\eta_{n+1}) + \sqrt{n}\,(\eta_{n+1} - 1)\, u. \tag{5}$$

It follows from $\mathbf{M}_{\beta+1} = (n+1)/n$, $\mathbf{D}\eta_{n+1} = (n+1)/n^2$ that $\eta_{n+1} \to 1$ in probability, and the variable $\sqrt{n}\,(\eta_{n+1} - 1)$ is bounded in probability. Taking account of this, we obtain the assertion of the lemma from (5).

Lemma 4. *For any fixed* $a > 0$ *and* $c > 0$

$$\lim_{n\to\infty} \mathbf{P}\left\{\sup_{0<u\leq\frac{a}{n}} \frac{|\gamma_n(u)|}{c\varphi(u)} \geq 1\right\} = 0.$$

Proof.

$$1-\mathbf{P}\left\{\sup_{0<u\leqslant\frac{a}{n}}\frac{|\gamma_n(u)|}{c\varphi(u)}\geqslant 1\right\}=\mathbf{P}\left\{|\gamma_n(u)|<c\sqrt{u}h(u),\ 0<u\leqslant\frac{a}{n}\right\}$$

$$=\mathbf{P}\left\{\left|\nu_n\left(\frac{v}{n}\right)-v\right|<c\sqrt{u}h\left(\frac{v}{n}\right),\ 0<v\leqslant a\right\}$$

$$\leqslant\mathbf{P}\{v-c_n\sqrt{u}<\nu(v)<v+c_n\sqrt{u},\ 0<v\leqslant a\}, \tag{6}$$

where $\nu(v)$ is a Poisson process with parameter 1, and $c_n = \inf_{0<u\leq a/n} ch(u) \longrightarrow \infty$ as $u \longrightarrow 0$ (on the strength of Lemma 1). The assertion of the lemma follows in an obvious way from (6).

Lemma 5. *If the random variable ν had Poisson distribution with parameter λ, then*

$$\mathbf{P}\{|\nu-\lambda|>z\}\leqslant 2\exp\left\{-\lambda\left[\left(1+\frac{z}{\lambda}\right)\ln\left(1+\frac{z}{\lambda}\right)-\frac{z}{\lambda}\right]\right\}. \tag{7}$$

Proof. For any $\vartheta > 0$

$$\mathbf{P}\{\nu-\lambda>z\}=\mathbf{P}\{e^{\vartheta(\nu-\lambda)}>e^{\vartheta z}\}\leqslant\frac{\mathbf{M}e^{\vartheta(\nu-\lambda)}}{e^{\vartheta z}}=\exp[\lambda(e^{\vartheta}-1-\vartheta)-\vartheta z].$$

Letting $\vartheta = \ln(1 + z/\lambda)$ (the value for which the right part attains a minimum) and combining with the analogous inequality for $\mathbf{P}\{-(\nu - \lambda) > z\}$, we obtain (7).

Estimating the right part of (7) from above, we derive a cruder but more convenient inequality. Using the relation

$$(1+y)\ln(1+y)-y=\int_0^y \ln(1+t)\,dt$$

and the estimate $\ln(1 + t) \geq (1 - \alpha)t$, $\alpha > 0$, $0 \leq t \leq t_\alpha$ (t_α is a constant depending on α), we obtain

$$\mathbf{P}\{|\nu-\lambda|>z\}\leqslant\begin{cases}2\exp\left[-\dfrac{(1-\alpha)^2z^2}{2\lambda}\right], & 0\leqslant z\leqslant\lambda t_\alpha,\\ 2\exp\left[-\dfrac{c_\alpha z}{2}\right], & z\geqslant\lambda t_\alpha,\end{cases}$$

where we can assume $c_\alpha = (1 - \alpha)^2 t_\alpha$. Moreover, for any $z > 0$,

$$\mathbf{P}\{|\nu-\lambda|>z\}\leqslant 2\exp\left[-\frac{(1-\alpha)^2z^2}{2\lambda}\right]+2\exp\left[-\frac{c_\alpha z}{2}\right]. \tag{8}$$

Applying (8) to the process $\gamma_n(u)$, we obtain ($u'' > u'$, $\Delta = u'' - u'$):

$$\mathbf{P}\{|\gamma_n(u'')-\gamma_n(u')|>z\}=\mathbf{P}\{|\nu_n(u'')-\nu_n(u')-n\Delta|>z\sqrt{n}\}$$

$$\leqslant 2\exp\left[-\frac{(1-\alpha)^2z^2}{2\Delta}\right]+2\exp\left[-\frac{c_\alpha z\sqrt{n}}{2}\right]. \tag{9}$$

Lemma 6. *Let (assuming* $\Delta = u'' - u' > 0$)

$$\sup_{u' \leqslant u \leqslant u''} \mathbf{P}\{|\gamma_n(u) - \gamma_n(u')| > z\} = \psi_z(\Delta).$$

Then

$$\mathbf{P}\{\sup_{u' \leqslant u \leqslant u''} |\gamma_n(u) - \gamma_n(u')| > z\} \leqslant 2\psi_{\frac{z}{2}}(\Delta).$$

Proof is analogous to the proof of Lemma 1 [6].

The right part of (9) is monotone increasing as Δ increases; therefore it can serve as the estimate from above for $\psi_z(\Delta)$, and we obtain

$$\mathbf{P}\left\{\sup_{u' \leqslant u \leqslant u''} |\gamma_n(u) - \gamma_n(u')| > z\right\} \leqslant 4\exp\left[-\frac{(1-\alpha)^2 z^2}{8\Delta}\right] + 4\exp\left[-\frac{c_\alpha z\sqrt{n}}{4}\right]. \tag{10}$$

We proceed with the proof of Theorem 1.

We take an arbitrary fixed number $a > 0$. We have:

$$\mathbf{P}\left\{\sup_{0<u\leqslant\tau} \frac{\gamma_n(u)}{(1+\delta)\varphi(u)} > 1\right\}$$

$$\leqslant \mathbf{P}\left\{\sup_{0<u\leqslant\frac{a}{n}} \frac{|\gamma_n(u)|}{\varphi(u)} > 1+\delta\right\} + \mathbf{P}\left\{\sup_{\frac{a}{n}<u\leqslant\tau} \frac{|\gamma_n(u)|}{\varphi(u)} > 1+\delta\right\}.$$

By Lemma 4, the first summand goes to 0 as $n \to \infty$. We estimate the second summand. According to Lemma 2 we choose values τ_0 and q_0 corresponding to $\alpha = \delta/4$. We take a number q such that $q_0 \leq q < 1$ and the inequality (15) (below) is satisfied. Let $u_k = q^k$, $\Delta_k = u_k - u_{k+1}$, $k(u)$ be the number for which $u_{k(u)+1} < u \leq u_{k(u)}$,

$$p_k = \mathbf{P}\left\{\sup_{u_{k+1}<u\leqslant u_k} \frac{|\gamma_n(u)|}{\varphi(u)} \geqslant 1+\delta\right\}$$

(the quantities p_k, and w_k and v_k introduced below, also depend on n, but for simplicity we do not introduce explicit notation for this). Then

$$\mathbf{P}\left\{\sup_{\frac{a}{n}\leqslant u\leqslant\tau} \frac{|\gamma_n(u)|}{\varphi(u)} > 1+\delta\right\} \leqslant \sum_{k=k(\tau)-1}^{k\left(\frac{a}{n}\right)} p_k.$$

Let $h_k = \max_{u_{k+1}<u\leq u_k} h(u)$, $h_k' = \min_{u_{k+1}<u\leq u_k} h(u)$. It follows from $q \geq q_0$ that $(1+\delta)\, h_k' > (1+\delta/2)\, h_k$. Let

$$w_k = \mathbf{P}\left\{|\gamma_n(u_{k+1})| > \left(1+\frac{\delta}{4}\right)\sqrt{u_{k+1}}\,h_k\right\},$$

$$v_k = \mathbf{P}\left\{\sup_{u_{k+1}\leqslant u\leqslant u_k} |\gamma_n(u) - \gamma_n(u_{k+1})| > \frac{\delta}{4}\sqrt{u_{k+1}}\,h_k\right\}.$$

Then $p_k \leq v_k + w_k$, and for proof of the theorem it suffices to show that

$$\lim_{\substack{\tau\to 0\\ n\to\infty}} \sum_{k=k(\tau)-1}^{k\left(\frac{a}{n}\right)} v_k = 0, \tag{11}$$

$$\lim_{\substack{\tau\to 0\\ n\to\infty}} \sum_{k=k(\tau)-1}^{k\left(\frac{a}{n}\right)} w_k = 0. \tag{12}$$

We now make use of the fact that from the convergence of the integral (1), assumed by the condition of the theorem, follows

$$\sum_{k=1}^{\infty} \exp\left[-\frac{h_k^2}{2}\right] < \infty. \tag{13}$$

Consider Σv_k. Using (9), we obtain

$$\sum v_k \leqslant 4 \sum \exp\left[-\frac{(1-\alpha)^2\delta^2 u_{k+1} h_k^2}{128\Delta_k}\right] + 4\sum \exp\left[-\frac{c_\alpha\delta\sqrt{nu_{k+1}}h_k}{16}\right]. \tag{14}$$

We assume that for some $\alpha > 0$,

$$\frac{(1-\alpha)^2\,\delta^2 q}{64\,(1-q)} > 1. \tag{15}$$

By (13),

$$\sum_{k=k(\tau)-1}^{k\left(\frac{a}{n}\right)} \exp\left[-\frac{(1-\alpha)^2\delta^2 u_{k+1}h_k^2}{128\Delta_k}\right] \leqslant \sum_{k=k(\tau)-1}^{\infty} \exp\left[-\frac{h_k^2}{2}\right] \to 0 \quad \text{as} \quad \tau\to 0.$$

Consider the second sum in (14). Let $b(\tau) = (c_\alpha\delta\sqrt{aq}/16)\min_{k\geq k(\tau)-1} h_k$. By Lemma 1 we have $b(\tau) \to \infty$ as $\tau \to 0$. Writing the terms of this sum in reverse order, we obtain

$$\sum_{k=k(\tau)-1}^{k\left(\frac{a}{n}\right)} \exp\left[-\frac{c_\alpha\delta\sqrt{nu_{k+1}}h_k}{16}\right] \leqslant \sum_{j=0}^{\infty} \exp\left[-b(\tau)\sqrt{\left(\frac{1}{q}\right)^j}\right].$$

Relation (11) is proved.

Consider Σw_k. We have:

$$w_k = \mathbf{P}\left\{|\nu_n(u_{k+1}) - nu_{k+1}| > \left(1+\frac{\delta}{4}\right)\sqrt{nu_{k+1}}h_k\right\},$$

where $\nu_n(u_{k+1})$ has a Poisson distribution with parameter nu_{k+1}. We take $\alpha > 0$ such that $(1 + \delta/4)\,(1 - \alpha) > 1$ and apply (8). Then

$$\sum_{k=k(\tau)-1}^{k\left(\frac{a}{n}\right)} w_k \leqslant 2\sum_{k=k(\tau)-1}^{\infty} \exp\left[-\frac{h_k^2}{2}\right] + 2\sum_{k=k(\tau)-1}^{k\left(\frac{a}{n}\right)} \exp\left[-\frac{c_\alpha\left(1+\frac{\delta}{4}\right)\sqrt{nu_{k+1}}h_k}{2}\right].$$

Estimating these sums like the sums in (14), we obtain (12). The theorem is proved.

Proof of Theorem 2. If $\phi(u)$ is a lower function, then for any $\tau > 0$ and $\epsilon > 0$ there exists a finite collection $0 < u_1 < u_2 < \cdots < u_s \leq \tau$ such that

$$\mathbf{P}\left\{\sup_{1\leq j\leq s} \frac{|\beta(u_j)|}{\varphi(u_j)} > 1\right\} > 1 - \varepsilon.$$

Indeed, let $r_1, r_2, \cdots$ be rational points from $(0, \tau]$. The events $B_k = \{\sup_{1\leq j\leq k} |\beta(r_j)|/\phi(r_j) > 1\}$ form a monotonically increasing sequence having as limit the event $B = \{\sup_{1\leq j<\infty} |\beta(r_j)|/\phi(r_j) > 1\}$, whose probability, because of the separability of the process $\beta(u)$ and the definition of lower function, is equal to 1, which proves our assertion. The assertion of the theorem now follows from the convergence of the finite-dimensional distributions of $\beta_n(u)$ to the finite-dimensional distributions of $\beta(u)$.

Proof of Theorem 3. The necessity of the condition of the theorem is shown by the following simple example. We assume that the integral (4) (for definiteness, for $i = 1$) converges when $l > l^* > 0$, and diverges when $0 < l < l^*$. Then the functional

$$f(x(u)) = \overline{\lim_{u\to 0}}\, \psi(u)\, x(u) \tag{16}$$

is continuous in the metric ρ_ψ; here we have $f(\beta(u)) = l^*$, $f(\beta_n(u)) = 0$ with probability 1 for each n. If the integral (4) diverges for any $l > 0$, then in (16) we take any function $\bar{\psi}(u)$ for which $l^* < \infty$ instead of $\psi(u)$.

Sufficiency. Let $\mathfrak{B}_\psi$ denote the smallest σ-algebra containing the open sets of the space R_ψ, let μ_n and μ be measures defined on $\mathfrak{B}_\psi$ induced by the processes $\beta_n(u)$ and $\beta(u)$. The sequence μ_n converges weakly to μ if for any continuous bounded functional f on R_ψ

$$\int f d\mu_n \to \int f d\mu \quad \text{as} \quad n \to \infty.$$

Each of the following conditions is necessary and sufficient for weak convergence of $\mu_n \Longrightarrow \mu$.

1°. For any closed $F \in R_\psi$ we have $\overline{\lim}\sup \mu_n(F) \leq \mu(F)$.

2°. For any continuity set $A \in R_\psi$ of the measure μ

$$\mu_n(A) \to \mu(A).$$

(The set A is called a continuity set of the measure μ if the μ-measure of its boundary equals zero.)

3°. For any functional f which is continuous on R_ψ,

$$\mu_n^f \to \mu^f.$$

As is shown by 3°, the assertion of the theorem is equivalent to weak convergence $\mu_n \Longrightarrow \mu$. For the proof we use the well-known convergence $\mu_n \Longrightarrow \mu$ in the case where $\psi(u) \equiv 1$ $(0 \le u \le 1)$ (uniform metric) (see [3]). In this case we write R, ρ, $\mathfrak{B}$ instead of R_ψ, ρ_ψ, $\mathfrak{B}_\psi$; also, we shall write ρ_ψ -(ρ-) continuity, -closed set, etc., to indicate which metric is being considered.

Without loss of generality, we can assume that $\psi(u) > 1$, $0 \le u \le 1$, so that $\rho(x, y) \le \rho_\psi(x, y)$. Then any ρ-open is ρ_ψ -open in $\mathfrak{B}_\psi \supseteq \mathfrak{B}$.

We take the ρ_ψ-closed set F and prove that it satisfies 1°. Let $H_\epsilon = \{x: \inf_{x_0 \in F} \rho_\psi(x, x_0) < \epsilon\}$. We note that $F = \lim_{\epsilon \to 0} H_\epsilon$, and therefore

$$\lim_{\varepsilon \to 0} \mu(H_\varepsilon) = \mu(F). \tag{17}$$

Take an arbitrary $\epsilon > 0$. For each $x(u)$ we let $x^\tau(u) = x(u)$ for $\tau < u \le 1 - \tau$, $x^\tau(u) = 0$ for $0 \le u \le \tau$, $1 - \tau < u \le 1$. Let $S = \{x: \rho_\psi(x, x^\tau) < \epsilon/3$, where τ is chosen (using 2.3 and Theorem 1) such that $\mu(S) > 1 - \epsilon$, $\mu_n(S) > 1 - \epsilon$ for sufficiently large n. We let $\psi_\tau = \max_{\tau \le u \le 1-\tau} \psi(u)$ and take $\alpha < \epsilon/3\psi_\tau$ such that the (ρ-open) set $G = \{x: \inf_{x_\rho \in F \cap S} \rho(x, x_0) < \alpha\}$ is a set of ρ-continuity of the measure μ (this can be done since the set of values α, for which G is not a ρ-continuity set of the measure μ, is at most countable). For any $x_1, x_2 \in S$ such that $\rho(x_1, x_2) < \alpha$,

$$\rho_\psi(x_1, x_2) \leqslant \rho_\psi(x_1, x^\tau) + \rho_\psi(x_1^\tau, x_2^\tau) + \rho_\psi(x_2^\tau, x_2) < \varepsilon. \tag{18}$$

For each n we have $\mu_n(F) \le \mu_n(F \cap S) + \epsilon \le \mu_n(G) + \epsilon$, and since G is a set of ρ-continuity of the measure μ, then

$$\overline{\lim_{n \to \infty}} \mu_n(F) \leqslant \mu(G) + \varepsilon. \tag{19}$$

As a result of (18), $G \cap S \subset H_\epsilon$, and therefore

$$\mu(G) \leqslant \mu(G \cap S) + \varepsilon \leqslant \mu(H_\varepsilon) + \varepsilon. \tag{20}$$

The assertion of the theorem follows from (17), (19), (20), and the arbitrariness of ϵ.

BIBLIOGRAPHY

[1] A. Ja. Hinčin, *Asymptotic laws of probability theory*, ONTI, Moscow, 1936.

[2] P. Levy, *Le mouvement brownien*, Mémor. Sci. Math. no. 126, Gauthier-Villars, Paris, 1954. MR 16, 601.

[3] M. D. Donsker, *Justification and extension of Doob's heuristic approach to the Kolmogorov-Smirnov theorems*, Ann. Math. Statistics 23 (1952), 277–281. MR 13, 853.

[4] T. W. Anderson and D. A. Darling, *Asymptotic theory of certain "goodness of fit" criteria based on stochastic processes*, Ann. Math. Statistics 23 (1952), 193–212.

[5] M. J. Karamata, *Sur un mode de croissance regulière*, Bull. Math. France 61 (1963), 55–62.

[6] E. B. Dynkin, *Criteria of continuity and of absense of discontinuities of the second kind for trajectories of a Markov process*, Izv. Akad. Nauk SSSR Ser. Mat. 16(1952), 563–572. MR 14, 567.

Translated by:
L. Rosenblatt

DETECTION OF A RANDOMLY APPEARING TARGET IN A MULTI-CHANNEL SYSTEM *

A. N. ŠIRJAEV

1. It is assumed that a device can in turn examine each of N paths. If the kth path is examined, the result of the observation is represented by the n-dimensional vector-process

$$\varphi^k(t) = \{\varphi_1^k(t), \ldots, \varphi_n^k(t)\},$$

where n is the number of channels of the device performing the scanning. The processes ϕ_i^k are mutually independent, if there is no target, ϕ_i^k satisfies the equation

$$d\varphi_i^k = -dt + \sqrt{2}\, d\psi_i^k. \tag{1}$$

Here ψ_i^k are standard Gaussian (Wiener) processes with

$$\mathbf{M}\Delta\psi_i^k = 0, \quad \mathbf{M}(\Delta\psi_i^k)^2 = \Delta t, \quad \psi_i^k(0) = 0. \tag{2}$$

In the case where there is a target, say in the kth path of the ith channel, we have

$$d\varphi_i^k = dt + \sqrt{2}\, d\psi_i^k, \tag{3}$$

where ψ_i^k satisfies the same conditions (2).

Below we assume that a target appears at a random moment of time, and the conditional distribution of the time of its appearance in any finite interval, under the condition that it appears there, is uniform. (For more details about this hypothesis see [1].) Also below we begin with an equiprobable distribution of appearance of a target in paths and channels. Here it is assumed that a target in some channel does not disappear later.

Let T be the mean time between two false alarms [1], and $\tau = \tau(\mathrm{T}, N, n)$ be the mean time of delay in detecting a target for given T, number of paths N and number of channels n in each path.

Our problem consists of finding the dependence $\tau = \tau(\mathrm{T}, N, n)$ for the case where the survey is made in cyclic order, while the determination of the presence or absence of a target is carried out in the following way (cf. the Neyman-Pearson method in [1]). On each path k a certain time m is spent in observation. At the end of this time we consider the quantity

$$\max_i \varphi_i^k \tag{4}$$

*The present article generalizes results of [1], whose notation and concepts are used throughout the present paper.

and if $\max_i \phi_i^k > h$ (h is some threshold of comparison), then we assume that a target is present on the given path. * Otherwise we assert absence of a target on this path and pass to observation of the following path.

The choice of the number of channels n depends, generally speaking, on the time of observation: $n = n_1(m)$. It will be convenient first to assume that the number of channels n is a function not of the observation time m, but of the time $\mathbf{T}$: $n = n_2(\mathbf{T})$. After obtaining the asymptotics for $\mathbf{T} \to \infty$ and $n_2(\mathbf{T}) = e^{o(\mathbf{T})}$ for the function $\tau(\mathbf{T}, N, n)$ (Theorem 1), we then easily obtain the dependence $\tau(\mathbf{T}, N, n_1)$ for $n_1 = e^{o(m)}$ and $\mathbf{T} \to \infty$ (Theorem 2).

2. We have the following theorems.

Theorem 1. *For* $\mathbf{T} \to \infty$ *and* $n_2(\mathbf{T}) = e^{o(\mathbf{T})}$

$$\tau(\mathbf{T},\ N,\ n_2(\mathbf{T})) \sim \frac{N+2}{2} \ln \mathbf{T} n_2(\mathbf{T}), \tag{5}$$

$$(N+2) f(\mathbf{T},\ n_2(\mathbf{T})) \geqslant \tau(\mathbf{T},\ N,\ n_2(\mathbf{T})) \geqslant N f(\mathbf{T},\ n_2(\mathbf{T})), \tag{6}$$

where

$$f(\mathbf{T},\ n_2(\mathbf{T})) = \frac{1}{2} \ln \mathbf{T} n_2(\mathbf{T}) \Bigg\{ 1 + \sqrt{\frac{2 \ln \ln \mathbf{T} n_2(\mathbf{T})}{\ln \mathbf{T} n_2(\mathbf{T})}} \times \left[1 - \frac{\pi - 2}{2} \frac{1}{\ln \ln \mathbf{T} n_2(\mathbf{T})} + O\left(\frac{1}{\ln \ln \mathbf{T} n_2(\mathbf{T})} \right)^2 \right] \Bigg\}. \tag{7}$$

Theorem 2. *For* $\mathbf{T} \to \infty$ *and* $n_1(m) = e^{o(m)}$

$$\tau(\mathbf{T},\ N,\ n_1) \sim \frac{N+2}{2} \ln \mathbf{T}, \tag{8}$$

i.e., the principal term under the condition that $n_1(m) = e^{o(m)}$ *is the same as in the case of one channel.*

Proof of Theorem 1. The following formulas (derived exactly like formulas (2.4), (2.30) in [1]) hold for $\mathbf{T}$ and τ: **

$$\mathbf{T} = \frac{m}{\mathfrak{A}}, \tag{9}$$

$$\tau(\mathbf{T},\ N,\ n,\ \omega) = \frac{m}{2} + \frac{m}{1-\mathfrak{B}} \left(\frac{1}{m} \int_0^m \mathfrak{B}(t)\, dt \right) + m(N-1)\left(\frac{1}{2} + \frac{\mathfrak{B}}{1-\mathfrak{B}} \right), \tag{10}$$

where

$$\mathfrak{A} = 1 - (1-\alpha)^n, \tag{11}$$

$$\mathfrak{B}(t) = \beta(t)(1-\alpha)^{n-1}, \quad \mathfrak{B} = \mathfrak{B}(0), \tag{12}$$

and α and $\beta(t)$ have the same meaning as in [1]. According to [1], the parameter ω takes on one of six values: $m, h, \alpha, \beta, x, y$. We introduce the function

* For criteria to determine in which channel to assume presence of a target, if $\phi_i^k > h$ simultaneously in several channels, see, for example, [2], §57.

** Computed for the given m and h.

$$f(T,\ n,\ \omega)=m\left(\frac{1}{2}+\frac{\mathfrak{B}}{1-\mathfrak{B}}\right)=m\left(\frac{1}{2}+\frac{\beta(1-\alpha)^{n-1}}{1-\beta(1-\alpha)^{n-1}}\right). \tag{13}$$

For

$$f(\mathrm{T},\ n)=\min_{\omega} f(\mathrm{T},\ n,\ \omega)$$

and

$$\tau(\mathrm{T},\ N,\ n)=\min_{\omega}\tau(\mathrm{T},\ N,\ n,\ \omega)$$

we have (see the lemma in [1]):

$$(N+2)f(\mathrm{T},\ n)\geqslant\tau(\mathrm{T},\ N,\ n)\geqslant Nf(\mathrm{T},\ n).$$

We begin the proof of the theorem by establishing formula (7). Direct verification shows that for $n_2(\mathrm{T}) = e^{o(\mathrm{T})}$ the root $\widetilde{x}_n$ of the equation

$$x_n=\sqrt{2\mathrm{T}[1-(1-\Phi(x_n))^n]}$$

for large T can be written, by analogy with [1], in the following form:

$$\frac{x_n^2}{2}=\ln\frac{\mathrm{T}n_2(\mathrm{T})}{(2\ln\mathrm{T}n_2(\mathrm{T}))^{3/2}}+O(1).$$

Now in order to show that the value x_{nf} which minimizes the function $f(\mathrm{T},\ n,\ x)$ lies to the left of $\widetilde{x}_n$, we consider $\partial f(\mathrm{T},\ n,\ x)/\partial x$. Computation shows that

$$\frac{1}{\mathrm{T}}\frac{\partial f(\mathrm{T},\ n,\ x)}{\partial x}=\frac{n}{2}\varphi(x)(1-\Phi(x))^{n-1}$$
$$+\frac{\varphi(x)\varphi(y)(1-\Phi(x))^{n-1}}{[1-(1-\Phi(y))(1-\Phi(x))^{n-2}]^2}\left\{\frac{1}{\varphi(x)}-\frac{1}{\varphi(y)}\right.$$
$$+(1-\Phi(x))^{n-1}\left[\frac{1}{\varphi(y)}-\frac{1}{\varphi(x)}+\frac{\Phi(x)}{\varphi(x)}-\frac{\Phi(y)}{\varphi(y)}+\frac{x-y}{2}\right]$$
$$\left.+(n-1)\frac{x-y}{2}(1-\Phi(x))^{n-1}-\frac{n-1}{\varphi(y)}+\frac{n-1}{\varphi(y)}\frac{1-\Phi(y)}{1-\Phi(x)}\right\}.$$

In [1] it is shown that

$$y=x-\sqrt{2\mathrm{T}[1-(1-\Phi(x)^n]} \tag{14}$$

and that for $\mathrm{T}\geq\pi/\Phi(\sqrt{2\pi})$ and $y\geq 0$

$$\frac{\Phi(x)}{\varphi(x)}-\frac{\Phi(y)}{\varphi(y)}+\frac{x-y}{2}\geqslant 0.$$

At the same time it is obvious that for $y\geq 0$

$$\frac{1}{\varphi(x)}-\frac{1}{\varphi(y)}-[1-\Phi(x)]^{n-1}\left(\frac{1}{\varphi(x)}-\frac{1}{\varphi(y)}\right)\geqslant 0.$$

Now consider the function

$$V(y)=\frac{x-y}{2}(1-\Phi(x))^{n-1}-\frac{1}{\varphi(y)}\left[1-\frac{1-\Phi(y)}{1-\Phi(x)}\right]$$

$$=\frac{x-y}{2}(1-\Phi(x))^{n-1}-\frac{1}{\varphi(y)}\left[\frac{\Phi(y)-\Phi(x)}{1-\Phi(x)}\right],$$

where x and y are connected by relation (14). It is obvious that for $y = C$, where C is a nonnegative constant, there exists $\mathrm{T}_C < \infty$ such that $V(C) > 0$ for all $\mathrm{T} \geq \mathrm{T}_C$. Moreover, for all $y \geq 0$ for which (for large T) $x \leq \bar{x} = \sqrt{2 \ln \mathrm{T} n_2(\mathrm{T})}$, we also have $V(y) \geq 0$. Indeed, for $\bar{x} = \sqrt{2 \ln \mathrm{T} n_2(\mathrm{T})}$,

$$\bar{y} = \bar{x} - \sqrt{2\mathrm{T}[1-(1-\Phi(x))^n]} = \sqrt{2 \ln \mathrm{T} n_2(\mathrm{T})} + O\left[\frac{1}{\ln \mathrm{T} n_2(\mathrm{T})}\right]^{1/4}.$$

Further, for large T, $n = n_2(\mathrm{T}) = e^{O(\mathrm{T})}$ and $x \leq \bar{x}$,

$$(1-\Phi(x))^n > \frac{1}{2}.$$

Therefore the given assertion follows from the fact that in the region $x \leq \bar{x}$,

$$\frac{x-y}{4} \geqslant \frac{1}{y}.$$

It remains only to show that y, determined by x_{nf} from relation (14), cannot be greater than $\bar{y}$. This is proved as follows. For all $x \geq \bar{x}$ and $\mathrm{T} \to \infty$

$$x(\mathrm{T}) \sim y(\mathrm{T})$$

and thus

$$\Phi(x(\mathrm{T})) \sim \Phi(y(\mathrm{T})).$$

Therefore, by formula (13), for $x \geq \bar{x}$,

$$f(\mathrm{T}, n, x) \sim \mathrm{T},$$

while for $y = 0$ (i.e., $x = \tilde{x}_n \sim \sqrt{2 \ln \mathrm{T} n_2(\mathrm{T})}$)

$$f(\mathrm{T}, n, \tilde{x}_n) \sim \frac{3}{2} \ln \mathrm{T} n_2(\mathrm{T}).$$

Thus the value x_{nf} must be sought in the class of those x_n for which $y_n \geq 0$, and $y_n(\mathrm{T}) \to -\infty$ as $\mathrm{T} \to \infty$. The last assertion is proved exactly as in Theorem 3 of [1]. Now it is easy to show that

$$y_{nf} \sim -\sqrt{\ln \ln \mathrm{T} n_2(\mathrm{T})},$$

$$x_{nf} \sim \sqrt{2 \ln \mathrm{T} n_2(\mathrm{T})}.$$

From this it is easy to establish (7) in the same way as in [1].

It is shown analogously that

$$y_{n\tau} \sim -\sqrt{\ln \ln \mathrm{T} n_2(\mathrm{T})},$$

$$x_{n\tau} \sim \sqrt{2 \ln \mathrm{T} n_2(\mathrm{T})},$$

and consequently the validity of formula (5) is established.

Proof of Theorem 2 now presents no difficulty. Let $n_1(m) = e^{o(m)}$. In Theorem 1 we take the value $m = m_{n_T}(\mathrm{T}) \sim \ln \mathrm{T} n_2(\mathrm{T})$ which minimizes $\tau(\mathrm{T}, N, n_2(\mathrm{T}))$, where $n_2(\mathrm{T})$ is chosen by $n_1(m_{n_T}(\mathrm{T})) = n_2(\mathrm{T})$. We easily find

$$\ln n_2(\mathrm{T}) \sim o(\ln \mathrm{T} n_2(\mathrm{T})),$$

whence $\ln n_2(\mathrm{T}) = o(\ln T)$. By Theorem 1, for such $n_2(\mathrm{T})$,

$$\tau(\mathrm{T},\ N,\ n_2(\mathrm{T})) \sim \frac{N+2}{2} \ln \mathrm{T}$$

and consequently

$$\tau(\mathrm{T},\ N,\ n_1) \sim \frac{N+2}{2} \ln \mathrm{T},$$

if $n_1(m) = e^{o(m)}$. The theorem is proved.

In conclusion we note that i follows from Theorems 1 and 2 that for the case of many channels with observation by the Neyman-Pearson method, the statistic $\max_i \phi_i^k$ is asymptotically optimal for $\mathrm{T} \to \infty$ (from the point of view of minimum value of time of delay for the given method), when $n_1(m) = e^{o(m)}$ or $n_2(\mathrm{T}) = e^{o(\ln \mathrm{T})}$.

BIBLIOGRAPHY

[1] A. N. Širjaev, *On finding disorder in a manufacturing process*. II, Teor. Verojatnost. i Primenen. **8** (1963), 431–443. MR **28** #649.

[2] L. A. Vaĭnšteĭn and V. D. Zudakov, *Extraction of signals from noise*, "Sovetskoe radio". 1960; English transl., International Series in Appl. Math., Prentice-Hall, Englewood Cliffs, N. J., 1962. MR 25 #5812.

Translated by:
L. Rosenblatt

ON THE THEORY OF DECISION FUNCTIONS AND CONTROL OF A PROCESS OF OBSERVATION BASED ON INCOMPLETE INFORMATION

A. N. ŠIRJAEV

§1. Introduction

It is well known that the apparatus for the theory of decision functions [1,2], in its application to the problems of testing hypotheses and estimation of an unknown parameter $\theta \in \Theta$ is developed on the assumption that this parameter does not change in the process of observation. In other words, if we consider the parameter θ as a function of time t, $\theta = \theta_t$, then the usual assumptions are that the process $\theta = \{\theta_t\}$ does not change with time. From a practical point of view in this connection, of course, the most interesting is that case when we observe not the process $\theta = \{\theta_t\}$ itself but some random process $\xi = \{\xi_t\}$, whose distribution depends on the values of $\theta = \{\theta_t\}$. In the simplest cases, the process $\xi = \{\xi_t\}$ (where the time t is discrete) consists of a sequence identically distributed independent random variables,[1)] whose distribution belongs to probability family $\mathcal{P} = \{P_\theta, \theta \in \Theta\}$, depending on the unknown parameter $\theta \in \Theta$, $\theta_t = \theta$, [3]. On the basis of the results of observations of ξ ("on the basis of incomplete information") we must take some decision or another concerning this parameter, assuming in the general case, that the observations can be performed in a sequential manner [1–4].

One of the basic problems of the present paper is in the presentation of the results which have to do with the case when the *parameter* $\theta = \{\theta_t\}$ *can change with time*. Complete results are obtained for the discrete time case under the assumption that the process $\theta = \{\theta_t\}$ is a *Markov process*, generally speaking, *controlled*, and the results of observations of ξ_t are conditionally independent. Let us point out here one (in principle) important result, consisting of this: in the situation considered, which generalizes the classical, (for details see the theorem in §2) the a posteriori probability often plays the role of a sufficient statistic from the results of the observations. Thereby the solution of a series of problems is simplified in that we can operate not with all past information, but only with the a posteriori measure or some characteristics of it. This is quite well illustrated in §4 in the example of the Markov process $\theta = \{\theta_t\}$ with two states, say 0 and 1, and with a unique transition from $0 \to 1$, accompanied by a change of the probability distribution of the process $\xi = \{\xi_t\}$ under observation. One of

---

1) In the continuous time case, the process $\xi = \{\xi_t\}$ with stationary independent increments plays an analogous role.

the main problems arising here is the problem of the fastest detection of the moment of (spontaneous) transition of the process $\theta = \{\theta_t\}$ from the zero state to unity. In the applications, this change can be treated as the beginning of a "disorder" in a production process, as a change in intensity of radioactive decay and the like.

The results treated here were first reported in our joint paper with A. N. Kolmogorov "Application of Markov Processes to the Detection of a Disorder in a Production Process" at the 6th All-Union Conference on Probability Theory and Mathematical Statistics (Vilno, September, 1960).

In [5,6] we indicated that for solutions of these new problems we make essential use of the possibility of forming recurrence relations for risks which generalize the well-known relations of Wald and Wolfowitz [7], Arrow, Blackwell, and Girshick [8] is essentially used. Strictly speaking, in the example of the solution of the given problem, it became clear that we could also apply the apparatus of the recurrence relations to the considerably more general situation dealt with in the present article. In [5] it is pointed out that the case of a Markov process with two states falls under the general scheme of controlled Markov processes proposed by A. N. Kolmogorov, to which we also confine attention in the present paper.

The approach which we are considering is Bayesian. But in subsection 3, §4, one nonparametric problem is cited, the solution of which makes use (in an essential manner) of the apparatus (developed here) of Bayes decisions (controls).

In §3 we consider the problem of control by a process of observation (the scanning problem), the solution of which utilizes essentially the recurrence relations constructed in §2. Let us discuss here the formulation of this problem.

Let us assume that there are N directions (channels) and one device which can examine them by turns. Let the state of these N directions be described by the N-dimensional process (not varying with time) $\theta = \{\theta_t\}$ with $N+1$ states $\theta^- = (0, \cdots, 0)$, $\theta^1 = (1, 0, \cdots, 0), \cdots, \theta^N = (0, \cdots, 0, 1)$. If at the moment t, the ith direction $(i = 1, \cdots, N)$ is observed, then it is assumed that

$$P\{\xi_t < x \mid \theta = \theta^j\} = \begin{cases} F_1(x) & i = j, \\ F_0(x) & \text{elsewhere,} \end{cases}$$

where the disorder functions F_0 and F_1 are assumed to be independent of i. In other words, if the observation is performed in the ith direction and there is a target in that direction, i.e., $\theta = \theta^i$, then the distribution function of the observed variable ξ_t is equal to $F_1(x)$. If there is not a target in that direction, i.e., $\theta \neq \theta^i$, then the distribution function of ξ_t is equal to $F_0(x)$.

Since at each of the moments of time t our device can investigate only one direction, there arises the problem (the scanning problem) of the search for the best control, in some sense, of the process of observation. Here, of course, in the general case, the observations are performed in a sequential manner. Clearly this problem is a *multidimensional* analogue of the Wald problem of testing two simple hypotheses ($N = 1$, hypothesis H_0: $\theta = \theta^-$, H_1: $\theta = \theta^1$).

Let us note that although in principle the solution of these and a number of other problems for the discrete time case can be obtained from the result of §2, a concrete solution is quite cumbersome. Going over to the continuous time case permits us to obtain complete solutions. It is necessary, of course, in this connection to mention that in the transition to continuous time, a difficulty arises in the determination of optimal decisions, controls, and the like. Here we shall follow an idea in [9], according to which the only solutions considered are those which can be obtained by passage to the limit from the discrete scheme. In this way we can make use of the recurrence relations of §2 for obtaining *parabolic differential equations* for the risks and for determining the boundaries; when the latter are reached, the process of observation is stopped and some decision is made. Let us remark further that the apparatus of differential equations for finding the risks and the boundaries is also used in [10, 11].

It is my pleasant duty to express my gratitude to A. N. Kolmogorov for posing the problems and for his assistance in the work on them.

§2. Recurrence relations for nonadditive functionals. Discrete time

1. Everywhere in this section it is assumed that the parameter t (time) takes discrete values, $0, \Delta, 2\Delta, \cdots, T < \infty$. Functions of $t \in [0, T]$ are denoted by $\phi = \{\phi_t\}$. With every function $\phi = \{\phi_t\}$ we associate functions ϕ^t, ϕ^t_s, defined respectively for values $\tau \leq t$ and $s < \tau \leq t$ only and coinciding on these sets with the function $\phi = \{\phi_t\}$:

$$\phi^t = \{\phi_\tau, \tau \leq t\}, \quad \phi^t_s = \{\phi_\tau, s < \tau \leq t\}.$$

By $(\Phi_t, \mathfrak{B}_{\Phi_t})$, $(\Phi^t, \mathfrak{B}_{\Phi^t})$, $(\Phi^t_s, \mathfrak{B}_{\Phi^t_s})$ and $(\Phi, \mathfrak{B}_\Phi)$ we will denote the measurable spaces generated by ϕ_t, ϕ^t, ϕ^t_s and $\phi = \phi^T$.

Let $(D, \mathfrak{B}_D)$, $(\Theta, \mathfrak{B}_\Theta)$, $(\Xi, \mathfrak{B}_\Xi)$ be the measurable spaces generated respectively by the functions $d = \{d_t\}$, $\theta = \{\theta_t\}$, $(t \geq 0)$, $\xi = \{\xi_t\}$, $(t \geq \Delta)$ which play a further role in the realization of a *decision rule* (*control*) of *controlled* and *observed processes*.[1)]

It is assumed that for every t there are given $\mathfrak{B}_D \times \mathfrak{B}_\Theta \times \mathfrak{B}_\Xi$-measurable functionals $W^t = W(t; \zeta^T)$, $\zeta^T = (d^T, \theta^T, \xi^T)$, $W^t \geq 0$, characterizing the "losses" which occur at the moment of time t if the decisions d^T led to the value ζ^T.

---

1) We shall apply these notations also for the same random processes.

Here, of course, in accordance with the principle of causality, we will assume that W^t is measurable relative to the σ-algebra of sets $\mathfrak{B}_{Dt} \times \mathfrak{B}_{\Theta t} \times \mathfrak{B}_{\Xi t}$. In other words, for two realizations ζ^T and $\tilde{\zeta}^T$ such that $\zeta^T = \tilde{\zeta}^T$, it is assumed that $W^t(\zeta^T) = W^t(\tilde{\zeta}^T)$. Clearly there will be no misunderstanding if in this case we will simply write $W^t = W(t; \zeta^t)$. For $t = 0$, by ζ^0 we understand (d^0, θ^0).

Let us turn now to the definition of the processes d, θ, ξ themselves. It is assumed that a process of observation is performed in a sequential manner; moreover at the moment t the decision d_t is chosen from the space D_t without delay, that is only on the basis of knowledge of preceding values $\eta^{t-\Delta} = (d^{t-\Delta}, \xi^{t-\Delta})$. More precisely, defining an arbitrary *decision rule* (*control*) δ is the same as defining a collection of functionals $\delta = \{\delta_t\}$ such that the functional

$$\delta_t = \delta_t(D_t;\, d^{t-\Delta},\, d_t^T,\, \xi^T), \qquad t > 0 \tag{1}$$

for each $D_t \in \mathfrak{B}_{D_t}$ is measurable with respect to the remaining variables and for fixed values of the latter represents a probability measure on $(D_t, \mathfrak{B}_{D_t})$. Here we again consider only functionals δ_t without delay; that is, only functionals which for each D_t are measurable relative to the σ-algebra $\mathfrak{B}_{Yt}$, generated by the functions $\eta^{t-\Delta} = (d^{t-\Delta}, \xi^{t-\Delta})$. Also, just as for W^t, we will use the notation

$$\delta_t = \delta_t(D_t;\, d^{t-\Delta},\, \xi^{t-\Delta}).$$

For the case $t = 0$, it represents a probability distribution

$$\delta_0 = \delta_0(D_0).$$

Analogously, it is assumed that a "*controlled*" process $\theta = \{\theta_t\}$ is defined by a system of functionals $\beta = \{\beta_t\}$ such that

$$\beta_t = \beta_t(\Theta_t;\, d_t,\, \zeta^{t-\Delta}), \quad t > \Delta, \quad \beta_0 = \beta_0(\Theta_0; d_0) \tag{2}$$

for each $\Theta_t \in \mathfrak{B}_{\Theta_t}$ measurable with respect to the remaining variables, and for fixed values of the latter, they represent probability measures on $(\Theta_t, \mathfrak{B}_{\Theta_t})$.

An "*observed*" process $\xi = \{\xi_t\}$, $t \geq \Delta$, usually producing "incomplete information"[1] relative to a process $\theta = \{\theta_t\}$ not subject to direct observation,[2] is given by a system of functionals $\gamma = \{\gamma_t\}$ where, as in the case of the functionals δ_t and β_t, it is assumed that $\gamma_t = \gamma_t(\Xi_t;\, d^T,\, \theta^T,\, \xi^{t-\Delta},\, \xi^T)$ are for fixed $\Xi_t \in \mathfrak{B}_{\Xi_t}$ measurable relative to the σ-algebra $\mathfrak{B}_{D_t} \times \mathfrak{B}_{\Theta_t} \times \mathfrak{B}_{\Xi^{t-\Delta}}$, and, for fixed

---

1) In the sense that the functionals γ_t, $t \geq \Delta$, (defined below) take values different from 0 and 1.

2) The case of the coincidence of the processes θ and ξ is not excluded in the following recurrence relations (10–12).

last arguments, are probability measures on $(\Xi_t, \mathfrak{B}_{\Xi_t})$. Again, we shall write this condition in the form:

$$\gamma_t = \gamma_t(\Xi_t;\ d_t, \theta_t,\ \xi^{t-\Delta}). \tag{3}$$

The systems of functionals $\beta = \{\beta_t\}$ and $\gamma = \{\gamma_t\}$, characterizing the evolution of the controlled and observed processes θ and ξ are given by the conditions of each concrete problem. The collection of functionals $\delta = \{\delta_t\}$ determining, as we say, an arbitrary rule (control) is required to be defined in such a way that it will will satisfy certain conditions of optimality. If some rule δ is given, then in the measurable space

$$(D \times \boldsymbol{\Theta} \times \Xi,\ \mathfrak{B}_D \times \mathfrak{B}_{\boldsymbol{\Theta}} \times \mathfrak{B}_{\Xi})$$

there is defined a probability measure which is naturally written with the subscript δ: $P_\delta(\cdot)$.

For each rule δ we will consider an average "risk"

$$r^T(\delta) = M_\delta W(T;\ \zeta^T), \tag{4}$$

where the subscript δ on the symbol for the mathematical expectation signifies that the mean is taken with respect to the measure P_δ.

We will call a decision rule $\tilde{\delta}$ *optimal* if

$$\rho^T = \inf_\delta r^T(\delta) = r^T(\tilde{\delta}). \tag{5}$$

Everywhere in the following we assume that such an optimal decision exists. It is not difficult to see that in these cases when a random process $\theta = \{\theta_t\}$ does not depend on the choice of a control $d = \{d_t\}$, the present definition of optimal decisions coincides with the concept of Bayes decisions.

Up to now we have not made any assumptions about the spaces D_t of the values of the decisions d_t at the moment t. In the general case, the typical situations are such that the preceding outcomes of observations $\eta^{t-\Delta} = (d^{t-\Delta},\ \xi^{t-\Delta})$ impose certain restrictions on the space from which these decisions can be selected. In other words, let $\mathfrak{D}_t = \mathfrak{D}_t(\eta^{t-\Delta}) \subseteq D_t$ be the maximal subspace of admissible values of the decision d_t, if the outcome of the observation is $\eta^{t-\Delta}$. Such restrictions, of course, are determined by the content of a concrete problem. If they arise, then the minimization in (5) is to be carried out only for admissible decisions, i.e., those such that for every (more precisely, almost every) $\eta^{t-\Delta}$, the supports of the measures δ_t belong to the spaces $\mathfrak{D}_t = \mathfrak{D}_t(\eta^{t-\Delta}) \subseteq D_t$.

2. We go over now to the construction of the recurrence relations enabling us, at least in principle, to find the optimal decisions.

Let

$$\Delta W_t^s = \Delta W_t^s(\zeta^s) = W(s;\ \zeta^s) - W(t;\ \zeta^t), \quad s > t, \tag{6}$$

and

$$\rho_t^T = \rho_t^T(\eta^t) = \inf M(\Delta W_t^T/\eta^t), \tag{7}$$

where inf is taken over all admissible rules δ_t^T, consistent with the outcome of preceding observations $\eta_t = (d^t,\ \xi^t)$. We observe that for every t

$$\begin{aligned}\rho^T &= \inf_{\delta} M_\delta W(T;\ \zeta^T) = \inf_{\delta} M_\delta\{W(t;\ \zeta^t) + \Delta W_t^T\}\\ &= \inf_{\delta^t}\inf_{\delta_t^T} M_\delta\{W(t;\ \zeta^t) + \Delta W_t^T\}\\ &= \inf_{\delta^t}\inf_{\delta_t^T} M_{\delta^t} M_{\delta_t^T}\{W(t;\ \zeta^t) + \Delta W_t^T/\eta^t\}\\ &= \inf_{\delta^t} M_{\delta^t}\{W(t;\ \zeta^t) + \rho_t^T(\eta^t)/\eta^t\}\\ &= \inf_{\delta^t} M_{\delta^t}\{M_{\delta^t}[W(t;\ \zeta^t)/\eta^t] + \rho_t^T(\eta^t)\}.\end{aligned} \tag{8}$$

Thus it is clear that for the chosen rule δ^t (with due regard for admissibility) which led to the outcome $\eta^t = (d^t,\ \xi^t)$, the optimal decision in the following stages coincides with the rule which minimizes $M(\Delta W_t^T/\eta^t)$. In order to find it we note that

$$\begin{aligned}\rho_t^T(\eta^t) &= \inf_{\delta_t^T} M_{\delta_t^T}(\Delta W_t^T/\eta^t) = \inf_{\delta_t^s}\inf_{\delta_s^T} M_{\delta_t^T}[\Delta W_t^s + \Delta W_s^T/\eta^t]\\ &= \inf_{\delta_t^s}\{M_{\delta_t^s}(\Delta W_t^s/\eta^t) + \inf_{\delta_t^T} M_{\delta_s^T}(\Delta W_s^T/\eta^t)\}\\ &= \inf_{\delta_t^s}\{M_{\delta_t^s}(\Delta W_t^s/\eta^t) + \inf_{\delta_s^T} M_{\delta_t^s}[M_{\delta_s^T}(\Delta W_s^T/\eta^s)/\eta^t]\}\\ &= \inf_{\delta_t^s}\{M_{\delta_t^s}(\Delta W_t^s/\eta^t) + M_{\delta_t^s}[\inf_{\delta_s^T} M_{\delta_s^T}(\Delta W_s^T/\eta^s)/\eta^t]\}\\ &= \inf_{\delta_t^s}\{M_{\delta_t^s}(\Delta W_t^s/\eta^t) + M_{\delta_t^s}[\rho_s^T(\eta^s)/\eta^t]\}\\ &= \inf_{\delta_t^s} M_{\delta_t^s}\{\Delta W_t^s + \rho_s^T(\eta^s)/\eta^t\}.\end{aligned} \tag{9}$$

In such manner for $0 \le t \le T - \Delta$

$$\rho_t^T(\eta^t) = \inf_{\delta_t^s} M_{\delta_t^s}\{\Delta W_t^s + \rho_s^T(\eta^s)/\eta^t\}, \tag{10}$$

and in the general case $s = t + \Delta$

$$\rho_t^T(\eta^t) = \inf_{\delta_{t+\Delta}} M_{\delta_{t+\Delta}}\{\Delta W_t + \rho_{t+\Delta}^T(\eta^{t+\Delta})/\eta^t\}, \tag{11}$$

where $\Delta W_t = \Delta W_t^{t+\Delta}$.

To the systems (10) and (11), which give the possibility of finding optimal rules $\delta_\Delta, \cdots, \delta_T$, it is necessary further to add the relation

$$\rho^T = \inf_{\delta_0} M_{\delta_0}\{W(0;\ \zeta_0) + \rho_0^T(\eta^0)/\eta^0\} \tag{12}$$

for finding δ_0.

These recurrence relations for "cut-off" functions of nonadditive risks ρ_t^T constitute a direct generalization of the relations considered in the works [7], [8], [5], [11]. The relations (11–12), considered successively for $t = T - \Delta$, $T - 2\Delta, \cdots$, in principle solve the problem of finding an optimal rule $\delta = \{\delta_t\}$, although the solution becomes apparent only in special cases, one of which we shall consider now.

3. Let

$$\text{A.}\quad W(t;\ \zeta^t) = \sum_{\tau \leq t} W(\tau,\ \zeta_\tau), \tag{13}$$

where the functions $W(\tau,\ \zeta_\tau)$ satisfy obvious conditions of measurability.

We will make the following simplifying assumptions concerning the processes $\theta = \{\theta_t\}$ and $\xi = \{\xi_t\}$:

B. The controlled process $\theta = \{\theta_t\}$ is Markov, i.e., the system of functionals $\beta = \{\beta_t\}$ describing its evolution is such that

$$\beta_t(\Theta_t;\ d^t,\ \theta^{t-\Delta},\ \xi^{t-\Delta}) = \beta_t(\Theta_t;\ d_t,\ \theta_{t-\Delta}).$$

In other words,

$$\beta_t(\Theta_t;\ d^t,\ \theta^{t-\Delta},\ \xi^{t-\Delta}) = \beta_t(\Theta_t;\ \tilde{d}^t,\ \tilde{\theta}^{t-\Delta},\ \tilde{\xi}^{t-\Delta}),$$

everytime when

$$d_t = \tilde{d}_t \text{ and } \theta_{t-\Delta} = \tilde{\theta}_{t-\Delta}.$$

C. The observed process $\xi = \{\xi_t\}$ is composed of a sequence of conditionally independent random variables, i.e., such that

$$\gamma_t(\Xi_t;\ d^t,\ \theta^t,\ \xi^{t-\Delta}) = \gamma_t(\Xi_t;\ \tilde{d}^t,\ \tilde{\theta}^t,\ \tilde{\xi}^{t-\Delta}),$$

if $\tilde{d}_t = d_t$, $\tilde{\theta}_t = \theta_t$.

D. Let us also assume that the spaces $\mathfrak{D}_t$ depend on the past data in the Markov manner: $\mathfrak{D}_t = \mathfrak{D}_t(\eta_{t-\Delta})$, $\eta_{t-\Delta} = (\xi_{t-\Delta},\ d_{t-\Delta})$.

Let us consider, in assumptions A–D, the recurrence relations (11).

In this case

$$\rho_t^T(\eta^t) = \inf_{\delta_{t+\Delta}} M\{W(t+\Delta, \zeta_{t+\Delta}) + \rho_{t+\Delta}^T(\eta^{t+\Delta})/\eta^t\}. \tag{14}$$

For $t = T - \Delta$

$$\rho_{T-\Delta}^T(\eta^{T-\Delta}) = \inf_{\delta_T} M\{W(T, \zeta_T)/\eta^{T-\Delta}\}.$$

Let an arbitrary decision $d_T \in \mathfrak{D}_T(\eta_{T-\Delta})$ be chosen. Under conditions B–C, it is easy to show, in view of $P(\Theta_T, \Xi_T/d_T, \eta^{T-\Delta})$, that $M[W(T, \zeta_T)/\eta^{T-\Delta}]$ is measurable relative to the σ-algebra $\mathfrak{B}_{\Pi_{T-\Delta}}$, where $\mathfrak{B}_{\Pi_t}$ denotes the smallest σ-algebra with respect to which, for each fixed $\Theta_t \in \mathfrak{B}_{\Theta_t}$, the conditional distributions $\pi_t(\Theta_t) = P(\Theta_t/\eta^t)$ are measurable.

Then from the above it follows that the choice of an optimal decision δ_T at moment T is defined, not by all "past" $\eta^{T-\Delta} = (d^{T-\Delta}, \xi^{T-\Delta})$, but only by the a posteriori measure $\pi_{T-\Delta} = \pi_{T-\Delta}(\cdot)$ and $\eta_{T-\Delta} = (d_{T-\Delta}, \xi_{T-\Delta})$. In the same way $\rho_{T-\Delta}^T(\eta^{T-\Delta})$ is also shown to be measurable relative to the smallest σ-algebra generated by $\mathfrak{B}_{\Pi_{T-\Delta}}$ and $\mathfrak{B}_{Y_{T-\Delta}}$. In order to emphasize this, we shall denote $\rho_{T-\Delta}^T(\eta^{T-\Delta})$ by $\rho_{T-\Delta}(\pi_{T-\Delta}, \eta_{T-\Delta})$. Similar notations are introduced also in the following, without explicit mention.

We will now consider the relations (14) for $t = T - 2\Delta$. By the calculation given above concerning the choice of an optimal decision, we obtain

$$\begin{aligned}\rho_{T-2\Delta}^T(\eta^{T-2\Delta}) &= \inf_{\delta_{T-\Delta}} \{MW(T-\Delta, \zeta_{T-\Delta}) + \int \rho_{T-\Delta}^T(\eta^{T-\Delta})\, dP(\eta_{T-\Delta}/\eta^{T-2\Delta})\} \\ &= \inf_{d_{T-\Delta}} \{M[W(T-\Delta, \zeta_{T-\Delta})/\eta^{T-2\Delta}, d_{T-\Delta}] \\ &\quad + \int \rho_{T-\Delta}^T(\pi_{T-\Delta}, \eta_{T-\Delta})\, dP(\eta_{T-\Delta}/\eta^{T-2\Delta}, d_{T-\Delta})\}. \end{aligned} \tag{15}$$

For every $d_{T-\Delta} \in \mathfrak{D}_{T-\Delta}(\eta_{T-2\Delta})$, $M[W(T-\Delta, \zeta_{T-\Delta})/\eta^{T-2\Delta}, d_{T-\Delta}]$ is $\mathfrak{B}_{\Pi_{T-2\Delta}}$ measurable. Further, for the computation of the conditional measure $\pi_{T-\Delta} = P(\cdot/\eta^{T-\Delta})$ it is sufficient to know from the "past" $\eta^{T-\Delta}$ only the measure $\pi_{T-2\Delta}$ and $\eta_{T-\Delta}$. This result is obtained by direct calculation in view of conditions B and C. We notice that essentially this fact lies at the basis of study of conditional Markov processes [12]. On the other hand, for $Y_{T-\Delta} = \Xi_{T-\Delta} \times \{d_{T-\Delta}\}$

$$\begin{aligned}P(Y_{T-\Delta}/\eta^{T-2\Delta}, d_{T-\Delta}) &= \int P(\Xi_{T-\Delta}/\theta_{T-\Delta}, \eta^{T-2\Delta}, d_{T-\Delta})\, dP(\theta_{T-\Delta}/\eta^{T-2\Delta}, d_{T-\Delta}) \\ &= \int P(\Xi_{T-\Delta}/\theta_{T-\Delta}, d_{T-\Delta})\, dP(\theta_{T-\Delta}/\eta^{T-2\Delta}, d_{T-\Delta}).\end{aligned}$$

Here

$$P(\Theta_{T-\Delta}/\eta^{T-2\Delta}, d_{T-\Delta}) = \int P(\Theta_{T-\Delta}/\theta_{T-2\Delta}, d_{T-\Delta})\, dP(\theta_{T-2\Delta}/\eta^{T-2\Delta}, d_{T-\Delta}).$$

But the measure $P(\Theta_{T-2\Delta}/\eta^{T-2\Delta}, d_{T-\Delta})$ does not depend on $d_{T-\Delta}$, since $P(\Theta_{T-2\Delta}, Y_{T-2\Delta}, Y_{T-2\Delta})$ does not depend on $d_{T-\Delta}$, in view of the fact that the functionals δ, β, γ, appear as functionals without delay. Thus for each fixed value of $d_{T-\Delta}$, the expressions in the braces of formula (15) depend only on $\pi_{T-2\Delta}$ and $\eta_{T-2\Delta}$. Consequently an optimal decision at moment of time $t-\Delta$ does not depend on the whole past course of the process η_t, $t \leq T-2\Delta$, but only on the a posteriori measure $\pi_{T-2\Delta}$ and $\eta_{T-2\Delta}$. Therefore also

$$\rho^T_{T-2\Delta}(\eta^{T-2\Delta}) = \rho^T_{T-2\Delta}(\pi_{T-2\Delta}, \eta_{T-2\Delta}).$$

The further argument for moments of time $t = T-3\Delta, T-4\Delta, \cdots$, is completely analogous. In the final calculation we come to the equation

$$\rho^T = \inf_{d_0} M\{W(0; \zeta_0) + \rho^T_0(\pi_0, d_0)\}, \tag{16}$$

[where π_0 is the measure $P(\cdot/d_0)$ defined on the measurable sets $\Theta_0 \in \mathfrak{B}_{\Theta_0}$] which gives the magnitude of the minimal possible risk.

Thus it is possible to formulate the following result:

Theorem. *If conditions* A–D *are satisfied, then the optimal decision* δ_t *at moment of time* t *depends on the past data* $\eta^{t-\Delta} = (\xi^{t-\Delta}, d^{t-\Delta})$ *only through an a posteriori measure* $\pi_{t-\Delta}$ *and* $\eta_{t-\Delta}$. *In other words it is sufficient to look for optimal decisions* δ_t *in the class of functionals* $\delta_t = \delta_t(D_t; \eta^{t-\Delta})$, *measurable with respect to the* σ*-algebra of sets* $\mathfrak{B}_{\Pi_{T-\Delta}} \times \mathfrak{B}_{Y_{T-\Delta}}$.

4. Let us consider especially the case that among the points of the spaces D_t there are decisions after which the process of observation can be terminated even before the end of the time allotted for observation. Let us assume that the spaces $\mathfrak{D}_t$ depend on $\xi_{T-\Delta}$ and let $D_t = D_t^0 \cup D_t' \cup \{\tilde{d}\}$, $t \geq \Delta$, $D_0 = D_0^0 \cup \{\bar{d}\}$. Here $\tilde{d}$ is the decision consisting of the fact that the process of observation is terminated and is accompanied by a "transition" of values of the processes θ and ξ into arbitrary auxiliary states $\tilde{\theta}$, $\tilde{\xi}$, i.e., we will suppose that $\beta_t(\tilde{\theta}; \theta_{t-\Delta}, \tilde{d}) = 1$, $\gamma_t(\tilde{\xi}, \tilde{\theta}, \tilde{d}) = 1$. By $d_t^0 \in D_t^0$, $t \geq 0$, we understand "terminal" decisions, after which it is possible to follow only decision $\tilde{d}$ concerning the cessation of observations. By $\bar{d}$ we will denote the decision consisting of the fact that it is necessary to carry out observation at least for the time Δ. By D_t' we will denote the remaining decisions, where $\mathfrak{D}_t(d'_{t-\Delta}) = D_t \setminus \{\tilde{d}\}$, $d'_{t-\Delta} \in D'_{t-\Delta}$. By analogy with the above, it follows that

$$\mathfrak{D}_\Delta(\bar{d}) = D'_\Delta, \quad \mathfrak{D}_t(d^0_{t-\Delta}) = \{\tilde{d}\}, \quad \mathfrak{D}_t(\{\tilde{d}\}) = \{\tilde{d}\}.$$

Here we always assume $W(t, \tilde{\xi}) \equiv 0$. Thus it is possible to say that d_t^0 corresponds to the taking of an arbitrary terminal decision, $\tilde{d}$ (corresponds) to the decision that it is not necessary to carry out the observations. By d_t', depending on the problems, we can mean different decisions connected with the carrying out of the observations. For instance, they can be decisions that the observation continues to be carried out, let us say in the ith direction as in the scanning problem about which we spoke in the introduction. Thus typical realizations of decision rules have the form

$$\begin{gathered} d^0, \tilde{d}, \tilde{d}, \cdots \\ \bar{d}, d', d^0, \tilde{d}, \cdots \\ \bar{d}, d', d', d^0, \tilde{d}, \cdots. \end{gathered} \tag{17}$$

For every rule δ let ν be a random variable equal to the duration of the observation process. Thus for the first of the indicated realizations $\nu = 0$, for the second $\nu = \Delta$, and so on. We shall restrict ourselves in the following to the case when the spaces D_t do not depend explicitly on the time t: $D = D_t$.

In view of the assumed condition $W(t, \tilde{\zeta}) \equiv 0$, it is useful to note that $\rho_t^T(\pi_t, d_t) = 0$, provided that $d_t \in D_t^0$.

Since $\rho_t(\pi_t, d_t')$ obviously does not depend on d_t', we shall write $\rho_t^T(\pi_t) = \rho_t^T(\pi_t, d_t')$. Then by (11)

$$\rho_t^T(\pi_t) = \min[\rho_{t,0}^T(\pi_t), \rho_{t,1}^T(\pi_t)], \tag{18}$$

where the minimal risk $\rho_{t,0}^T(\pi_t)$ obtained for the terminal decision taken at the moment $t + \Delta$ is

$$\rho_{t,0}^T(\pi_t) = \inf_{d_t + \Delta \in D_t^0} M[W(t + \Delta, \zeta_{t+\Delta})/\pi_t], \tag{19}$$

and

$$\rho_{t,1}^T(\pi_t) = \inf_{d_t + \Delta \in D'} M[W(t + \Delta, \zeta_{t+\Delta}) + \rho_{t+\Delta}^T(\pi_{t+\Delta})/\pi_t] \tag{20}$$

is the minimal risk for continuation of observation.

5. Thus far we have considered the case $T < \infty$; while in the derivation of relations (11) essentially we make use of the existence of the last moment of time $T < \infty$. By an argument analogous to that which in the classical case $(\theta_t \equiv \theta)$ is applied in [7], [8], [9] (see also [13] where the case of a Markov process $\{\theta_t\}$ with two states considered), it can be shown that, for $\rho_t(\pi_t) = \rho_t^\infty(\pi_t)$, equations (18–20) also remain valid if, let us say, the optimal control is determined only in the class of those controls δ for which

$$M_\delta \sum_\tau V(\tau, \Theta_\tau, \xi_\tau) < \infty,$$

where

$$|W(\tau, \zeta_t)| = |W(\tau, \Theta_\tau, \xi_\tau, u_\tau)| \leq V(\tau, \Theta_\tau, \xi_\tau).$$

This fact is valid also in a more general case, but we shall not discuss this point further just now.

If we omit the index $T = \infty$ in the formulas, then we obtain by analogy with (18–20),

$$\rho_t(\pi_t) = \min[\rho_{t,0}(\pi_t), \rho_{t,1}(\pi_t)], \tag{21}$$

where

$$\rho_{t,0}(\pi_t) = \inf_{d_t+\Delta \in D^0} M[W(t+\Delta, \zeta_{t+\Delta})\pi/_t], \tag{22}$$

$$\rho_{t,1}(\pi_t) = \inf_{d_t+\Delta \in D'} M[W(t+\Delta, \zeta_{t+\Delta}) + \rho_{t+\Delta}(\pi_{t+\Delta})/\pi_t]. \tag{23}$$

In the stationary problems in which the functions $W(t, \zeta_t)$, the spaces $\mathfrak{D}_t$, the functionals β_t, γ_t do not depend explicitly on t, time t also does not enter into the risks $\rho_{t,0}(\pi_t)$, $\rho_{t,1}(\pi_t)$ and $\rho_t(\pi_t)$ in an explicit manner. Therefore we will simply denote these risks by $\rho_0(\pi_t)$, $\rho_1(\pi_t)$, $\rho(\pi_t)$.

§3. Transition to continuous observation. Determination of optimal decisions. A special case

1. The preceding section was devoted entirely to the discrete time case and from the logical point of view the definition of optimal decisions did not cause any difficulty. The situation is otherwise in those problems when observations are made in a continuous manner. It is necessary to state that a general theory of decision functions for that case has not been constructed (see also [9]). But the existence of recurrence relations for the discrete time case leads to the idea of constructing a theory in the general case by passage to the limit from the discrete scheme.

Let us assume for simplicity that the spaces D_t, $t \neq 0$ do not depend on t, and $T = \infty$. Let the space D of functions $d = \{d_t\}$ contain piecewise-constant left-continuous functions $d^\Delta = \{d_t^\Delta\}$ with points of discontinuity only at multiples of the time interval $\Delta \in T^2$, where T^2 is the set of dyadic rational numbers.

Let us consider an arbitrary control $\delta = \{\delta_t\}$, the realizations of which are functions d^Δ. We assume that with the realizations of the observed process $\xi = \{\xi_t\}$ and the controlled process $\theta = \{\theta_t\}$, for every $d^\Delta = \{d_t^\Delta\}$, $\xi^\Delta = \{\xi_t^\Delta\}$ and $\theta^\Delta = \{\theta_t^\Delta\}$ are also piecewise-constant left-continuous functions. Therefore it is sufficient to define the functionals $\{\gamma_t\}$ and $\{\beta_t\}$ only for the d^Δ at multiples of the time interval $\Delta \in T^2$.

By R^Δ we denote the collection of decision rules which have as their realizations the functions $d^\Delta = \{d_t^\Delta\}$, $\Delta \in T^2$. Following [9] we shall call a rule δ *regular* if there exists a sequence of decision functions $\{\delta^{\Delta_i}\}$, $\Delta_i \in T^2$, $\Delta_i \to 0$, such that

$$r(\delta^{\Delta_i}) \to r(\delta), \quad (r(\delta) = r^\infty(\delta)). \tag{24}$$

Let us denote the set of all regular decision functions by $\widetilde{R}$. A rule $\delta^* \in \widetilde{R}$ for which $r(\delta^*) \leq r(\widetilde{\delta})$ for all $\widetilde{\delta} \in \widetilde{R}$ will be called *optimal* (with respect to the given system of functionals β, γ).

Such a procedure for the definition of optimal decisions makes it possible to use the recurrence relations of the discrete schemes for their construction. But the question of how general are these definitions of optimality and the methods of finding out about them remains unclear.

We shall illustrate with two problems the procedure stated for finding optimal decisions. We turn to the first of them.[1)]

2. Let the space $D_0^0 = D^0 \cup \{\bar{d}\}$. We assume that the process $\theta = \{\theta_t\}$ is neither dependent on the controls nor varying with time. Let θ_t take $N+1$ values $\theta^- = (0, \cdots, 0)$, $\theta^1 = (1, 0, \cdots, 0), \cdots, \theta^N = (0, \cdots, 0, 1)$. Moreover, let $P(\theta = \theta^i) = \beta_0(\theta^i, \bar{d}) = \pi_i$, $\Sigma \pi_i \leq 1$. With the process θ, as we have already mentioned, it is convenient to associate the following model. There are N directions (channels). The state θ^i, $i \geq 1$ is interpreted as the presence of a "target" in the ith direction. Its absence is identical with the state θ^-. Further let the space of decisions D' be composed of N points $(d'_1, \cdots, d'_N)$ where decision d'_i is identified with observation of the ith direction. Thus it is possible for us to investigate all N directions.

Let us consider the observed process $\xi = \{\xi_t\}$. Let us assume that in the course of time $[t_1, t_2)$, decision d'_i is taken. Then we shall assume that in this interval the process $\xi = \{\xi_t\}$ becomes a Wiener process, for which, if there is a target in the ith direction, i.e. $\theta = \theta^i$, we have

$$M\,\Delta\xi = \Delta, \quad D\,\Delta\xi = \Delta.$$

In the case $\theta \neq \theta^i$, we assume

$$M\,\Delta\xi = 0, \quad D\,\Delta\xi = \Delta.$$

Moreover, everytime we assume that if $d'_{t_1-0} = d'_j$ where $j \neq i$, then $\xi(t_1) = 0$. In

---

1) The second problem is considered in §4.

other words, observation in each direction begins from the value of ξ equal to zero. In this connection it is assumed that the processes being observed in different directions are independent.

Further let $\mathfrak{D}^0$, the space of terminal decisions, be composed only of the two points d_+ and d_-, where decision d_+ is identified with the decision concerning the existence of a target in general (i.e., without reference to the number of the of the direction where it is), and d_- is the decision concerning its absence. Of more interest is the "information" problem in which the space $\mathfrak{D}^0$ is composed of $N+1$ points $(d_0^0, d_1^0, \cdots, d_N^0)$ where d_0^0 denotes the absence of a target in general and d_i^0, $i \geq 1$ denotes its presence, namely in the ith direction. For simplicity we will consider only the problem where $D^0 = \{d_+\} \cup \{d_-\}$.

So it is necessary for us to construct an optimal control. For this we consider at first the discrete scheme in which the realizations of the decisions coincide with the functions $d^\Delta = \{d_t^\Delta\}$. Thus we decide to take observations only at multiples of the time interval Δ. Let $\xi_\Delta, \xi_{2\Delta}, \cdots, \xi_{k\Delta}$ be the outcomes of the observations obtained at the times $\Delta, 2\Delta, \cdots, k\Delta$. In order to solve the question of what decision should be taken at time $(k+1)\Delta$, it obviously makes no difference whether we consider the variables $\xi_\Delta, \xi_{2\Delta}, \cdots, \xi_{k\Delta}$ or the mutually independent variables $\xi'_\Delta, \xi'_{2\Delta}, \cdots, \xi'_{k\Delta}$, where

$$\xi'_\Delta = \xi_\Delta,\ \xi'_{2\Delta} = \xi_{2\Delta} - \xi_\Delta, \cdots,\ \xi'_{k\Delta} = \xi_{k\Delta} - \xi_{(k-1)\Delta}.$$

Thus by taking the increments of the process ξ as observations, we obtain the process ξ', consisting of a sequence of independent random variables which leads to the satisfying of condition C.

If we choose the functions $W(t, \theta, d)$ to be the functions (not depending on t)

$$W(\theta, d) = \begin{cases} c\Delta, & d = d'_j, \\ 1, & d = d_+,\ \theta = \theta_0, \\ 1, & d = d_-,\ \theta \neq \theta_0, \end{cases}$$

and equal to zero in the remaining cases, then

$$r(\delta^\Delta) = M \sum_{\tau=0}^{\infty} W(\theta_\tau, d_\tau) = \pi_-[cM_-\nu + P(d_+/\theta^-)]$$
$$+ \pi_+[cM_+\nu + P(d_-/\theta \neq \theta^-)],$$

where $\nu \geq 0$ is the time of observation until taking one of the decisions d_+ or d_- and $M_-\nu$, $M_+\nu$ are the expectations of ν under the hypotheses of the absence or presence of a target respectively; $\pi_+ = \pi_1 + \cdots + \pi_N$, $\pi_- = 1 - \pi_+ > 0$.

Let [1] $\pi(t) = (\pi_-(t), \pi, (t), \cdots, \pi_N(t))$ be the vector of a posteriori probabilities of states $(\theta^-, \theta^1, \cdots, \theta^N)$ at moment of time t under some rule of observation or other (t is a multiple of Δ). Obviously

$$\rho_0^{(\Delta)}(\pi(t)) = \min\{\pi_+(t), \pi_-(t)\}, \tag{25}$$

$\pi_+(t) = \pi_1(t) + \cdots + \pi_N(t)$, $\pi_-(t) = 1 - \pi_+(t)$. The index (Δ) indicates that time t is a multiple of Δ. It is clear that

$$\rho_1^{(\Delta)}(\pi(t)) = \min\{\rho_{1i}^{(\Delta)}(\pi(t)), \cdots, \rho_{1N}^{(\Delta)}(\pi(t))\}, \tag{26}$$

where $\rho_{1i}^{(\Delta)}(\pi(t))$ is the minimal risk with the assumption that at the moment $t + \Delta$, decision d_i' is taken.

If the decision to continue the observation is taken, then by (26) it is obviously necessary to take this observation in the ith direction, for which $\rho_1^{(\Delta)}(\pi(t)) = \rho_{1i}^{(\Delta)}(\pi(t))$.

The risks $\rho_{1i}^{(\Delta)}(\pi(t)) = \rho_{1i}^{(\Delta)}(\pi_0(t), \cdots, \pi_N(t))$, considered as functions of N variables $\pi_1(t), \cdots, \pi_N(t)$ are convex in the range $0 \le \pi_1(t) + \cdots + \pi_N(t) \le 1$. Taking this fact into account, it is not difficult to show that we must take that decision d_i' for which $\pi_i(t) = \max(\pi_1(t), \cdots, \pi_N(t))$. In other words, it is necessary to continue observation in the most "dangerous" direction, a result which is intuitively clear.

With the aim of deciding the question concerning whether it is necessary to continue the observations or to terminate them, we must be able to compare the functions $\rho_0^{(\Delta)}(\pi(t))$ and $\rho_1^{(\Delta)}(\pi(t))$.

It follows that it is sufficient to consider the problem only for $N = 2$ since, as will be seen subsequently, the general case is easily reduced to this. Since $\pi_0(t) + \pi_1(t) + \pi_2(t) = 1$, in the case $N = 2$, only two independent variables remain. In is convenient to take the variables

$$T_i = T_i(t) = \frac{\pi_i(t)}{\pi_0(t)}, \quad i = 1, 2,$$

as such. Considering the risks $\rho_0(\pi(t))$ and $\rho_1(\pi(t))$ as functions of T_1 and T_2, we will denote them by $\rho_0(T_1, T_2)$, $\rho_1(T_1, T_2)$.

Let $T_2 > T_1$; then if the decision is taken to continue observation, we must carry it out in the second direction. Here (and this is essential) in view of the assumptions concerning the character of the observed process, the statistic T_1 does not change during the whole course of observation in the second direction.

---

[1] For convenience of writing, we sometimes use the notation $\phi(t)$ in place of ϕ_t.

Thus by (23)

$$\rho_1^{(\Delta)}(T_1(t),\ T_2(t)) = c\Delta + \int \rho^{(\Delta)}(T_1(t),\ T_2(t) + \Delta T_2)\, dP\,(\Delta T_2/T_1(t),\ T_2(t)), \tag{27}$$

where $\Delta T_2 = T_2(t + \Delta) - T_2(t)$. Let us now observe that under the assumption that at the moment t we began observation in the second direction and continued in that direction until the moment $t + \tau$:

$$T_2(t + \tau) = T_2(t)\, e^{\xi(\tau) - \frac{1}{2}\tau}, \quad \xi(t) = 0,$$
$$T_1(t + \tau) = T_1(t).$$

Let us consider the statistic $(T_1(t),\ T_2(t))$ as the coordinate of a moving point in the plane $(T_1,\ T_2)$ (T_1 the abscissa axis, T_2 the ordinate axis). Then the movement of the point $(T_1(t),\ T_2(t))$ is degenerate: for $T_2 > T_1$ the point moves along the vertical line with abscissa T_1, and for $T_1 > T_2$ the movement is along the horizontal line with ordinate T_2.

Let us consider equation (27) in that part of the region $T_2 > T_1$ where in fact it is necessary to continue taking observations. The region of values of T_1 can be broken up into two: $[0,\ T)$ and $[T,\ \infty)$ so that:

1. For $T_1 < T$ two points $\gamma_2(T_1), \gamma_1(T_1)$, $\gamma_2(T_1) \geq \gamma_1(T_1) \geq T_1$ exist such that if $\gamma_2 > T_2 > \gamma_1$, then observation is performed in the second direction until $T_2 \geq \gamma_2$ or $T_2 \leq \gamma_1$. Also in the first case, decision d_+ concerning the presence of a target is taken; in the second, d_- concerning its absence;

2. For $T_1 \geq T$ a point $\tilde{\gamma}_2(T_1)$ exists such that for $\tilde{\gamma}_2(T_1) > T_2 > T_1$, observation is continued in the second direction. If $T_2 \geq \tilde{\gamma}_2(T_1)$ then decision d_+ is taken; if $T_2 \leq T_1$ then transition to observation in the first direction takes place.

Since the problem is symmetric, the boundaries in the region $T_1 > T_2$ can be obtained by a reflection of the boundaries of the region $T_2 > T_1$ relative to the diagonal $T_1 = T_2$. Schematic representations of these boundaries are given in Figure 1.

Of course, for every choice of distance Δ, the boundaries $\gamma_1, \gamma_2, \tilde{\gamma}_2$ depend on Δ: $\gamma_1 = \gamma_1^\Delta, \gamma_2 = \gamma_2^\Delta, \tilde{\gamma}_2 = \tilde{\gamma}_2^\Delta$. We will denote by γ^Δ the part of the space $T_2 \geq T$ where observations are made. Let

$$\gamma_1(T_1) = \lim_{\Delta \to 0,\ \Delta \in T^2} \gamma_1^\Delta(T_1), \quad \gamma_2(T_1) = \lim_{\Delta \to 0,\ \Delta \in T^2} \gamma_2^\Delta(T_1).$$

The stated limits actually exist since for every T_1, $\gamma_1^{\Delta_1}(T_1) \leq \gamma_1^{\Delta_2}(T_1)$, $\gamma_2^{\Delta_1}(T_1) \geq \gamma_1^{\Delta_2}(T_1)$, $\Delta_1 \geq \Delta_2$ which can be seen by considering the risks $\rho_0^{(\Delta)}(\pi(t))$ and $\rho_1^{(\Delta)}(\pi(t))$.

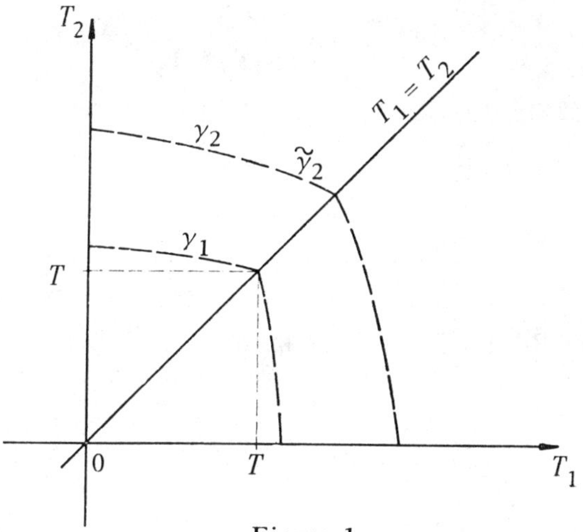

Figure 1.

Let us consider that part of the region $T_2 > T_1$ where observations are made. Then in this region

$$\rho_1^{(\Delta)}(T_1(t),\ T_2(t)) = \rho^{(\Delta)}(T_1(t),\ T_2(t)).$$

By (27), in the region γ^Δ

$$\rho^{(\Delta)}(T_1(t),\ T_2(t)) = c\Delta + \int \rho^{(\Delta)}(T_1(t),\ T_2(t) + \Delta T_2)\, d\,P(\Delta T_2/T_1,\ T_2).$$

Further let

$$\rho(T_1,\ T_2) = \lim_{\Delta \to 0,\ \Delta \in T^2} \rho^{(\Delta)}(T_1,\ T_2).$$

It is not difficult to show that this limit really exists and for arbitrary Δ, the following recurrence relation for $\rho(T_1,\ T_2)$ $(T_2 > T_1)$ is valid

$$\rho(T_1(t),\ T_2(t)) = c\Delta + \int \rho(T_1(t),\ T_2(t) + \Delta T_2)\, d\,P(\Delta T_2/T_1,\ T_2) + o\,(\Delta). \qquad (29)$$

Let us assume that the risk $\rho(T_1,\ T_2)$ is twice continuously differentiable (this, of course, is a restriction on the class of decision rules considered). Then by expanding $\rho(T_1,\ T_2 + \Delta T_2)$ in a Taylor series in the neighborhood of the point T_2 we obtain

$$-c\Delta = \frac{\partial \rho}{\partial T_2} \int \Delta T_2\, dP\,(\Delta T_2/T_1,\ T_2) + \frac{1}{2}\frac{\partial^2 \rho}{\partial T_2^2} \int (\Delta T_2)^2\, dP\,(\Delta T_2/T_1,\ T_2) + o\,(\Delta). \qquad (30)$$

By formula (28)

$$\int \Delta T_2 \, dP(\Delta T_2/T_1, T_2) = \frac{T_2^2(t)}{1 + T_1 + T_2} (e^{\Delta} - 1),$$

$$\int (\Delta T_2)^2 dP(\Delta T_2/T_1, T_2) = T_2^2(t)\,\Delta.$$

Thus passing to the limit with $\Delta \to 0$ in (30) we get

$$-c = \frac{\partial \rho}{\partial T_2} \frac{T_2^2}{1 + T_1 + T_2} + \frac{1}{2} \frac{\partial^2 \rho}{\partial T_2^2} T_2^2. \tag{31}$$

Putting $C - 2c$, $B - 1 + T_1$, we then have

$$-C = \frac{\partial \rho}{\partial T_2} \frac{2T_2^2}{B + T_2} + \frac{\partial^2 \rho}{\partial T_2^2} T_2^2.$$

This equation, considered in that part of the region $T_2 > T_1$ where observation is continued, for every fixed $T_1 = B - 1$ is in fact an ordinary differential equation, the general solution of which is given by the formula

$$\rho(T_1, T_2) = 2C \ln T_2 \frac{B - T_2}{B + T_2} + \frac{c_1(B)}{B + T_2} + c_2(B), \tag{32}$$

where $c_1(B)$ and $c_2(B)$ depend on $B = T_1 + 1$. Let us recall that in the assumed notation

$$\rho_0(T_1, T_2) = \min[\rho_{0\gamma_1}, \rho_{0\gamma_2}], \tag{33}$$

where

$$\rho_{0\gamma_1} = 1 - \frac{1}{B + T_2}, \quad \rho_{0\gamma_2} = \frac{1}{B + T_2}.$$

Let us consider the region $T_2 \geq T_1$, where $T_1 < T$, i.e., where there are two boundaries: $\gamma_1 = \gamma_1(T_1)$ and $\gamma_2 = \gamma_2(T_1)$. We shall find these boundaries γ_1 and γ_2, by using the solution (32) and the following boundary conditions:

$$\rho_0(T_1, \gamma_1(T_1)) = \rho(T_1, \gamma_1(T_1)), \tag{34}$$

$$\rho_1(T_1, \gamma_2(T_1)) = \rho(T_1, \gamma_2(T_1)), \tag{35}$$

$$\left. \frac{\partial \rho_0(T_1, T_2)}{\partial T_2} \right|_{T_2 = \gamma_1(T_1)} = \left. \frac{\partial \rho(T_1, T_2)}{\partial T_2} \right|_{T_2 = \gamma_1(T_1)}, \tag{36}$$

$$\left. \frac{\partial \rho_0(T_1, T_2)}{\partial T_2} \right|_{T_2 = \gamma_2(T_1)} = \left. \frac{\partial \rho(T_1, T_2)}{\partial T_2} \right|_{T_2 = \gamma_2(T_1)} \tag{37}$$

The first two conditions are clear. In the application to the special case of one direction $N = 1$, the last two conditions were derived by V. S. Mihalevič [14].

Let us give a brief outline of the argument for $N \geq 1$. Let us consider, for example, the derivation of condition (36), we have

$$\rho_1(T_1, T_2) = \int_{-\infty}^{\infty} \rho(T_1, T_2 + \Delta T_2)\, dP(\Delta T_2/T_1, T_2) + O(\Delta).$$

On the other hand, it is easily seen that

$$\rho_{0\gamma_1}(T_1, T_2) = \int_{-\infty}^{\infty} \rho_{0\gamma_1}(T_1, T_2, + \Delta T_2)\, dP(\Delta T_2/T_1, T_2) + O(\Delta).$$

At the point $T_2 = \gamma_1(T_1)$

$$\rho_1(T_1, \gamma_1(T_1)) = \rho_{0\gamma_1}(T_1, \gamma_1(T_1)),$$

therefore

$$0 = \int_{-\infty}^{\infty} [\rho_{0\gamma_1}(T_1, \gamma_1(T_1) + \Delta T_2) - \rho(T_1, \gamma_1(T_1) + \Delta T_2)]\, dP(\Delta T_2/T_1, T_2) + O(\Delta)$$

$$= \int_{-\infty}^{0} + \int_0^{\epsilon} + \int_{\epsilon}^{\infty} \{[\rho_{0\gamma_1}(T_1, \gamma_1(T_1) + \Delta T_2) - \rho(T_1, \gamma_1(T_1) + \Delta T_2)]$$
$$\cdot\, dP(\Delta T_2/T_1, T_2)\} + O(\Delta),$$

where $\epsilon > 0$.

Let us note that the first integral is equal to zero, since in the region $\Delta T_2 \leq 0$ the minimal risk coincides with the risk of terminating the observations and taking decision d_-. In the second integral, let us expand $\rho_{0\gamma_1}$ and ρ in the neighborhood of the point $\gamma_1(T_1)$. We obtain

$$\int_0^{\epsilon} [\rho_{0\gamma_1}(T_1, \gamma_1(T_1) + \Delta T_2) - \rho(T_1, \gamma_1(T_1) + \Delta T_2)]\, dP(\Delta T_2/T_1, T_2)$$

$$= \left(\frac{\partial \rho_{0\gamma_1}}{\partial T_2} - \frac{\partial \rho}{\partial T_2}\right)\Bigg|_{T_2=\gamma_1(T_1)} \int_0^{\epsilon} \Delta T_2 dP(\Delta T_2/T_1, T_2)$$

$$+ \frac{1}{2}\left(\frac{\partial^2 \rho_{0\gamma_1}}{\partial T_2^2} - \frac{\partial^2 \rho}{\partial T_2^2}\right)_{T_2=\gamma_1(T_1)} \int_0^{\epsilon} (\Delta T_2)^2 dP(\Delta T_2/T_1, T_2) + O(\Delta).$$

It is not difficult to see that the integral $\int_0^{\infty} (\Delta T_2)\, dP(\Delta T_2/T_1, T_2)$ for every $\epsilon > 0$ is of the order $\sqrt{\Delta}$, while the second integral

$$\int_0^{\epsilon} (\Delta T_2)^2 dP(\Delta T_2/T_1, T_2) = O(\Delta).$$

The integral

$$\int_{\epsilon}^{\infty} \{\cdot\} dP\,(\Delta T_2/T_1,\ T_2) = o\,(\Delta)$$

because of the diffusion character of the process $T_2(t)$, ([15], §52).

Thus

$$0 = \left\{\frac{\partial \rho_{0\gamma_1}}{\partial T_2} - \frac{\partial \rho}{\partial T_2}\right\}\Bigg|_{T_2=\gamma_1(T_1)} o\,(\sqrt{\Delta}) + \frac{1}{2}\left\{\frac{\partial^2 \rho_{0\gamma_1}}{\partial T_2^2} - \frac{\partial^2 \rho}{\partial T_2^2}\right\}_{T_2=\gamma_1(T_1)} o\,(\Delta) + o\,(\Delta).$$

It follows that

$$\left\{\frac{\partial \rho_{0\gamma_1}}{\partial T_2} - \frac{\partial \rho}{\partial T_2}\right\}\Bigg|_{T_2=\gamma_1(T_1)} = 0.$$

A Remark. If we consider the "truncated" decision rules ($T < \infty$), then the risks considered depend on t: $\rho(t,\ T_1(t),\ T_2(t))$ and in place of equation (31) we will get

$$-C - \frac{\partial \rho}{\partial t} = \frac{\partial \rho}{\partial T_2}\,\frac{T_2^2}{1 + T_1 + T_2} + \frac{1}{2}\,\frac{\partial^2 \rho}{\partial T_2^2}\,T_2^2.$$

In this case the boundaries γ_1 and γ_2 already depend not only on T_1 but also on time t: $\gamma_1(t,\ T_1(t))$, $\gamma_2 = \gamma_2(t,\ T_1(t))$. Here for every t, conditions (34)–(35) are again satisfied. The last two, under the assumption that for fixed T_1, $\gamma_1 = \gamma_1(t,\ T_1)$, $\gamma_2 = \gamma_2(t,\ T_1)$, have finite derivatives with respect to t.

Thus we have four supplementary conditions which, for every B, make it possible to determine completely the unknown points γ_1, γ_2, and the constants c_1 and c_2:

$$2C \ln \gamma_1 \frac{B - \gamma_1}{B + \gamma_1} + \frac{c_1}{B + \gamma_1} + c_2 = 1 - \frac{1}{B + \gamma_1},$$

$$2C \ln \gamma_2 \frac{B - \gamma_2}{B + \gamma_2} + \frac{c_1}{B + \gamma_2} + c_2 = \frac{1}{B + \gamma_2},$$

$$\left[2C \ln \gamma \frac{B - \gamma}{B + \gamma} + \frac{c_1}{B + \gamma}\right]'_{\gamma=\gamma_1} = \left[1 - \frac{1}{B + \gamma}\right]'_{\gamma=\gamma_2},$$

$$\left[2C \ln \gamma \frac{B - \gamma}{B + \gamma} + \frac{c_1}{B + \gamma}\right]'_{\gamma=\gamma_2} = \left[\frac{1}{B + \gamma}\right]'_{\gamma=\gamma_2}.$$

Eliminating c_1 and c_2 from these equations, we obtain the following system of equations for determining γ_1 and γ_2:

$$C(\gamma_2 - \gamma_1)\left[\frac{B^2}{\gamma_1\gamma_2} - 1\right] = B - 1, \tag{38}$$

$$C(\gamma_2 - \gamma_1)\left[\frac{B^2}{\gamma_1\gamma_2} + 1\right] + 2CB\ln\frac{\gamma_2}{\gamma_1} = 1, \tag{39}$$

where, we recall, $B = T_1 + 1$.

Let $\gamma_1 = \gamma_1(T_1)$, $\gamma_2 = \gamma_2(T_1) > \gamma_1$. We must solve the system (38)–(39) for all of those values of T_1 for which $\gamma_1(T_1) > T_1$. Thus in the region $T_1 < T$; there are two boundaries and they are determined by (38)–(39). In this case observation is carried out the whole time in one (the second) direction with the adoption of d_+ or d_-, depending on the attainment of the upper or the lower boundary. If $T_1 = 0$, i.e., $B = 1$, we have the Wald case of the Bayes problem concerning two hypotheses. In this case, from (38) it follows that $\gamma_1\gamma_2 = 1$ and equation (39) for $\gamma_1 = \gamma_2$ becomes

$$\left(\gamma - \frac{1}{\gamma}\right) + 2\ln\gamma = \frac{1}{2C}$$

or, letting $\gamma = e^{\psi}$,

$$e^{\psi} - e^{-\psi} + 2\psi = \frac{1}{2C}.$$

This result was obtained as long ago as 1956 by V. S. Mihalevič, [14].

Let us consider now the region $T_1 \geq T$. It is possible to take here only one decision concerning the termination of observations: d^+. Therefore there is only one boundary $\gamma_2 = \gamma_2(T_1)$ on which conditions (35) and (37) are again satisfied. But here it is still necessary to add conditions on the diagonal $T_1 = T_2$ for $T_1 \geq T$. By the same argument as for the deduction of conditions (36), (37), it can be demonstrated that on the diagonal the "reflecting barrier" condition should be satisfied:

$$\left.\frac{\partial\rho(T_1, T_2)}{\partial T_1} = \frac{\partial\rho(T_1, T_2)}{\partial T_2}\right|_{T_1 = T_2 \geq T}. \tag{40}$$

By the symmetry of the problem (and consequently also the symmetry of the boundaries with respect to the straight line $T_1 = T_2$) it is sufficient from the point of view of analysis to look at the motion of the point $(T_1(t),\ T_2(t))$ only for $T_2 \geq T_1$ under the condition that the boundary $T_2 = T_1 \geq T$ is a *reflecting* barrier. In this connection, note that if the motion begins from the value $T_1 > T$, decision d_- can be taken only through the attaining of the point (T, T). The existence of drift at the boundary is an interesting peculiarity of this motion. Such processes are investigated in a more general situation by A. N. Kolmogorov in his paper "On the physical meaning of boundary conditions for diffusion processes", given on March 22, 1961 at the Section on Theory of Probability and Mathematical Statistics of the Moscow Mathematical Society.

Conditions (35), (37), (40) are changed into the following in the present case:

$$2C \ln \tilde{y}_2(1 + T_1 - \tilde{y}_2) + \hat{c}_1 + \hat{c}_2(1 + T_1 + \tilde{y}_2) = 0,$$

$$2C\left[\frac{(1+T_1)^2}{\tilde{y}_2} - \tilde{y}_2 - 2(1+T_1)\ln \tilde{y}_2\right] = \hat{c}_1,$$

$$2C\frac{1}{T_1}\frac{1}{1+2T_1} - 4C\ln T_1\frac{1}{1+2T_1} = \frac{\hat{c}_1'}{1+2T_1} + \hat{c}_2'$$

$$(\hat{c}_1(T_1) = c_1(B),\ \hat{c}_2(T_1) = c_2(B),\ B = 1 + T_1).$$

Hence, eliminating $\hat{c}_1$, $\hat{c}_2$ and their derivatives, we get for $\tilde{y} = \tilde{y}_2(T_1)$ the differential equation:

$$\frac{\dot{\tilde{y}}}{\tilde{y}^2} = \frac{\left[\frac{1}{T_1} - 2\ln T_1\right] - \left[\frac{1}{\tilde{y}} - 2\ln \tilde{y}\right]}{(T_1 - \tilde{y})(1 + T_1 + \tilde{y})}, \tag{41}$$

with the initial condition $\tilde{y}_2(T) = y_2(T)$ where we recall that T is such that $y_1(T) = T$.

Let us now examine the changes that must be made if $N > 2$. Let $T_i = \max(T_1, \cdots, T_N)$ and $T_j = \max(T_1, \cdots, T_{i-1}, T_{i+1}, \cdots, T_N)$. Let us set $\bar{B} = 1 + (T_1) + \cdots + T_N) - T_i - T_j$ and $B = 1 + (T_1 + \cdots + T_N) - T_i$. Then the process of observation is accomplished in the following manner. As long as $T_i > T_j$, the ith direction is observed. As soon as $T_i \leq T_j$, we transfer to observation of the jth direction and so on. For the determination of the boundaries y_1, y_2 it is necessary here to keep in mind the value of T^j as well as the value of $\bar{B}$. Here the boundaries $y_1 = y_1(T_j, \bar{B})$, $y_2 = y_2(T_j, \bar{B})$ are determined from equations (38), (39), which have to be solved in that region of values of $(\bar{B}, T_j)$ where $y_1(T_j, \bar{B}) \geq T_j$. In the remaining region of values of $(\bar{B}, T_j)$, for every $\bar{B}$ we obtain an analogous (to (41)) differential (with respect to T_j) equation for $\tilde{y} = \tilde{y}_2(\bar{B}, T_j)$:

$$\frac{\dot{\tilde{y}}}{\tilde{y}^2} = \frac{\left[\frac{\bar{B}}{T_j} - 2\ln T_j\right] - \left[\frac{\bar{B}}{y} - 2\ln \tilde{y}\right]}{(T_j - \tilde{y})(\bar{B} + T_j + \tilde{y})}, \tag{42}$$

where $\tilde{y}_2(\bar{B}, T_j') = y_2(\bar{B}, T_j')$, and T_j' such that $y_1(\bar{B}, T_j') = T_j'$.

Thus it is possible to draw the following conclusion: in the class of regular decision rules having twice continuously differentiable risks, the optimal rule (for $N = 2$) is based on the observation of the statistics T_1, T_2. If $T_2 > T_1$, then the second direction is observed; if $T_1 > T_2$, then the first. The boundaries on the attainment of which observations are stopped by taking decisions d_+ or d_- determined by equations (38), (39), (41).

§4. Problems connected with the detection of the moment of the probabilistic characteristics of the observed process

1. In this section various formulations of problems mentioned in the title are considered. The results referred to here were published by us in [5,6]. There is a detailed presentation in [13] and in the journal "Theory of Probability and its Applications", Volume VII 1, 3, 4, (1963). In the present paper we discuss only the fundamental aspects of the matter, referring to the above-mentioned literature for the details.

Thus let the process $\theta = \{\theta_t\}$, characterizing the state of the system (of interest to us), which is inaccessible to direct observation, be a Markov process with two states 0 and 1 and with a unique transition $0 \longrightarrow 1$ at moment of time θ^*. Let us assume that the time $t \geq 0$ is continuous and also that observation may be carried out in a continuous manner.

The Markov character of the process $\theta = \{\theta_t\}$ is equivalent to the fact that

$$P(\theta^* > t) = e^{-\lambda t}, \tag{43}$$

where λ is a known constant.

Concerning the observed process $\xi = \{\xi_t\}$, let us assume that it satisfies the following stochastic differential equation:

$$d\xi = \chi(t - \theta^*)\, dt + dw,$$

where $w = \{w_t\}$ is a Wiener process $w_0 = 0$,

$$\boldsymbol{M}\, \Delta w = 0, \quad \boldsymbol{D}\, \Delta w = \Delta t,$$

and[1])

$$\chi(s) = \begin{cases} 0, & s \leq 0, \\ \sqrt{2}, & s > 0. \end{cases}$$

Thus the moment θ^* describes the change of the probabilistic characteristics, namely of the average value of the observed process ξ. Let us note that in the case of the Wiener process, a change in dispersion can be detected over as small an interval of time as desired.

Let the space of decisions $\boldsymbol{D}$ consist of the points: d' meaning to continue

---

1) The value $\sqrt{2}$ is taken only for simplicity of the final formulas (46–47).

observation, decision d^0 meaning the presence of a "disorder", $\widetilde{d}$ meaning to terminate observation. Let us denote by ν the moment of taking decision d^0. Then it is natural to call the probability

$$\omega = P(\nu < \theta^*)$$

the probability of "false alarm". Further let

$$\tau = M(\nu - \theta^* | \nu \geq \theta^*) \tag{44}$$

be the mathematical expectation of the time delay in the detection of a disorder, i.e., in taking decision d^0. The following variational problem of detection is of interest: for a given probability ω, to find the rule of observation minimizing $\tau = \tau(\omega)$. It is not difficult to see that for every rule δ the risk

$$r(\delta) = c\omega + (1 - \omega)\,\tau(\omega), \quad c \geq 0$$

is an additive functional of type (13).

Applying the apparatus which was presented in §3 we find, by passage to the limit from the discrete scheme that in the class of regular decision rules the optimal decision is characterized in the following way: we must observe the a posteriori probability

$$\pi(t) = P(\theta^* \leq t | \xi'), \quad \pi(0) = 0,$$

until the first achievement of some level π' which is so chosen that the probability $P(\nu < \theta^*)$ is equal to the given number ω. The following stochastic differential equation

$$d\pi = (1 - \pi)(\lambda - 2\pi^2) + \sqrt{2} \cdot \pi(1 - \pi)\,d\xi$$

which we can use for the actual model holds for the a posteriori probability $\pi(t)$.

The recent article [16] borders on this range of problems. In it are considered cases when the "penalty" for the delay in the detection is characterized by more complicated functionals than (44).

2. The following "multistage" problem also can be of interest. Let us assume that after the decision about the presence of a disorder, we "verify" whether in fact a disorder took place. Formally this means taking some new decision d_{np}. If verification showed the presence of a disorder (i.e., established that at the moment of verification $\theta_t = 1$), then observation is stopped. Otherwise it is resumed and is carried out according to exactly the same rule as the first stage of observation. The latter suggestion does not seem strange if we note that the process $\theta = \{\theta_t\}$ is a strong Markov process and thereby for any moment of time s (it can be random but not depending on the future), its conditional distribution given the condition $\theta_s = 0$ coincides with the unconditional.

Then the following statement of the problem is natural. Let $\mathfrak{N} = M\kappa$ be the mathematical expectation of the number κ of "false alarms" (i.e., of verifications by which the absence of disorders is established). Also let us assume that these verifications are accomplished at moments of time $\nu_1, \nu_1 + \nu_2, \cdots, \nu_1 + \cdots + \nu_\kappa$. Denoting $\tau = M(\nu_1 + \cdots + \nu_{\kappa+1} - \theta^*)$ we set up the problem thus: for a given $\mathfrak{N}$, to find the rule of observation which minimizes τ. It turns out that the solution of this problem is reduced to the preceding one since $\mathfrak{N} = \omega/(1-\omega)$.

3. More interesting and natural from the point of view of applications is the problem of the fastest detection in the case where no a priori hypotheses are made concerning the character of the distribution of the process $\theta = \{\theta_t\}$.

Let us suppose that an arbitrary rule δ for making the decision d^0 is chosen. Let us denote by ν the moment of taking d^0 on the assumption of the absence of a disorder. Let all the rules δ for which $M\nu = T$, where T is fixed, be considered. Also, after each decision d_0 let the process of observation be resumed and carried out exactly in the same way as the first stage of observation. If for the rules δ in question, the random variable ν is nonlattice, then a stationary regime is established in the sense that

$$\lim_{t\to\infty} P\{t - (\nu_1 + \cdots + \nu_{\kappa_t}) \geq u\} = \frac{1}{T}\int_u^\infty [1 - F(x)]\,dx,$$

where $\kappa_t = \max\{k\colon \nu_1 + \cdots + \nu_k < t\}$, and $F(x)$ is the distribution function of the random variable $\nu = \nu_i$, $i = 1, 2, \cdots$ on the assumption of the absence of a disorder in general.

Let us assume that the beginning of the disorder takes place under the assumption of the existence of an established stationary regime of observation. Let us denote by $\tau = \tau(T)$ the minimal possible average time of delay. The problem arises of finding the decision rule for which $\tau = \tau(T)$ is attained.

It is shown that this *nonparametric* problem can be considered as a limiting case, when $\lambda \to 0$, of the problem in which the a priori distribution of θ^* is given by formula (43). Here the optimal decision rule (in the class of regular decision rules) is based on the observation of the process ϕ^* controlled by the stochastic equation

$$d\phi^* = \left[1 - \frac{1}{T}e^{\phi^*}\right]dt - \sqrt{2}\cdot d\xi$$

with conditions $\phi^*(0) = \infty$ and with the declaration of "alarm" at the attainment of the zero level. After this the process of observation is resumed and continues indefinitely, according to the same rule as the first stage of observation.

The explicit formula of dependence obtained: (see [6])

$$\tau(T) = e^b(-\mathrm{Ei}(-b)) - 1 + b\int_0^\infty e^{-bu}\frac{\ln(1+u)}{u}\,du, \qquad (46)$$

where $b = 1/T$,

$$-\mathrm{Ei}(-x) = \int_x^\infty \frac{e^{-t}}{t}\,dt, \quad x \geq 0,$$

is the integral exponential function. It is interesting to consider the special cases $T \to 0$, $T \to \infty$. From (46) we get immediately:

$$\tau(T) = \begin{cases} \ln T - 1 - C + O\left(\frac{1}{T}\right), & T \to \infty, \\ \frac{T}{2} + O(T^2), & T \to 0, \end{cases} \qquad (47)$$

where $C = 0.577\cdots$ is Euler's constant.

Of course the most interesting case is for large T. Here in the calculation of the integral

$$I(x) = \int_0^\infty e^{-xu}\frac{\ln(1+u)}{u}\,du$$

for small x we can make use of the asymptotic expansion which A. F. Nikiforov and V. B. Uvarov pointed out to me:

$$I(x) = \frac{\pi^2}{4} + E_1(x)\left[\frac{1}{2}E_1(x) - f(x)\right] - 2e^{-x}\psi(x),$$

where

$$E_1(x) = -\mathrm{Ei}(-x) = -C - \ln x + \sum_{n=1}^\infty \frac{(-1)^{n+1}x^n}{n!\,n},$$

$$f(x) = \int_0^\infty \frac{e^s - e^{-s}}{s} = 2\sum_{n=0}^\infty \frac{x^{2n+1}}{(2n+1)!(2n+1)},$$

$$\psi(x) = \sum_{k=1}^\infty \left\{\frac{x^{2k-1}}{(2k-1)!} + \frac{x^{2k}}{(2k)!}\right\}\sum_{s=0}^{k-1}\frac{1}{(2s+1)^2}.$$

It would be interesting to obtain formula (46) by a direct method and to determine the degree of its applicability when the observed process is different from a Wiener process.

4. As is seen from (45) an optimal decision rule contains in it a *nonlinear* reduction of the information ξ being received. We investigated several *linear* systems of observation containing elements of the sequential analysis of Wald and of the method of Neyman-Pearson. A short presentation of the results relevant here is given in [6], and a detailed exposition in [13]. We will indicate here only that the so-called *degenerate sequential analysis*, [6], gives a good approximation

to the optimal. For it, for large T

$$\tau_B(T) = \ln T - 1,5 + O\left[\frac{\ln^2 T}{T}\right].$$

(Compare with (47).)

5. In [13] are also investigated the cases when the disorder can appear in one of N directions against a background of an established stationary regime of observation. There is proposed one procedure of observation, containing elements of sequential analysis, applied to the problem of distinguishing between two hypotheses for which the average time of delay $\tau = \tau_B(T; N)$ for given T and N, when T is great, is

$$\tau_B(T; N) = \ln T + N - 1,5 - \frac{1}{N} + O\left\{N\frac{\ln^2 T}{T}\right\}.$$

How far this method of observation is from the optimal is unknown to us. For a discussion of these questions, see not only [13] but also [17]. A complete exposition of the results relevant to the case of N directions appeared in the journal "Probability Theory and its Applications" in Volume VIII.

BIBLIOGRAPHY

[1] A. Wald, *Statistical decision functions*, Wiley, New York and Chapman & Hall, London, 1950. MR **12**, 193.

[2] D. Blackwell and M. A. Girshick, *Theory of games and statistical decisions*, Wiley, New York and Chapman & Hall, London, 1954; Russian transl., IL, Moscow, 1958. MR **16**, 1135.

[3] E. L. Lehmann, *Testing statistical hypotheses*, Wiley, New York and Chapman & Hall, London, 1959; Russian transl., "Nauka", 1964. MR **21** #6654.

[4] A. Wald, *Sequential analysis*, Russian transl., Fizmatgiz, Moscow, 1960. MR **22** #7228.

[5] A. N. Širjaev, *Detection of spontaneous effects*, Dokl. Akad. Nauk SSSR **138** (1961), 799–801 = Soviet Math Dokl. **2** (1961), 740–743. MR **24** #A2504.

[6] ———, *The problem of quickest detection of the destruction of a stationary regime*, Dokl. Akad. Nauk SSSR **138** (1961), 1039–1042 = Soviet Math Dokl. **2** (1961), 795–799. MR **24** #A2505.

[7] A. Wald and J. Wolfowitz, *Bayes solutions of sequential decision problems*, Ann. Math. Statist. **21** (1950), 82–99. MR **11**, 529.

[8] K. J. Arrow, D. Blackwell and M. A. Girshick, *Bayes and minimax solutions of sequential decision problems*, Econometrica **17** (1949), 213–244. MR **11**, 261.

[9] V. S. Mihalevič, *Consecutive Bayes' solutions and optimal methods of statistical acceptance control,* Teor. Verojatnost. i Primenen. 1 (1956), 437–465. MR 19, 694.

[10] H. Chernoff, *Sequential tests for the mean of a normal distribution,* Proc. 4th Berkeley Sympos. Math. Statist. and Prob., Vol. I, Univ. of California Press, Berkeley, Calif., 1961, 79–91. MR 24 #A1788.

[11] R. L. Stratonovič, *Some extremal problems of mathematical statistics and conditional Markov processes,* Teor. Verojatnost. i Primenen. 7 (1962), 226–229.

[12] V. S. Mihalevič, *Sequential Bayes solutions and optimal methods of statistical acceptance control,* Dissertation, Moscow State University, Moscow, 1956.

[13] A. N. Širjaev, *The problem of quickest detection of the destruction of a stationary regime,* Dissertation, Mathematical Institute of the Academy of Sciences, USSR, Moscow, 1961.

[14] V. S. Mihalevič, *Sequential Bayes solutions and optimal methods of statistical acceptance control,* Dissertation, Moscow State University, Moscow 1956.

[15] B. V. Gnedenko, *A course in the theory of probability,* GITTL, Moscow, 1954.

[16] R. L. Stratonovič, *On optimal detection of disorder in a manufacturing process,* Vestnik Moskov. Univ. Ser. I. Math. Meh. 2 (1962), 63–71.

[17] A. N. Širjaev, *On problems of the most rapid detection of randomly arising effects,* All-Union Mathematical Congress, Leningrad, July 1961.

Translated by:
Margaret Stein

OPTIMUM CONTROL IN LINEAR SYSTEMS WITH TIME LAG

N. N. KRASOVSKIĬ

The paper deals with the problem of forming the stabilizing controller signal that minimizes the integral square error in a linear control system with small aftereffect. The problem is solved by the method of the small parameter.

1. **Statement of the problem.** Let us consider a system described by the equation

$$\frac{dx}{dt} = Ax(t) + \mu Bx(t-h) + m\xi, \tag{1}$$

where x is an n-dimensional vector specifying the deviation of the actual motion $(\{x_i(t)\},\ i = 1, \cdots, n)$ from the assigned motion $(\{x_i = 0\},\ i = 1, \cdots, n)$; ξ is a scalar denoting the stabilizing action of the controller (the controller, in brief); $A = (a_{ij})$ and $B = (b_{ij})$ are $(n \times n)$-matrices; $m = (m_i)$ is a vector; $h > 0$ is the signal lag in the channels of the system; μ is a parameter that characterizes the smallness of the aftereffect. The controller ξ is formed at each instant of time t on the basis of information about the actual deviation $x(t + \vartheta)$ for $-h \leq \vartheta \leq 0$. In other words, in equation (1) the quantity ξ is a functional $\xi[x(t+\vartheta); -h \leq \vartheta \leq 0]$. For $t > t_0$, the motion of system (1) is determined [1], [2] by the initial state $x^0(t_0 + \vartheta)$ $(-h \leq \vartheta \leq 0)$; therefore it is convenient to consider as an element of motion, corresponding to some instant t, the section of trajectory $x(t+\vartheta)$ $(-h \leq \vartheta \leq 0)$. The functions $x(\vartheta)$ $(-h \leq \vartheta \leq 0)$ will be denoted by the symbol $x(\cdot)$. The motion of system (1) for $\xi = \zeta$, generated by the initial condition $x^0(\cdot)$ (for $t_0 = 0$), will be denoted by the symbol $x(x^0(\cdot), \zeta, t)$ if we are referring to the point $\{x_i(t)\}$, and by the symbol $x(x^0(\cdot), \zeta, t, \cdot)$ if we were referring to the section of the curve $\{x_i(t+\vartheta)\}$ $(-h \leq \vartheta \leq 0)$. (The symbols $x^0(\cdot)$, ζ, will be dropped whenever this can be done without leading to confusion.)

The problem consists in determining the optimum controller, i. e. the functional $\xi^0[x(\cdot)]$, that satisfies the following conditions.

1) For $\xi = \xi^0[x(t, \cdot)]$, the motion $x = 0$ of system (1) must be asymptotically stable for any initial deviations $x^0(\cdot)$.

2) For $\xi = \xi^0[x(t, \cdot)]$, the integral

$$I_\xi[x^0(\cdot)] = \int_0^\infty \left(\sum_{i=1}^n x_i^2(x^0(\cdot), \xi, t) + \xi^2[x(x^0(\cdot), \xi, t, \cdot)] \right) dt \tag{2}$$

must be finite and minimal with respect to $\xi[x(\cdot)]$ belonging to the set of continuous functionals $\{\xi[x(\cdot)]\}$ defined on the continuous curves $x(\cdot) = \{x_i(\vartheta)\}$

$(-h \leq \vartheta \leq 0,\ i = 1, \cdots, n$; the norm $\|x(\cdot)\| = \sup[x_1^2 + \cdots + x_n^2)^{1/2}, -h \leq \vartheta \leq 0])$.

The object of the present paper is to describe and to prove the validity of the method of construction of the optimum controller ξ^0 in the form of a power series in the parameter μ.

2. **Optimality criterion.** Let us formulate the optimality criterion based on the method of Ljapunov functions [3], [4], modified for equations with aftereffect [5]. This criterion is also based on the reasoning characteristic of the method of dynamic programming of R. Bellman [6]. The theorem proved in this section is an improved version of Theorem 3.1 of reference [7].

Definition 1. The functional $v[x(\cdot)]$ is said to be positive definite with respect to $\|x(0)\| = (x_1^2 + \cdots + x_n^2)^{1/2}$ if $v[x(\cdot)] \geq 0$ and if it is possible to find a continuous function $w(r)$ that satisfies the conditions

$$w(r) > 0 \quad \text{for } r > 0, \ \lim w(r) = \infty \quad \text{for } r \to \infty, \tag{3}$$

$$v[x(\cdot)] \geqslant w(\|x(0)\|) \text{ on the curves } \|x(\cdot)\| = \|x(0)\|. \tag{4}$$

Definition 2. We shall say that the functional $v[x(\cdot)]$ admits an upper bound with respect to $\|x(\cdot)\|$ if there exists a continuous function $W(r)$ that satisfies the conditions

$$W(0) = 0, \ |v[x(\cdot)]| \leqslant W(\|x(\cdot)\|). \tag{5}$$

Theorem 1. *Suppose that the functionals $v[x(\cdot)]$ and $\xi^0[x(\cdot)]$ satisfy the following conditions:*

a) *The functional $v[x(\cdot)]$ is positive definite with respect to $\|x(0)\|$ and admits an upper bound with respect to $\|x(\cdot)\|$;*

b) *by equations* (1) *for $\xi = \xi^0[x(t, \cdot)]$, we have for the derivative $(dv/dt)_{\xi^0}$ (see [5]) of the functional v (for $t = 0$) the formula*

$$\min\left[\left(\frac{dv}{dt}\right)_\xi + \sum_{i=1}^n x_i^2(0) + \xi^2\right] = \left(\frac{dv}{dt}\right)_{\xi^0} + \sum_{i=1}^n x_i^2(0) + (\xi^0)^2 = 0. \tag{6}$$

Thereupon ξ^0 will be an optimum controller and

$$v[x^0(\cdot)] = I_{\xi^0}[x^0(\cdot)]. \tag{7}$$

Let us prove that Condition 1 is satisfied. We first show that under the conditions of the theorem the motion $x = 0$ is stable in the sense of Ljapunov, i. e., for any $\epsilon > 0$ it is possible to find a $\delta > 0$ such that for $t > 0$ we have the inequality $\|x(x^0(\cdot), \xi^0, t)\| < \epsilon$ if $\|x^0(\cdot)\| \leq \delta$. Suppose that the number δ satisfies the condition $\sup[W(r)$ for $0 \leq r \leq \delta] < w(\epsilon)$. Thereupon, by (5), we have $v[x^0(\cdot)] < w(\epsilon)$ for $\|x^0(\cdot)\| \leq \delta$, and therefore (by (6))

$$v[x(x^0(\cdot), \xi^0, t, \cdot)] < w(\varepsilon) \quad \text{for} \quad t > 0. \tag{8}$$

If we were to assume that at the instant $t = t^*$ the trajectory $x(x^0(\cdot), \xi^0, t)$

emerges for the first time onto the plane $x_1^2 + \cdots + x_n^2 = \epsilon^2$, then the equation $\|x(\cdot)\| = \|x(0)\| = \epsilon$ would be satisfied at this instant, i.e., according to (4) we would have $v[x(x^0(\cdot), \xi^0, t^*, \cdot)] \geq w(\epsilon)$, which contradicts (8). The contradiction proves the stability of the motion $x = 0$ in the sense of Ljapunov.

Now, by reasoning in the same way as before and by using the second condition of (3), it is possible to verify that the trajectory $x(x^0(\cdot), \xi^0, t)$ remains in a bounded region of the space $\{x\}$. But then the relation

$$\lim x(x^0(\cdot), \xi^0, t) = 0 \qquad \text{for } t \to \infty,$$

must be satisfied, since otherwise

$$\lim (v[x(x^0(\cdot), \xi^0, T, \cdot)] - v[x^0(\cdot)]) = \lim\left(\int_0^T (dv/dt)_{\xi^0} dt\right) = -\infty \quad \text{for } T \to \infty,$$

which is impossible in view of $v \geq 0$. Thus Condition 1 is fulfilled. Condition 2 can be verified in the same way as in the analogous case that occurs in the proof of Theorem 3.1 of [7]. Here we leave out this proof.

Thus the problem reduces to the determination of the functionals $v[x(\cdot)]$ and $\xi^0[x(\cdot)]$ that satisfy the conditions of Theorem 1.

3. **Construction of optimum controller.** In reference [7] it is shown that the functional $v[x(\cdot)]$ must be sought in the form

$$v[x(\cdot)] = \sum_{i=1}^n \sum_{j=1}^n \alpha_{ij} x_i(0) x_j(0) + 2 \sum_{i=1}^n \sum_{j=1} x_i(0) \int_{-h}^0 \beta_{ij}(\vartheta) x_j(\vartheta) d\vartheta$$

$$+ \sum_{i=1}^n \sum_{j=1}^n \left[\int_{-h}^0 \int_{-h}^0 \gamma_{ij}(\vartheta, \tau) x_i(\vartheta) x_j(\tau) d\vartheta\, d\tau\right] \tag{9}$$

$$(\alpha_{ij} = \alpha_{ji} = \text{const}; \ \gamma_{ij}(\vartheta, \tau) = \gamma_{ji}(\tau, \vartheta);$$
$$x(\cdot) = \{x_i(\vartheta)\}, -h \leqslant \vartheta \leqslant 0, \ i = 1, \ldots, n).$$

The actual construction of the functional (9) is in general beset by great computational difficulties. For the system (1) we shall seek the functional (9) in the form of a power series in μ, i.e.,

$$v[x(\cdot)] = \sum_{k=0}^{\infty} \mu^k v_k[x(\cdot)], \tag{10}$$

$$v_k[x(\cdot)] = \sum_{i=1}^n \sum_{j=1}^n \alpha_{ij}(k) x_i(0) x_j(0) + 2 \sum_{i=1}^n \sum_{j=1}^n x_i(0) \int_{-h}^0 \beta_{ij}(k; \vartheta) x_j(\vartheta) d\vartheta$$

$$+ \sum_{i=1}^n \sum_{j=1}^n \left[\int_{-h}^0 \int_{-h}^0 \gamma_{ij}(k; \vartheta, \tau) x_i(\vartheta) x_j(\tau) d\vartheta\, d\tau\right]. \tag{11}$$

Let us set up for the functional (10), (11) the equations that follow from condition (6). The derivative (dv/dt) of the functional (9) along the trajectories of system (1) has the form (for $t = 0$)

$$\left(\frac{dv}{dt}\right)_\xi = \sum_{i=1}^{n}\left\{\left[\frac{\partial}{\partial x_i(0)}\left(\sum_{i=1}^{n}\sum_{j=1}^{n}\alpha_{ij}x_i(0)\,x_j(0)\right.\right.\right.$$

$$\left.\left.+2\sum_{i=1}^{n}\sum_{j=1}^{n}x_i(0)\int_{-h}^{0}\beta_{ij}(\vartheta)\,x_j(\vartheta)\,d\vartheta\right)\right]\left[\sum_{l=1}^{n}a_{il}x_l(0)+\mu\sum_{l=1}^{n}b_{il}x_l(-h)+m_i\xi\right]\Bigg\}$$

$$+2\sum_{i=1}^{n}\sum_{j=1}^{n}x_i(0)\int_{-h}^{0}\beta_{ij}(\vartheta)\,[dx_j(\vartheta)/d\vartheta]\,d\vartheta$$

$$+2\sum_{i=1}^{n}\sum_{j=1}^{n}\left[\int_{-h}^{0}\int_{-h}^{0}\gamma_{ij}(\vartheta,\tau)\,x_j(\tau)\,[dx_i(\vartheta)/d\vartheta]\,d\vartheta\,d\tau\right].$$

By substituting this expression for (dv/dt) in the right equation (6), we obtain the first of the desired equations for v and ξ^0. The second equation for these quantities can be obtained by differentiation of the first equation with respect to ξ, since by virtue of the left equation (6) we have a minimum for $\zeta = \zeta^0$. By eliminating ξ from the equations thus obtained and by substituting $\alpha_{ij} = \Sigma_{k=0}^{\infty}\alpha_{ij}(k)\,\mu^k$,

$$\beta_{ij}(\vartheta)=\sum_{k=0}^{\infty}\beta_{ij}(k;\vartheta),\mu^k,\ \gamma_{ij}(\vartheta,\tau)=\sum_{k=0}^{\infty}\gamma_{ij}(k;\vartheta,\tau)\,\mu^k,$$

we obtain the equation

$$\sum_{k=0}^{\infty}\mu^k\left\{\sum_{i=1}^{n}\sum_{j=1}^{n}\sum_{l=1}^{n}[(\alpha_{il}(k)\,a_{lj}+\alpha_{jl}(k)\,a_{li})\,x_i(0)\,x_j(0)\right.$$

$$+2\alpha_{il}(k-1)\,b_{lj}x_l(0)\,x_j(-h)+2x_i(0)\int_{-h}^{0}\beta_{ij}(\vartheta)\,[dx_j(\vartheta)/d\vartheta]\,d\vartheta$$

$$+2\,(a_{li}x_i(0)+\mu b_{li}x_l(-h))\int_{-h}^{0}\beta_{lj}(\vartheta)\,x_j(\vartheta)\,d\vartheta \tag{12}$$

$$\left.+2\int_{-h}^{0}\int_{-h}^{0}\gamma_{ij}(\vartheta,\tau)\,x_j(\tau)\,[dx_i(\vartheta)/d\vartheta]\,d\vartheta]\right\}$$

$$-\left[\sum_{k=0}^{\infty}\mu^k\left\{\sum_{i=1}^{n}\sum_{j=1}^{n}\sum_{l=1}^{n}\alpha_{ij}(k)\,m_ix_j(0)+m_i\int_{-h}^{0}\beta_{ij}(k;\vartheta)\,x_j(\vartheta)\,d\vartheta\right\}\right]^2$$

$$+\sum_{i=1}^{n}x_i^2(0)=0.$$

By equating terms with equal powers of μ, we obtain the equations for $\alpha_{ij}(k)$, $\beta_{ij}(k, \vartheta)$ and $\gamma_{ij}(k; \vartheta, \tau)$. Let us first consider the equation for $v_0[x(\cdot)]$, i. e., equation (12) for $\mu = 0$. This equation can be satisfied by setting $\beta_{ij}(0, \vartheta) = 0$ and $\gamma_{ij}(0; \vartheta, \tau) = 0$, i. e., $v_0[x(\cdot)]$ reduces simply to the function $v_0(x(0)) = \Sigma\alpha_{ij}(0)x_i(0)x_j(0)$, whose coefficients must satisfy the equation

$$\sum_{i=1}^{n}\sum_{j=1}^{n}\sum_{l=1}^{n}(\alpha_{il}(0)\alpha_{lj} + \alpha_{jl}(0)\alpha_{li})x_i(0)x_j(0)$$
$$-\left[\sum_{i=1}^{n}\sum_{j=1}^{n}\alpha_{ij}(0)m_i x_j(0)\right]^2 + \sum_{i=1}^{n}x_i^2(0) = 0. \tag{13}$$

This equation coincides naturally with the equation obtained (see [8]) in solving the problem of optimum stabilization of the system

$$\frac{dx}{dt} = Ax + m\xi, \tag{14}$$

$$I_\xi^0[x^0(0)] = \int_0^\infty\left[\sum_{i=1}^{n}x_i^2(x^0(0), \xi, t) + \xi^2[x(x^0(0), \xi, t)]\right]dt = \min. \tag{15}$$

A sufficient condition [9] for the existence of a solution of (13) in the form of a positive-definite form $v_0 = \Sigma\alpha_{ij}(0)x_i(0)x_j(0)$ is the independence of the vectors $m, Am, \cdots, A^{n-1}m$. Suppose that this condition is satisfied. The optimum controller $\xi_0^0[x(0)]$ for the problems (14), (15) is a linear function

$$\xi_0^0 = \sum_{j=1}^{n}\omega_j x_j(0) = -\sum_{i=1}^{n}\sum_{j=1}^{n}\alpha_{ij}(0)m_i x_j(0). \tag{16}$$

By substituting ξ_0^0 into equation (14) we obtain an asymptotically stable linear system

$$\frac{dx}{dt} = Hx = Ax + m\xi_0^0[x]. \tag{17}$$

Now the equations for the functionals $v_k[x(\cdot)]$, obtained from (12), can be written as follows:

$$\left(\frac{dv_k[x(x^0(0), t, \cdot)]}{dt}\right)_{(17)} = -2\sum_{i=1}^{n}\sum_{j=1}^{n}\sum_{l=1}^{n}\Big\{\alpha_{ij}(k-1)b_{lj}x_i(t)x_j(t-h)$$
$$-b_{li}x_i(t-h)\int_{-h}^{0}\beta_{lj}(k-1;\vartheta)x_j(t+\vartheta)\,d\vartheta\Big\} \tag{18}$$
$$+\sum_{r+s=k}L_rL_s = F_k[t; x(x^0(0), t, \cdot)],$$

where the symbol $(dv_k/dt)_{(17)}$ denotes the derivative of the functional v_k along the trajectory $x(x^0(0), t)$ of system (17), and

$$L_r=\sum_{i=1}^{n}\sum_{j=1}^{n}\left[\alpha_{ij}(r)\,m_ix_j(t)+m_i\int_{-h}^{0}\beta_{ij}(r,\vartheta)\,x_j(t+\vartheta)\,d\vartheta\right]\quad(r\geqslant 1).$$

System (17) is asymptotically stable; therefore equations (18) can be solved successively. The solution of equations (18) has the form

$$v_k[x(\cdot)]=-\int_0^{\infty}F_k[t;x(x(0),t,\cdot)]\,dt. \tag{19}$$

After constructing the series (10), it is possible to find $\xi^0[x(\cdot)]$ from the condition

$$\frac{\partial}{\partial\xi}\left[\sum_{i=1}^{n}\frac{\partial v[x(\cdot)]}{\partial x_i(0)}m_i\xi+\xi^2\right]=0.$$

Hence the series for ξ^0 will have the form

$$\xi^0[x(\cdot)]=\sum_{k=0}^{\infty}\mu^k\xi_k^0[x(\cdot)], \tag{20}$$

$$\xi_k^0[x(\cdot)]=-\sum_{i=1}^{n}\sum_{j=1}^{n}\left[\alpha_{ij}(k)\,m_ix_j(0)+m_i\int_{-h}^{0}\beta_{ij}(k;\vartheta)\,x_j(\vartheta)\,d\vartheta\right]. \tag{21}$$

4. **Proof of validity of method of construction of ξ^0.** Let us show that the series (10), (19) constructed in the preceding section converges at least for small values of μ. We shall denote by N_k the upper bound of the quantities $|\alpha_{ij}(k)|$, $|\beta_{ij}(k;\vartheta)|$, $|\gamma_{ij}(k;\vartheta,\tau)|$ $(i,j=1,\cdots,n;\ -h\le\vartheta\le 0,\ -h\le\tau\le 0)$. Owing to the asymptotic stability of system (17), the solutions $x(x^0(0),t)$ of equation (17) will satisfy the inequality

$$\|x(x^0(0),t)\|\leqslant N\|x^0(0)\|\exp(-\lambda t)\ (N,\lambda>0-\text{const}),$$

therefore we derive from (18) and (19) the estimates

$$N_k\leqslant MN_{k-1}+P\sum_{r+s=k}N_sN_r\quad(r\geqslant 0,\ s\geqslant 0), \tag{22}$$

where the constants M and P do not depend on the number k.

Using the estimate (22), let us construct a majorant for the series (10). Let us consider the equation

$$\rho^2-2(a+\mu)\rho+b=0, \tag{23}$$

whose root

$$\rho=a-\sqrt{(a+\mu)^2-b}$$

can be expanded, under the conditions $a^2>b>0$ and $|\mu|<a-b^{1/2}=\omega$, in a power series

$$\rho(\mu)=\sum_{k=0}^{\infty}\mu^k\rho^k. \tag{24}$$

After substituting (24) into equation (23), the coefficients ρ_k can be found from the relations

$$\rho_0^2 - 2a\rho_0 + b = 0, \quad \rho_0 = a - \sqrt{a^2 - b},$$

$$2(\rho_0 - a)\rho_k = -2\rho_{k-1} - 2\sum_{s+r=k} \rho_s\rho_r \quad (s > 0,\ r > 0),$$

or

$$\rho_k = \frac{1}{\varepsilon}\rho_{k-1} + \frac{1}{\varepsilon}\sum_{s+r=k} \rho_r\rho_s \quad (r \geqslant 0,\ s \geqslant 0;\ \varepsilon = \sqrt{a^2 - b}).$$

If a and b satisfy the condition $1/\epsilon > \max[M, P]$, $\rho_0 > N_0$, then the series (24) will be (by virtue of (22)) a majorant for the series (11); this proves the convergence of series (11), and hence also of series (20), (21) for ξ^0. Thus we have obtained the following result.

Theorem 2. *If the vectors $m, Am, \cdots, A^{n-1}m$ are linearly independent, then there exists for sufficiently small values of μ an optimum controller ξ^0 that satisfies the Conditions 1 and 2. This controller has the form (20), (21). The coefficients of system (17) can be determined by the solution of the quadratic equations (13).*

5. **Example.** Let us consider the equation

$$\frac{dx}{dt} = \mu x(t-1) + \xi, \tag{25}$$

where it is required to determine the functional $\xi[x(\cdot)]$ $(x(\cdot) = x(\vartheta)\ (-1 \leq \vartheta \leq 0))$ that minimizes the integral

$$I_\xi[x^0(\cdot)] = \int_0^\infty [x^2(x^0(\cdot), \xi, t) + \xi^2[x(x^0(\cdot), \xi, t, \cdot)]\, dt. \tag{26}$$

Let us find the system of first approximation (17) for the problem (25), (26). Equations (13) have in our case the form

$$-\alpha_{11}^2(0)\, x^2(0) + x^2(0) = 0,$$

i.e., $\alpha_{11}(0) = 1$. Thus the terms with μ^0 in the series for the functionals v and ξ^0 have here the form $v_0 = x^2(0)$, $\xi_0^0 = -(1/2)\,\partial v_0/\partial x(0) = -x(0)$. Hence the system of first approximation will be

$$\frac{dx}{dt} = -x. \tag{27}$$

The solutions of equation (27) are

$$x(x^0(0), t) = x^0(0)\, e^{-t}. \tag{28}$$

Let us write down equation (18) for our case.

For $k = 1$ we have

$$\left(\frac{dv_1[x(x^0(0), t, \cdot)]}{dt}\right) = -2x(t)\, x(t-1), \tag{29}$$

i. e., by (19),

$$v_1[x(\cdot)] = 2\int_0^\infty x(t)\,x(t-1)\,dt. \tag{30}$$

In evaluating the integral (30), it must be taken into account that $x(t) = x(0)e^{-t}$ for any $t \geq 0$, $x(t-1) = x(0)e^{-(t-1)}$ for $t \geq 1$, and that $x(t-1)$ for $0 \leq t < 1$ is determined by the initial curve $x(\cdot) = \{x(\vartheta),\ -h \leq \vartheta \leq 0\}$, whereby $x(t-1) = x(\vartheta)$ $(\vartheta = t-1)$ (see figure).

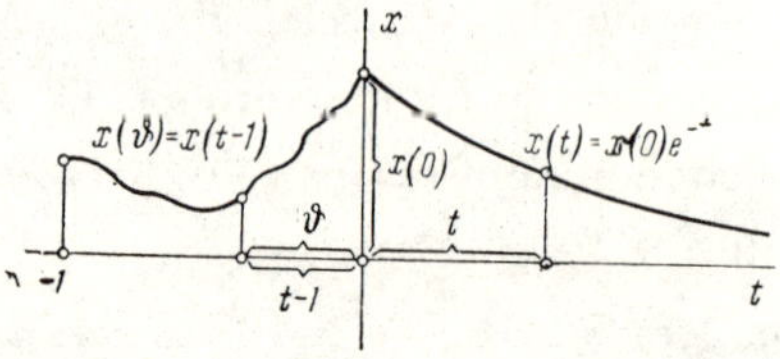

Fig. 1

We have

$$\int_0^\infty x(t)\,x(t-1)\,dt = x(0)\,e^{-1}\int_{-1}^0 e^{-\vartheta}x(\vartheta)\,d\vartheta + \frac{e^{-1}}{2}x^2(0),$$

and hence

$$v_1[x(\cdot)] = e^{-1}x^2(0) + 2x(0)\int_{-1}^0 e^{-(\vartheta+1)}x(\vartheta)\,d\vartheta,$$

i. e., to an accuracy of terms of the order of μ^2 the optimum controller ξ^0 for the problem (25), (26) has the form

$$\xi^0[x(\cdot)] = -x(0) - \mu\left[e^{-1}x(0) + \int_{-1}^0 e^{-(\vartheta+1)}x(\vartheta)\,d\vartheta\right] + \ldots.$$

BIBLIOGRAPHY

[1] L. È. Èl'sgol'c, *Qualitative methods in mathematical analysis*, GITTL, Moscow, 1955; English transl., Transl. Math. Monographs, Vol. 12, Amer. Math. Soc., Providence, R. I., 1964. MR **17**, 847.

[2] A. D. Myškis, *Linear differential equations with retarded argument*, GITTL, Moscow, 1951. MR **14**, 52.

[3] A. M. Ljapunov, *General problem of stability of motion*, GITTL, Moscow, 1950; photographic reproduction of French transl., Ann of Math. Studies, No. 17, Princeton Univ. Press, Princeton, N. J. and Oxford Univ. Press, London, 1947. MR **9**, 34; MR **12**, 612.

[4] N. G. Četaev, *Stability of motion*, GITTL, Moscow, 1955. MR **17**, 1087.

[5] N. N. Krasovskiĭ, *Certain problems in the stability of motion*, Fizmatgiz, Moscow, 1959; English transl., *Stability of motion. Application of*

Lyapunov's second method to differential systems and equations with delay, Stanford Univ. Press, Stanford, Calif., 1963. MR **21** #5047; MR **26** #5258.

[6] R. Bellman, *Dynamic programming*, Princeton Univ. Press, Princeton, N. J., 1957; Russian transl., IL, Moscow, 1960. MR **19**, 820; MR **22** #4898.

[7] N. N. Krasovskiĭ, *On the analytic construction of an optimum control in a system with time lags*, Prikl. Mat. Meh. **26** (1962), 39–51; English transl., J. Appl. Math. Mech. **26** (1962), 50–67. MR **26** #2706.

[8] A. M. Letov, *The analytic design of control systems*, Avtomat. i Telemeh. **22** (1961), 425–435; translated at Automat. Remote Control **22** (1961), 363–372. MR **26** #1228.

[9] F. M. Kirillova, *Contribution to the problem on the analytic construction of regulators*, Prikl. Mat. Meh. **25** (1961), 433–439; English transl., J. Appl. Math. Mech. **25** (1961), 644–653. MR **24** # B2517.

Translated by L. Ebner

ON THE PROBLEM OF THE EXISTENCE OF OPTIMAL CONTROL FOR A LINEAR SYSTEM WITH RANDOM PERTURBATION

F. M. KIRILLOVA

In this note we consider linear controlled processes under random perturbations without aftereffect and solve the problem of the possibility of the choice of a law for the variation of the controling action which ensures the damping of a transient process to zero with maximal statistical high speed.

This problem is a generalization of the well-known problem of optimal control described in [1–4], and it differs from the usual well studied statistical problems of control [5, 6] in that here the question concerns the choice of a controlling action and not the choice of an optimal operator. An analogous problem was considered earlier in [7]. Statistical problems of optimal control where there are possibilities of a choice of control were considered, for example, in [8, 9] with a different formulation.

1. Let there be given a system of differential equations

$$\frac{dx}{dt} = Ax + bu + c\eta, \tag{1}$$

where $x = (x_1, \cdots, x_n)$, $x \in X$ is a representing vector in the phase space X, $b = (b_1, \cdots, b_n)$, $c = (c_1, \cdots, c_n)$ are constant vectors, $A = (a_{ij})$ is a constant matrix of order n, $u(t)$ is a piecewise continuous controlling action, constrained by the condition

$$|u(\tau)| \leq 1, \quad t_0 \leq \tau \leq t. \tag{2}$$

We assume that the system (1) is nonsingular; then the vectors

$$b, Ab, \cdots, A^{n-1}b$$

are linearly independent, and the characteristic numbers p_i of the matrix A satisfy the inequality

$$\operatorname{Re} p_i < 0, \quad i = 1, \cdots, n. \tag{3}$$

Let the function $\eta(\tau)$ define a Markov random process [11]. In this note we confine ourselves to one special case, namely: we will assume that at each given moment $\tau = t$ the values of the function $\eta(\tau)$ may be one of the numbers η_m, $m = 1, 2, \cdots, l$, and a change of the values η_m is possible only at $t = k\tau_0$, $k = 1, 2, \cdots$, where τ_0 = a positive constant, and also, the probability of the transition $\eta_{m1} \rightarrow \eta_{m2}$ is a fixed number. We suppose also that the realizations of the function $\eta(\tau)$ are functions that are continuous from the right.

We denote by $\|x\|$ the norm of the vector $x(\tau)$:

$$\|x\| = (x_1^2 + \cdots + x_n^2)^{\frac{1}{2}},$$

and by $g(t)$ a realization of the random function $\eta(\theta)$ for $t_0 \leq \theta < t$. Let the symbol $p(x_0, t_0, \eta_0, u, t)$ denote the probability of the inequality $\|x(x_0, t_0, \eta_0, u, \eta, t)\| > 0$ for $t > t_0$, where $x(x_0, t_0, \eta_0, u, \eta, t)$ is a random solution of the equation (1) generated by the equation $u = u(t)$ and the function $\eta(t)$ with initial value $\eta = \eta_0$ (t_0 is the initial moment, $x(t_0) = x_0$). Everywhere in what follows we consider the process of control for the realization $g(\theta)$ complete at the moment $\tau = t$, when the point $x(x_0, t_0, \eta_0, u, g, \tau)$ hits the origin of coordinates $x = 0$ for the first time.

We consider the following problem: given $t_0 = 0$, $x(0) = x_0$, η_0, at any moment of time t, there enters information about the value of $\eta(\tau)$ for $0 \leq \tau < t$; it is required to choose a control $u_t(g)$ such that for condition (2) the minimum of the functional

$$T(u_t) = \int_0^\infty p(x_0, t_0, \eta_0, u_t, t)\, dt \tag{4}$$

is ensured. It is easy to see that the integral (4) expresses the mathematical expectation over time of a hit at the origin of coordinates of the point $x(x_0, t_0, \eta_0, u_t, \eta, t)$.

Definition 1. The functional $u_t(g)$ will be called an admissible control if $u_t(g)$ satisfies condition (2) and the inequality $T(u_t) < +\infty$ is correct.

Definition 2. The functional $u_t^0(g)$ will be called an optimal control if u_t^0 satisfies condition (2) and

$$T(u_t^0) = \min_{u_t} T(u_t).$$

In the following section it will be proved that for the system (1), for any initial conditions x_0, η_0, there exists an admissible piecewise-continuous control. This statement, together with the results of [7], where the existence of an optimal control is proved provided there exists an admissible control, enables us to state that for the given system, for any initial conditions x_0, η_0, there exists an optimal piecewise-constant control. Here we assume that a random perturbation $\eta(\tau)$ satisfies definite restraints, concerning which more will follow below.

Let $\Lambda(t, \beta_g)$ be the minimum of the values [1)]

$$\int_{t_0}^{t} |[l \cdot F(t-\tau)\, b]|\, dt \tag{5}$$

---

1) We use the symbols $(a \cdot b)$ and $[a \cdot b]$ to designate the scalar product of the vectors a and b.

on a set of vectors l, satisfying the equation

$$(l \cdot \int_{t_0}^{t} F(t-\tau)\, c g(\tau)\, d\tau) = (l \cdot \beta_g) = -1, \quad \|\beta_g\| \neq 0,$$

where $F(t) = \phi_{ij}(t)$ is the fundamental matrix of solutions of the homogeneous part of the system (1).

Since the system (1) is nonsingular, the function $\Lambda(t, \beta_g)$ is well defined and positive (cf. [10], p. 173) for all $t > t_0$ and any realization $g(\theta)$. For fixed t, the function $\Lambda(t, \beta_g)$ is continuous in β_g.

We will say that the function $\eta(\tau)$ fulfills condition B, if for arbitrary initial condition $t_0 \in [0, \tau_0]$, $\eta(t_0) = \eta_0$, there exists an increasing sequence of numbers $S^{(n)}(\eta_0, t_0, g)$, $n = 1, 2, \cdots$, and a number $\alpha > 0$ such that

$$\Lambda(t_0 + S^{(n)}, \beta_g) \geq 1 + \alpha \tag{6}$$

and

$$S^{(1)} \geq M_1(t_0, \eta_0),\ S^{(n)} - S^{(n-1)} \leq M_2(\eta_0),\ S^{(n)} \longrightarrow \infty$$

for $n \longrightarrow \infty$. Here M_1 and M_2 are constants, depending on t_0, η_0 and η_0, respectively. Condition B for $\eta(\tau)$ is satisfied, for example, when the realization $g(\theta)$ is sufficiently small in modulus, as we shall now prove.

We consider the function (5) on the sphere $\|l\| = 1$. On this sphere, the variable (5) attains a positive minimum $\mu(t_0, t)$. It is easy to see that the inequality

$$\Lambda(t, \beta_g) = \min_{(l \cdot \beta_g) = -1} \int_{t_0}^{t} |[l \cdot F(t-\tau)\, b]|\, d\tau$$

$$\geqslant \min_{\|l\| \geqslant \frac{1}{\|\beta_g\|}} \int_{t_0}^{t} |[l \cdot F(t-\tau)\, b]|\, d\tau = \frac{\mu(t_0, t)}{\|\beta_g\|} \quad (t_0 \geqslant 0) \tag{7}$$

is correct.

Since the characteristic numbers of the matrix A satisfy condition (3), there exists a positive number M_3 such that the inequality

$$\max_{t} \left(\sum_{i=1}^{n} \left(\sum_{j=1}^{n} \int_{t_0}^{t} |\varphi_{ij}(t-\tau)|\, d\tau \right)^2 \right)^{\frac{1}{2}} \leqslant M_3, \quad t_0 \geqslant 0, \tag{8}$$

will be fulfilled, where the constant M_3 does not depend on t_0. Consequently, it is possible to strengthen the inequality (7):

$$\Lambda(t, \beta_g) \geqslant \frac{\mu(t_0, t)}{\|\beta_g\|} \geqslant \frac{\mu(t_0, t)}{M_3 \cdot \max\limits_{t \geqslant 0, i} |c_i g(t)|}. \tag{9}$$

Let $t = t^{(1)}$, $t^{(1)} > \tau_0$, be fixed, and let t_0 vary within the bounds of the interval $[0, \tau_0]$. The function $\mu(t_0, t^{(1)})$ is continuous in t_0 for fixed $t^{(1)}$ (cf. [3]). Hence

$$\min_{0 \le t_0 \le \tau_0} \mu(t_0, t^{(1)}) \ge \mu_0(t^{(1)}) > 0.$$

Therefore it follows from (9) that:

$$\Lambda(t, \beta_g) \geqslant \frac{\mu_0(t^{(1)})}{M_3 \cdot \max\limits_{t \geqslant 0, i} |c_i g(t)|}, \tag{10}$$

where the function $\mu_0(t^{(1)})$, by virtue of the boundedness of the vectors of l and the inequality (8), has the property:

$$0 < \mu_0(t^{(1)}) \le \mu_0^0 = \text{const}, \quad \tau_0 < t^{(1)} < +\infty.$$

Put $S^{(1)}(\eta_0, t_0, g) = t_0^{(1)} > \tau_0$. If $\tilde{\mu} = \min_{t_0^{(1)} < t^{(1)}} \mu_0(t^{(1)})$, then for an arbitrary sequence

$$S^{(k)}(S^{(1)} = t_0^{(1)}), \; S^{(k)} - S^{(k-1)} \le M = \text{const} > 0, \; S^{(k)} \longrightarrow \infty \text{ for } k \longrightarrow \infty,$$

and an arbitrary number $\alpha > 0$ with

$$\max_{t \geqslant 0, i} |c_i g(t)| \leqslant \frac{\tilde{\mu}}{M_3(1+\alpha)}$$

we have

$$\Lambda(t, \beta_g) \ge 1 + \alpha, \; t = t_0 + S^{(k)}, \quad k = 1, 2, \cdots.$$

The assertion is proved.

2. In this section, the proof is given for the existence of an admissible control for the system (1). We prove first one auxiliary proposition.

Lemma. *Let the random function $\eta(\tau)$ satisfy condition* B. *For arbitrary initial conditions t_0, $x(t_0)$, η_0, and for any realization $g(\theta)$, $t_0 \le \theta \le t$, there exists a piecewise-continuous control $u_t(g)$ which transfers the point $x_0 = x(t_0)$ into the origin of coordinates in a finite time $T(g)$. If $x_0 \in G$, where G is a bounded region, then the inequality*

$$T(g) \le M_4 = \text{const} \quad \textit{for all } t_0 < +\infty, \tag{11}$$

is fulfilled for the number $T(g)$.

Proof. We write the solution to the system (1):

$$x(t) = F(t)x_0 + \int_{t_0}^{t} F(t-\tau)[bu(\tau) + cg(\tau)]\, d\tau.$$

If for $t = t_1$ the point $x(t)$ attains the origin of coordinates, we obtain

$$-F(t_1)x_0 - \int_{t_0}^{t_1} F(t_1 - \tau)\, cg(\tau)\, d\tau = \int_{t_0}^{t_1} F(t_1 - \tau)\, bu(\tau)\, d\tau,$$

or if we denote the left-hand part of the identity by $-\beta$,

$$-\tilde{\beta} = \int_{t_0}^{t_1} F(t_1 - \tau)\, bu(\tau)\, d\tau. \tag{12}$$

By the inequality (3) there exists a t_2 such that for $t_1 > t_2$ we have

$$\|\tilde{\beta} - \beta_g\| < \epsilon, \quad \beta_g = \int_{t_0}^{t_1} F(t_1 - \tau)\, cg(\tau)\, d\tau,$$

where ϵ is an arbitrarily small number. Therefore, taking into account the inequality (6) and the continuous dependence of the function $\Lambda(t, \beta_g)$ on β_g (t and g are fixed), we conclude: for all $t_k > \tilde{t}_2 \geq t_2$ the inequality

$$\Lambda(t_k, \beta_g) \geq 1, \quad t_k = t_0 + S^{(k)},$$

will be fulfilled. This means [10] that the system (12) is solved by the piecewise-continuous functions $u_t(g)$, which satisfy conditions (2).

We assume that G is a finite region of the space X. Then $\max_{x_0 \in G} \|x_0\| \leq K < \infty$. Therefore the system (12) has a piecewise-continuous solution $u_t(g)$; moreover the transition time of the process depends only on the realization $g(\theta)$.

The validity of the inequality (11) follows from condition (8). The lemma is proved.

Let g_k^m, $k = 1, 2, \cdots, m = 1, 2, \cdots, l$, be realizations of the function $\eta(\tau)$ for $0 \leq \theta \leq k\tau_0$. We designate by p_k^m the probability that the realizations g_k^m will actually occur and we order these realizations in the following way. If $k_1 < k_2$, we will say that the realization $g_{k_1}^m$ precedes the realization $g_{k_2}^m$; if $k_1 = k_2$, we will order the realizations in the order of the change in m; for example, we will consider that the realization $g_{k_1}^{m1}$ precedes the realization $g_{k_1}^{m2}$ if $m_1 < m_2$ at $k = k_3 < k_1$, where k_3 is the smallest of the numbers $k = 1, 2, \cdots$, when $m_1 \neq m_2$. Then we have the following theorem.

Theorem 1. *If the random function $\eta(\tau)$ satisfies condition* B, *then for arbitrary initial conditions $x(0) = x_0$, $\eta(0) = \eta_0$, there exists an admissible piecewise-continuous control $u_t(g)$.*

Proof. We consider the triple $\{g_k^m, u_k^m, T_k^m\} = \{g, u, T\}_k^m$, composed of the realizations g_k^m, the controls u_k^m, taken from the domain (2), and the numbers T_k^m

for which the inequality $(k-1)\tau_0 \leq T_k^m \leq k\tau_0$ is true. We denote by $\Psi(x_0, \eta_0)$ the set of triples $\{g, u, T\}_k^m$ at which the controls u_k^m have the property: (a) under the piecewise-continuous control u_k^m, the point $x(x_0, t_0, \eta_0, u_k^m, g_k^m, t)$ first hits the origin of coordinates at $t = T_k^m$.

By virtue of the lemma, the set is not empty. Therefore there exists a number [1] $M^{(1)}$ such that at least one T_k^m does not exceed $M^{(1)}$.

We consider the realizations g_k^m for which

$$T_k^m \leq M^{(1)}. \tag{13}$$

Let $\tau = T_{k_1}^{m_1}$ be the smallest of the numbers (13). In the equation (1), we put $u(t) = u_{k_1}^{m_1}$ and investigate the random solution $x(x_0, t_0, \eta_0, u_{k_1}^{m_1}, \eta, t)$. It is easy to see that the inequality,

$$p(x_0, t_0, \eta_0, u_{k_1}^{m_1}, T_{k_1}^{m_1}) \leq 1 - p^{(1)} \quad (\text{here } t_0 = 0), \tag{14}$$

holds, where $p^{(1)}$ is the smallest of the numbers p_k^m for the realizations g_k^m defined on the interval $t_0 \leq t \leq T_{k_1}^{m_1}$.

For convenience of notation, we will assume everywhere in the sequel that $T_{k_1}^{m_1} = k_1\tau_0$. As analysis of condition B shows that such an assumption does not reduce the generality of the arguments.

We establish the correspondence $u_{k_1}^{m_1} \longleftrightarrow g_{k_1}^{m_1}$. If a realization $g_{k_1}^m$, different from $g_{k_1}^{m_1}$ was realized in the interval $0 \leq t \leq k_1\tau_0$, we act in the following way.

We renumber all realizations which are possible under the initial condition $\eta^{(1)} = g_{k_1}^{m_1}(k_1\tau_0)$, leaving the previous form of the entry g_k^m.

Let $x^{(1)} = x(x_0, t_0, \eta_0, u_{k_1}^{m_1}, g_{k_1}^{m_1}, k_1\tau_0)$. We consider the set $\Psi(x^{(1)}, \eta^{(1)})$ of triples $\{g, u, T\}_k^m$, where the piecewise-continuous controls u_k^m satisfy condition (a). By virtue of the lemma, the set $\Psi(x^{(1)}, \eta^{(1)})$ is not empty. We designate by $T_{k_2}^{m_2} = k_2\tau_0$ the smallest of the numbers T_k^m:

$$T_k^m \leq M^{(2)} = \text{const} > 0.$$

If u is set equal to $u_{k_2}^{m_2}$, then obviously

1) If for all realizations $g_k^m(\eta(t_0) = \eta_0)$ the inequality $T_k^m \leq M^{(1)}$ holds, then the functional $u_t(g)$ defined by the correspondence $u_k^m \longleftrightarrow g_k^m$ is an admissible control, since in this case the inequality $T(u_t(g)) \leq M^{(1)}$ obviously holds.

$$p(x^{(1)}, k_1\tau_0, \eta^{(1)}, u_{k_2}^{m_2}, k_2\tau_0) \leq 1 - p^{(2)}. \tag{15}$$

Here $p^{(2)}$ is the smallest of the numbers p_k^m for the realizations g_k^m of the triples $\Psi(x^{(1)}, \eta^{(1)})$, $k_1\tau_0 \leq t < k_2\tau_0$.

We estimate the probability $p(x_0, t_0, \eta_0, u^{(2)}, k_2\tau_0)$, where

$$u^{(2)}(t) = \begin{cases} u_{k_1}^{m_1}, & 0 \leqslant t < k_1\tau_0, \\ u_{k_2}^{m_2}, & k_1\tau_0 \leqslant t < k_2\tau_0. \end{cases}$$

Using the inequalities (14) and (15), we have

$$p(x_0, t_0, \eta_0, u^{(2)}, k_2\tau_0) \leq (1 - p_2)^2,$$

$$p_2 = \min p^{(i)}, \quad i = 1, 2.$$

We assume $u_{k_2}^{m_2} \longleftrightarrow g_{k_2}^{m_2}$. Continuing this argument and applying the method of induction, on the sth step we construct a piecewise-continuous control $u^{(s)}$, defined by the formulas

$$u^{(s)}(t) = \begin{cases} u_{k_1}^{m_1}, 0 \leqslant t < k_1\tau_0, \\ \dots\dots\dots\dots \\ u_{k_s}^{m_s}, k_{s-1}\tau_0 \leqslant t < k_s\tau_0, \end{cases}$$

where $u_{k_s}^{m_s}$ belongs to that triple $\{g, u, T\}_k^m$ of the set $\Psi(x^{(s-1)}, \eta^{(s-1)})$ for which T_k^m is smallest. We observe that for the construction of the sets $\Psi(x^{(s-1)}, \eta^{(s-1)})$ we carry out the ordering of the realizations $g(t)$ every time, leaving the old form of the notation.

By construction, the control $u^{(s)}$ has these properties:

$$p(x^{(s-1)}, k_{s-1}\tau_0, \eta^{(s-1)}, u_{k_s}^{m_s}, k_s\tau_0) \leq 1 - p^{(s)},$$

$$p^{(s)} = \min p_k^m, \ k_{s-1}\tau_0 \leq t < k_s\tau_0$$

and

$$p(x_0, t_0, \eta_0, u^{(s)}, k_s\tau_0) \leq (1 - p_s)^s,$$

where

$$p_s = \min p^{(i)}, \quad i = 1, \cdots, s.$$

We denote the functional given by the correspondence [1)] $u_{k_i}^{m_i} \longleftrightarrow g_{k_i}^{m_i}$ by $u_t(g)$

---

1) We assume for this that $u_t(g) = u^{(i-1)}(t)$, for $0 \leq \theta < t = k_{i-1}\tau_0$.

and estimate $T(u_t(g))$:

$$T(u_t) \leq k_1 \tau_0 p_{k_1}^{m_1} + \sum_{s=2}^{\infty} (1 - p_s)^{s-1} k_s \tau_0. \tag{16}$$

By virtue of the inequality (8) and the boundedness of the functions $u^{(s)}(t)$, $g(t)$, the random solution $x(x_0, t_0, \eta_0, u, \eta, t)$ does not leave some finite region $G(x_0, \eta_0)$. Applying the lemma, we conclude: it is possible to choose a number $T_{k_s}^{m_s}$ such that the inequality

$$T_{k_s}^{m_s} - T_{k_{s-1}}^{m_{s-1}} \leq M^0 = \text{const} > 0$$

will be fulfilled. Hence among the numbers p_s, $s = 1, \cdots,$ there is a smallest, which we denote by p_0^0.

We transform the inequality (16) in accordance with the remarks made above:

$$T(u_t) \leq k_1 \tau_0 p_{k_1}^{m_1} + M^0 \sum_{s=2}^{\infty} (1 - p_0^0)^{s-1} s.$$

The series $\sum_{s=2}^{\infty} (1 - p_0^0)^{s-1} s$ converges. Hence $u_t(g)$ is an admissible piecewise-continuous control. The theorem is proved.

Theorem 2. *If the random function $\eta(\tau)$ satisfies condition* B, *then for arbitrary initial conditions $x(0)$, $\eta(0)$, there exists, for the system* (1), *at least one optimal piecewise-constant control $u_t^0(g)$, which has the form*

$$u_t^0(g) = \text{sign}\,(l \cdot \psi(g, t)),$$

where $\psi(g, t)$ is (for a fixed realization $g(t)$) a continuous function which vanishes at separate isolated points t.

We do not give the proof of this proposition here: in our case, for the initial conditions $x(0)$, $\eta(0)$ there exists an admissible control; therefore it is possible to carry out the subsequent argument according to the plan explicitly described in [7].

Remark. After substitution of $u_t^0(g)$ into the right-hand part of the system (1), solutions of this equation turn out to be random functions of time. For the whole synthesis of the system it is important that this process be a Markov process without aftereffect.

We show that it is possible to choose the functional $u_t^0(g)$ such that the value of the control will be defined in a unique manner by the values of $x(t)$, $\eta(t)$ at a given time. In the construction of the optimal control $u_t^0(g)$ we will hold to the following rule: if under the initial conditions $x(t_1)$, $\eta(t_1)$ and realizations $g_1(t)$, $g_2(t)$, $g_1(t) = g_2(t)$, $t_1 \leq t \leq t_1 + T_1$, it is possible to find controls $u(g_1)$, $u(g_2)$ which transfer the point $x(t_1)$ into the origin of coordinates at time T_1 and T_2 ($T_1 < T_2$) respectively, then as the optimal value of the functional $u_t^0(g)$ we will put the control $u(g_1)$.

In conclusion, the author gratefully acknowledges that N. N. Krasovskiĭ acquainted him with problems of optimal control.

BIBLIOGRAPHY

[1] L. C. Pontrjagin, *Optimal regulation processes*, Uspehi Mat. Nauk 14 (1959), no. 1(85), 3–20; English transl., Amer. Math. Soc. Transl. (2) **18** (1961), 321–339. MR 22 #11189.

[2] R. V. Gamkrelidze, *Theory of processes in linear systems which are optimal with respect to rapidity of action*, Izv. Akad. Nauk. SSSR Ser. Mat. 22 (1958), 449–474. MR 20 #4039.

[3] N. N. Krasovskiĭ, *Concerning the theory of optimal control*, Avtomat, i Telemeh. **18** (1957), 960–970. MR 20 #803.

[4] R. Bellman, I. Glicksberg and O. Gross, *On the "bang-bang" control problem*, Quart. Appl. Math. 14 (1956), 11–18. MR 17, 1206.

[5] V. S. Pugačev, *Theory of random functions and its application to problems of automatic control*, GITTL, Moscow, 1957. MR **20** #359.

[6] Dzh. Kh. Lening and R. G. Bettin, *Random processes in problems of automatic control*, IL, Moscow, 1958. (Russian)

[7] N. N. Krasovskiĭ, *On optimum control in the presence of random distrubances*, Prikl. Mat. Meh. 24 (1960), 64–79; Translated as J. Appl. Math. 24, 82–102. MR 22 #4607.

[8] E. F. Miščenko and L. S. Pontrjagin, *A statistical problem on optimal control*, Dokl. Akad. Nauk SSSR **128** (1959), 890–892. MR 23 #B3085.

[9] R. Bellman, *Dynamic programming and stochastic control processes*, Information and Control **1** (1958), 228–239. MR 20 #3747.

[10] N. I. Akhiezer and M. G. Kreĭn, *On some questions in the theory of moments*, GONTI USSR, Kharkov, 1938, Article IV (Russian).

[11] B. V. Gnedenko, *Course in the theory of probability*, GITTL, Moscow, 1950. MR 13, 565.

Translated by:
R. H. Rodine

RANDOMIZED HOMOGENEOUS TESTS FOR THE BEHRENS-FISHER PROBLEM

JU. V. LINNIK

It is shown that the property of similarity (and *a fortiori* nonbias) for homogeneous randomized tests in the Behrens-Fisher problem is incompatible with the statistically expedient requirement of "null-regularity".

Let $x_1, \cdots, x_{n1} \in N(a_1, \sigma_1)$, $y_1, \cdots, y_{n2} \in N(a_2, \sigma_2)$ be two independent repeated samples from normal populations. The Behrens-Fisher problem is to design tests for testing the hypothesis H_0: $a_1 = a_2$ in the presence of the unknown nuisance parameters σ_1 and σ_2.

We shall consider randomized tests for testing the null hypothesis H_0: with each point $M(x_1, \cdots, x_{n1}; y_1, \cdots, y_{n2})$ of the Cartesian product of the sample spaces there will be associated a quantity $\Phi(M)$ such that $0 \leq \Phi(M) \leq 1$, representing the probability of rejecting H_0 using a subsidiary random choice (e.g., a table of random numbers); $\Phi(M)$ is called *the critical test-function.* In the sequel we shall call $\Phi(M)$ simply *the test.*

We shall consider tests which depend only on the sufficient statistics

$$\bar{x} = \frac{1}{n_1} \sum_{i=1}^{n_1} x_i, \quad \bar{y} = \frac{1}{n_2} \sum y_j,$$

$$s_1^2 = \frac{1}{n_1} \sum_{i=1}^{n_1} (x_i - \bar{x})^2, \quad s_2^2 = \frac{1}{n_2} \sum (y_j - \bar{y})^2$$

and which moreover, are invariant under translation of all the observations by one and the same quantity and under rescaling by an equal factor. Such tests we shall call *homogeneous*; to them there correspond Lebesgue-measurable critical functions Φ, depending only on two variables:

$$\xi = \frac{\bar{x} - \bar{y}}{s_2}, \quad \eta = \frac{s_1}{s_2},$$

as can easily be deduced from the mentioned homogeneity property.

The *level* of the randomized test $\Phi(M)$ is the quantity

$$\alpha = \sup E_{H_{ou}} \Phi(M), \tag{0.1}$$

where H_{ou} runs through the simple hypotheses which make up H_0: $a_1 = a_2 = u$; σ_1 and σ_2 assume given values.

If in addition we have for some simple H_{ou}

$$E_{H_{ou}}(\Phi(M)) < \alpha, \tag{0.2}$$

then the test is biased in favor of the alternative hypotheses (see [1], Chapter IV). For the test $\Phi(M)$ to be unbiased it is necessary that

$$E_{H_{0u}}(\Phi(M)) = \alpha \tag{0.2'}$$

for all the simple hypotheses H_{ou} which make up H_0. A test $\Phi(M)$ satisfying such a condition (0.2′) will be called a *similar* test.

Consider a homogeneous similar test

$$\Phi(\xi, \eta), \quad \xi = \frac{\bar{x} - \bar{y}}{s_2}, \quad \eta = \frac{s_1}{s_2}. \tag{0.3}$$

If the magnitude

$$\frac{\bar{x} - \bar{y}}{\sqrt{n_1 s_1^2 + n_2 s_2^2}} = \frac{\xi}{\sqrt{n_1 \eta + n_2}} \tag{0.4}$$

vanishes or is very small, then it is natural to accept the null hypothesis H_0; a test not possessing such a property would be, as it were, "conditionally" biased in favor of the alternatives to H_0. These considerations lead us to introduce a new natural property of a test, which we shall term the property of *null-regularity*.

The property of null-regularity of a test $\Phi(\xi, \eta)$: there exists a number $\eta_0 = 1 + \eta_1 > 1$ such that the segment $0 \le \eta \le \eta_0$ of the axis $\xi = 0$ has the following property: around each point of this segment one can describe a circle such that the conditional probability that the hypothesis H_0 will be rejected by the test $\Phi(M)$, subject to the point (ξ, η) falling within this circle, is equal to zero.

1. We can now formulate the principal theorem of the present paper. In what follows we shall assume that $\max(n_1, n_2) \ge 3$.

Theorem. *There do not exist randomized and homogeneous (and a fortiori unbiased) tests for the Behrens-Fisher problem which possess the property of null-regularity.*

Let us make some observations regarding this theorem. It is best possible in the sense that if we were to abandon the requirement of null-regularity the theorem would no longer be true: there do exist similar randomized tests without that property. Namely, the density of the distribution of the statistics (ξ, η), which we shall write out below, is an even function of ξ and η under the null hypothesis H_0. Let $\Phi_1(\xi, \eta)$ be a continuous function of ξ, η in the half-plane Ω: $-\infty < \xi < \infty$, $0 \le \eta < \infty$, which for all values of η is an odd function of ξ and satisfies the condition

$$|\Phi_1(\xi, \eta)| \leqslant \frac{1}{3}.$$

Put

$$\Phi(\xi, \eta) = \frac{1}{2} + \Phi_1(\xi, \eta).$$

Then by what has been said above we have:

$$E\,\Phi_1(\xi, \eta) = 0, \quad E\,\Phi(\xi, \eta) = \frac{1}{2}$$

under the hypothesis H_0, so that $\Phi(\xi, \eta)$ is a randomized similar test. However, the requirement of null-regularity is obviously violated here.

Let us further observe that from the analytic point of view our theorem can be strengthened further. Instead of the similarity of the test $\Phi(\xi, \eta)$ with respect to all values of σ_1 and σ_2 it is sufficient to require that it be similar with respect to an arbitrary denumerable set of distinct values contained in some finite segment.

Furthermore, if the test is nonrandomized, i.e., if the function $\Phi(\xi, \eta)$ always equals 0 or 1, and if the boundary of the critical region (the region where $\Phi(\xi, \eta) = 1$) consists of a finite number of continuous curves, then the property of null-regularity simply means that the test does not reject the null hypothesis if $\bar{x} - \bar{y} = 0$ and $s_1^2 + s_2^2 > 0$, so that our theorem holds for it under a very simple and natural condition.

2. Let us work out the joint distribution of the variables $(\bar{x} - \bar{y})/s_2$, s_1/s_2 under the hypothesis H_0. Setting

$$\sigma_1^2 = \theta_1, \quad \sigma_2^2 = \theta_2, \quad X = (\bar{x} - \bar{y})\left(\frac{n_1\theta_2 + n_2\theta_1}{n_1 n_2}\right)^{\frac{1}{2}},$$

$$U = \left(\frac{n_1 s_2^2}{\theta_1}\right)^{\frac{1}{2}}, \quad V = \left(\frac{n_2 s_2}{\theta_2}\right)^{\frac{1}{2}}, \quad T = \frac{X}{V}, \quad W = \frac{U}{V},$$

we obtain, after a computation of the standard kind, the joint probability density

$$\rho(T, W) = C_{n_1 n_2} \frac{W^{n_1 - 2}}{(1 + T^2 + W^2)^{\frac{n_1 + n_2}{2} - \frac{1}{2}}}, \tag{2.1}$$

where $C_{n_1 n_2}$ depends only on n_1 and n_2.

Further, let us put $n_2\theta_1/n_1\theta_2 = \theta$, so that

$$\frac{\bar{x} - \bar{y}}{s_2} = T(1 + \theta)^{\frac{1}{2}}, \quad \frac{s_1}{s_2} = \frac{U}{V}\theta^{\frac{1}{2}},$$

$$\Phi(\xi, \eta) = \Phi\left(T(1 + \theta)^{\frac{1}{2}}, W\theta^{\frac{1}{2}}\right).$$

Thus, under the hypothesis H_0 we obtain from (0.2′):

$$\iint_{\Omega'} \Phi\left(T\,(1+\theta)^{\frac{1}{2}},\; W\theta^{\frac{1}{2}}\right) p\,(T,\,W)\,dT\,dW = \alpha, \tag{2.2}$$

where $\alpha > 0$ is the level of the similar test and Ω' is the half-plane

$$-\infty < T < \infty, \quad 0 \leqslant W < \infty.$$

Passing from the variables T, W in (2.2) to the variables ξ, η, we find

$$\iint_{\Omega} \Phi\,(\xi,\,\eta) \frac{\eta^{n_1-2}\,d\xi\,d\eta}{(\theta^2 + \theta\,(1 + \xi^2 + \eta^2) + \eta^2)^N} = \alpha\,\theta^{-\frac{n_2}{2}}\,(1+\theta)^{-N+\frac{1}{2}}, \tag{2.3}$$

where $N = (n_1 + n_2)/2 - 1/2$.

Equality (2.3) holds for all $\theta > 0$.

Following the basic idea of the author's notes [2], [3], let us consider the analytic continuation of both sides of equality (2.3) in the plane of the complex values of the parameter $\theta = \sigma + i\tau$. The right-hand side of (2.3) is continuable in the domain Λ consisting of the complex plane slit along the negative axis: $\tau = 0$, $-\infty < \sigma < 0$. The left-hand side is also continuable in that domain; the integral in (2.3) converges absolutely and uniformly in every closed subdomain lying in Λ (the proof of this assertion follows easily from the consideration of the critical curves $A\,(\xi,\,\eta) = D$ and $B\,(\xi,\,\eta) = D$, to be introduced below, and from the condition $\max(n_1,\,n_2) \geq 3$).

Thus equality (2.3) holds throughout the domain Λ of the values of the parameter θ.

In the denominator of the integrand in (2.3) there occurs a quadratic polynomial in θ:

$$P\,(\theta,\,\xi,\,\eta) = \theta^2 + \theta(1 + \xi^2 + \eta^2) + \eta^2. \tag{2.4}$$

We shall consider its values in the whole (unslit) complex θ-plane. For given ξ, η the roots of $P\,(\theta,\,\xi,\,\eta)$ are negative or zero. Putting

$$P\,(\theta,\,\xi,\,\eta) = (\theta + A\,(\xi,\,\eta))\,(\theta + B\,(\xi,\,\eta)), \tag{2.5}$$

we may assume that always

$$0 \leqslant A\,(\xi,\,\eta) \leqslant 1, \quad B\,(\xi,\,\eta) \geqslant 1. \tag{2.6}$$

The locus given by the equations

$$A\,(\xi,\,\eta) = D, \quad B\,(\xi,\,\eta) = D,$$

will be called, following notes [1] and [2], the *critical curves*, or the "*critics*" of the family of measures induced by the probability densities (2.1) and by the parameter θ. For $D < 1$, $A(\xi,\,\eta) = D$ we obtain a family of hyperbolas:

$$\frac{\eta^2}{D} - \frac{\xi^2}{1-D} = 1. \tag{2.7}$$

For $D = 1$ the hyperbola degenerates into the doubly described interval $\xi = 0$, $1 \leq \eta < \infty$; for $D = 0$ it degenerates into the axis $\eta = 0$. For $D > 1$, $B(\xi, \eta) = D$ we obtain a family of confocal semiellipses:

$$\frac{\xi^2}{D-1} + \frac{\eta^2}{D} = 1. \qquad (2.8)$$

For $D = 1$ they degenerate into the doubly described interval $\xi = 0$, $0 \leq \eta \leq 1$.

We can write the fundamental equation (2.3) in the form:

$$\iint_{\Omega} \Phi(\xi, \eta) \frac{\eta^{n_1-2}\, d\xi\, d\eta}{[(\theta + A(\xi, \eta))(\theta + B(\xi, \eta))]^N} = \alpha\theta^{-\frac{n_2}{2}}(1+\theta)^{-N+\frac{1}{2}}. \qquad (2.9)$$

By virtue of the null-regularity of the test $\Phi(\xi, \eta)$, there exists a number $\eta_0 = 1 + \eta_1 > 1$ such that around each point of the segment $\xi = 0$, $0 \leq \eta \leq \eta_0$ we can describe a circle within which $\Phi(\xi, \eta) = 0$ almost everywhere in the sense of Lebesgue measure. By the Heine-Borel theorem, we can select a finite covering out of these circles, and obtain an open domain Γ containing our segment such that $\Phi(\xi, \eta) = 0$ for almost all points $(\xi, \eta) \in \Gamma$.

Let us now consider the family (2.8) of confocal ellipses. We choose a number $D > 1$ so close to 1 that the corresponding ellipse $B(\xi, \eta) = D$ lies entirely inside the domain Γ (for smaller D this will clearly hold as well). Then we begin to increase D. Let $D_0 > 1$ be the least upper bound of the values of D for which $\Phi(\xi, \eta) = 0$ almost everywhere in the interior of the ellipse (2.8).

If $D_0 = \infty$, then $\Phi(\xi, \eta) = 0$ almost everywhere in Ω; the test is then a trivial similar test. We dismiss this case and shall assume that D_0 is a finite number.

3. We introduce notations which will be needed later on. We denote with the symbol H a magnitude, not always the same one, which is under the variation of the various parameters of the problem. By $\zeta_0, \zeta_1, \cdots$ we denote small positive constants, each to be chosen in dependence upon the preceding ones; by $L_\theta^{(k)}$ we denote the operator of differentiation k times with respect to θ at a point $\theta \in \Lambda$ ($k \geq 0$ is an integer).

We have $D_0 > 1$; let us put

$$D_0 - 1 = \zeta_0, \qquad \zeta_1 = \min\left(\frac{\zeta_0}{2}, \frac{1}{2}\right).$$

Let θ vary in the rectangle $Q \subset \Lambda$:

$$\theta = -D_0 + \delta + i\zeta, \quad 0 \leqslant \delta \leqslant \zeta_1, \quad 0 < \zeta < \zeta_1. \qquad (3.1)$$

The arguments which are to follow are based on applying the operator $L_\theta^{(k)}$ to both sides of equality (2.9) for $\theta \in Q$. We shall discover that, for a suitable choice of $\theta \in Q$ and sufficiently great k, the orders of magnitude of the right-hand and left-hand sides of (2.9) are different; this leads to a contradiction.

By the definition of D_0 we have $\Phi(\xi, \eta) = 0$ almost everywhere in the interior of the semiellipse $B(\xi, \eta) < D_0$. Let us denote with Ω_0 the half-plane Ω from which the semiellipse has been removed. We denote by $I_0(\theta)$ the corresponding integral, which coincides with the left-hand side of (2.9). We have:

$$I_0(\theta) = \iint\limits_{\Omega_0} \Phi(\xi, \eta) \frac{\eta^{n_1-2}\, d\xi\, d\eta}{[(\theta + A(\xi, \eta))(\theta + B(\xi, \eta))]^N} d\theta^{-\frac{n_2}{2}} (1+\theta)^{-N+\frac{1}{2}}. \tag{3.2}$$

We further note that

$$0 \leqslant A(\xi, \eta) \leqslant 1, \tag{3.3}$$

whence, for $\theta \in Q$,

$$|\theta + A(\xi, \eta)| \geqslant D_0 - \zeta_1 - 1 \geqslant \zeta_1 \tag{3.4}$$

by the definition of ζ_1.

4. Let us first consider the case that the sample sizes n_1 and n_2 are of different parity; then $N = (n_1 + n_2 - 1)/2$ is an integer and the intergrand in (3.2) is rational and does not have branchings.

From the integral $I_0(\theta)$ we separate the integral

$$I_1(\theta) = \iint\limits_{B(\xi, \eta) \geqslant D_0+1} \tag{4.1}$$

and estimate from above the quantity

$$L_\theta^{(k)} I_1(\theta) \tag{4.2}$$

for $\theta \in Q$ and large values of k. By (3.4) we obtain

$$L_\theta^{(m)} \frac{1}{(\theta + A)^N} = H \frac{\Gamma(N+m)}{\zeta_1^m} \tag{4.3}$$

(the meaning of the symbol H has been explained above; in particular, $H \cdot H = H$; $\zeta_1^{-N} = H$).

Further, let $B = B(\xi, \eta)$ vary in the region

$$D \leqslant B(\xi, \eta) < D + 1, \tag{4.4}$$

where $D \geq D_0 + 1$. Then we have

$$|\theta + B(\xi, \eta)| \geqslant D - D_0 \geqslant 1, \tag{4.5}$$

whence

$$L_\theta^{(m)} \frac{1}{(\theta + B)^N} = H \frac{\Gamma(N+m)}{(D - D_0)^{m+N}}. \tag{4.6}$$

Let us now estimate the quantity

$$L_\theta^{(k)} \frac{1}{(\theta + A)^N (\theta + B)^N} \tag{4.7}$$

for $\theta \in Q$ and $B = B(\xi, \eta)$ belonging to the region (4.5). For this we make use of (4.3), (4.6) and the standard formula for differentiating the product of two functions. We then find:

$$L_\theta^{(k)} \frac{1}{(\theta + A)^N (\theta + B)^N} = H \sum_{m=0}^{k} C_k^m \frac{\Gamma(N+m)}{\zeta_1^m} \frac{\Gamma(N+k-m)}{(D-D_0)^{k-m+N}}. \tag{4.8}$$

Since $C_k^m = k!/m!\,(k-m)!$, we obtain after a simple calculation

$$L_\theta^{(k)} \frac{1}{(\theta + A)^N (\theta + B)^N}$$

$$= H\Gamma(k+1)(D-D_0)^{-N} \zeta_1^{-k} k^{2N} = H\Gamma(k+1)(D-D_0)^{-N} \zeta_2^{-k}, \tag{4.9}$$

where $\zeta_2 = \zeta_1/2$.

From $I_1(\theta)$ we separate (see (4.1)) a part subject to the condition (4.4) and denote the latter by $I_{1D}(\theta)$. In the region (4.4), a horseshoe between the semi-ellipses, we have:

$$\eta = H(D-D_0)^{\frac{1}{2}};$$

the area of the horseshoe in $H(D-D_0)^{1/2}$. Taking (4.9) into account, we find that for $\theta \in Q$

$$L_\theta^{(k)} I_{1D}(\theta) = H\Gamma(k+1)\zeta_2^{-k}(D-D_0)^{\frac{n_1}{2}-N-\frac{1}{2}}. \tag{4.10}$$

In view of the condition $\max(n_1, n_2) \geq 3$, we obtain:

$$n_1 \leq n_2, \quad \frac{n_1}{2} - N - \frac{1}{2} = \frac{n_2}{2} \geq \frac{3}{2}.$$

In that case we may sum the estimates (4.10) for the values $D = D_0 + 1$, $D_0 + 2, \cdots$, as a result of which we obtain:

$$L_0^{(k)} I_1(\theta) = H\Gamma(k+1)\zeta_2^{-k}. \tag{4.11}$$

5. Put

$$I_{00}(\theta) = I_0(\theta) - I_1(\theta) = \iint\limits_{D_0 \leqslant B(\xi,\eta) \leqslant D_0+1}. \tag{5.1}$$

Let $R > 100$ be a large number, to be fixed later on, such that $M = R\zeta_2^{-1}$ is an integer. Let

$$\Delta = \zeta_2 R^{-1},$$

$$F_{0,r}(\theta) = \iint\limits_{D_0 + r\Delta \leqslant B(\xi,\eta) < D_0 + (r+1)\Delta}, \tag{5.2}$$

so that

$$I_{00}(\theta) = F_{00}(\theta) + F_{01}(\theta) + \ldots + F_{0,M-1}(\theta). \tag{5.3}$$

We have to study the behavior of

$$L_\theta^{(k)} F_{0r}(\theta) \tag{5.4}$$

in the region

$$D_0 + r\Delta \leqslant B(\xi, \eta) \leqslant D_0 + (r+1)\Delta \tag{5.5}$$

for $\theta \in Q$. We have:

$$\theta + B = r\Delta + \delta + i\zeta. \tag{5.6}$$

Let us put $\delta = \Delta/2$; thus the value of the number $\theta = -D_0 + \delta + i\zeta$ will depend on the numbers R and ζ which we shall determine later on. Now from (5.6) we obtain:

$$|\theta + B| = \left(\left(r+\frac{1}{2}\right)^2 \Delta^2 + \zeta^2\right)^{\frac{1}{2}}.$$

Let us also assume that $0 < \zeta \leq \Delta$. Then in the region (5.5) and for the given values of θ we have:

$$\left(r+\frac{1}{2}\right)\Delta \leqslant |\theta + B| \leqslant \Delta\left(\left(r+\frac{1}{2}\right)^2 + 1\right)^{\frac{1}{2}}. \tag{5.7}$$

Let us now return to (5.4). By (5.7), expression (4.6) is replaced by

$$L_\theta^{(m)} \frac{1}{(\theta+B)^N} = H \frac{\Gamma(N+m)}{\left(\left(r+\frac{1}{2}\right)\Delta\right)^{m+N}}, \tag{5.8}$$

and estimate (4.3) is preserved. Further, for $k > 0$ we have:

$$L_\theta^{(k)} \frac{1}{(\theta+A)^N(\theta+B)^N} = \frac{(-1)^k N(N+1)\ldots(N+k-1)}{(\theta+B)^{N+k}(\theta+A)^N} + \Upsilon_k, \tag{5.9}$$

where

$$\Upsilon_k = H\sum_{m=1}^{k} C_k^m \frac{\Gamma(N+m)}{\zeta_1^{m+N}} \cdot \frac{\Gamma(N+k-m)}{|\theta+B|^{k-m+N}} = H\frac{\Gamma(N+k)}{|\theta+B|^{k+N-1}} \frac{\left(r+\frac{1}{2}\right)\Delta}{\zeta_1}, \tag{5.10}$$

if $(r+1/2)\Delta < \zeta_1/2$. But if $(r+1/2)\Delta \geq \zeta_1/2$, then from (5.9) and the second of the three expressions (5.10) we deduce:

$$L_\theta^{(k)} \frac{1}{(\theta+B)^N(\theta+A)^N} = H\frac{\Gamma(N+k)}{\zeta_3^k}, \tag{5.11}$$

where $\zeta_3 = \zeta_1/2$.

Further, returning to (5.9) and using (5.7), we find:

$$\left| L_\theta^{(k)} \frac{1}{(\theta+A)^N(\theta+B)^N} \right| > c_0 \frac{\Gamma(N+k)}{|\theta+B|^{N+k}} \tag{5.12}$$

for

$$\left(r+\frac{1}{2}\right)\Delta < \zeta_4 \xi_1, \tag{5.13}$$

where $c_0 > 0$ is a constant.

In addition, if the number $i\zeta$ (see (3.1)) is sufficiently small, for instance $0 < \zeta < \Delta/10k^2$, then taking account of (3.3) we obtain for even $k+N$:

$$\operatorname{Re} L_\theta^{(k)} \frac{1}{(\theta+A)^N(\theta+B)^N} > \frac{1}{2} c_0 \frac{\Gamma(N+k)}{|\theta+B|^{N+k}}, \tag{5.14}$$

and if $(r+1/2)\Delta \geq \zeta_4\zeta_2$, then, analogously to expression (5.11), we get:

$$L_\theta^{(k)} \frac{1}{(\theta+A)^N(\theta+B)^N} = H\frac{\Gamma(N+k)}{(\zeta_4\zeta_1)^k}. \tag{5.15}$$

Let us now return to quantity (5.2). By the definition of number D_0 we must have:

$$\iint\limits_{D_0\leqslant B(\xi,\eta)\leqslant D_0+\Delta} \eta^{n_1-2}\Phi(\xi,\eta)\,d\xi\,d\eta > \beta_0(\Delta) > 0, \tag{5.16}$$

where $\beta_0(\Delta)$ is a positive number depending on Δ. We choose R so large – and therefore Δ so small – that $\Delta/2 < \zeta_4\zeta_1$, so that condition (5.13) is fulfilled for $r = 0$. Then from (5.12), (5.14) and (5.16) we obtain:

$$\operatorname{Re} L_\theta^{(k)} F_{00}(\theta) > \frac{c_0\beta_0(\Delta)}{2}\frac{\Gamma(N+k)}{\left(\Delta\sqrt{\frac{5}{4}}\right)^{N+k}}. \tag{5.17}$$

Taking into account (5.17), (5.14) and (5.13) and assuming r_0 to be the greatest integer subject to condition (5.13) , we find:

$$\operatorname{Re} L_\theta^{(k)}(F_{00}(\theta) + F_{01}(\theta) + \ldots + F_{0,r_0}(\theta)) > \frac{c_0\beta_0(\Delta)}{2}\frac{\Gamma(N+k)}{\left(\Delta\sqrt{\frac{5}{4}}\right)^{N+k}}. \tag{5.18}$$

By (5.15), we have

$$\operatorname{Re} L_\theta^{(k)}(F_{0,r_0+1}(\theta) + F_{0,r_0+2}(\theta) + \ldots + F_{0,M-1}(\theta)) = H\frac{\Gamma(N+k)}{(\zeta_4\zeta_1)^k}. \tag{5.19}$$

Taking account also of (4.11) and comparing (5.9), (4.11) and (5.18), we conclude that if Δ is chosen sufficiently small compared with $\zeta_4\zeta_1$ and ζ_2 (which implies choosing a sufficiently large R)– for instance, if

$$\Delta \leqslant \frac{1}{1000}\min(\zeta_4, \zeta_1, \zeta_2),$$

and then, Δ having been fixed, k is chosen sufficiently large and with the same parity as N, then we get:

$$\operatorname{Re} L_\theta^{(k)} I_0(\theta) > \frac{1}{4}c_0\beta_0(\Delta)\frac{\Gamma(N+k)}{\left(\Delta\sqrt{\frac{5}{4}}\right)^{N+k}} \tag{5.20}$$

for $\theta = D_0 + \Delta 2 + i\Delta/10k^2$, and this inequality, as we shall see later, leads to a contradiction.

6. Let us now consider the case that the numbers n_1 and n_2 have the same parity, and show that an inequality analogous to (5.20) holds. In this case $N = (n_1+n_2-1)/2$ is half of an odd number, and the expression

$$\frac{1}{((\theta+A)(\theta+B))^N},$$

occuring in (3.2) will have branch-points.

Retaining our previous notation, let us consider a sequence of large integers $\{k\}$ and for each k let us choose, as before,

$$\theta = -D_0 + \frac{\Delta}{2} + i\frac{\Delta}{10k^2}. \tag{6.1}$$

For $\theta \in \Lambda$, assuming $\operatorname{arc}\theta = 0$ for $\theta > 0$, we obtain using (3.3) and (5.6):

$$\frac{1}{[(\theta + A)(\theta + B)]^N} = i^N \frac{1}{[(-\theta - A)(\theta + B)]^N}. \tag{6.2}$$

In addition, for the values of θ determined by (6.1),

$$\operatorname{arc}(-\theta - A) = O\left(\frac{1}{k^2}\right), \tag{6.3}$$

and for the values of $B = B(\xi, \eta)$ for $(\xi, \eta) \in \Omega_0$

$$\operatorname{arc}(\theta + B) = O\left(\frac{1}{k^2}\right). \tag{6.4}$$

Thus for sufficiently large k we have:

$$\frac{1}{[(\theta + A)(\theta + B)]^N} = i^N \frac{1}{(-\theta - A)^N} \frac{1}{(\theta + B)^N}, \tag{6.5}$$

and we can apply here the arguments of §§ 4 and 5 with the following minor changes: instead of (4.3) we write the estimate

$$L_\theta^{(m)} \frac{1}{(-\theta - A)^N} = H \frac{\Gamma(N+m)}{\zeta_1^m}, \tag{6.6}$$

which follows from the well-known formula

$$\frac{\Gamma(N+m)}{\Gamma(N)} = N(N+1)\ldots(N+m-1).$$

Formula (4. 6) remains unchanged; the following arguments proceed also in the same way, where $\theta + A$ should be replaced by $-\theta - A$. Instead of formula (5.9) we obtain:

$$i^{-N} L_\theta^{(k)} \frac{1}{(-\theta - A)^N (\theta + B)^N} = \frac{(-1)^k N(N+1)\ldots(N+k-1)}{(\theta + B)^{N+k}(-\theta - A)^N} + \gamma_k \tag{6.7}$$

with the same estimate (5.10) for γ_k. Consequently,

$$(-1)^k i^{-N} \operatorname{Re} \frac{1}{(-\theta - A)^N (\theta + B)^N} > \frac{c_0}{2} \frac{\Gamma(N+k)}{|\theta + B|^{N+k}} \tag{6.8}$$

for $(r + 1/2)\Delta < \zeta_4 \zeta_1$.

The following arguments proceed as in § 5, except that in inequalities (5.17), (5.18) and (5.20) one has to replace the quantity

$$\operatorname{Re} L_\theta^{(k)}(\ldots) \quad \text{by} \quad \operatorname{Re}(-1)^k i^{-N} L_\theta^{(k)}(\ldots).$$

As a result we arrive once again at a formula of the type (5.20):

$$(-1)^k i^{-N} \operatorname{Re} L_\theta^{(k)} I_0(\theta) > \frac{c_0 \beta_0(\Delta)}{4} \frac{\Gamma(N+k)}{\left(\Delta\sqrt{\frac{5}{4}}\right)^{N+k}} \tag{6.9}$$

for

$$\theta = -D_0 + \frac{\Delta}{2} + \frac{i\Delta}{10k^2}. \tag{6.10}$$

From (5.20) and (6.9) we get for the values (6.10):

$$|L_\theta^{(k)} I_0(\theta)| > \frac{c_0\beta_0(\Delta)}{4} \frac{\Gamma(N+k)}{\left(\Delta\sqrt{\frac{5}{4}}\right)^{N+k}} \tag{6.11}$$

this time already for arbitrary values of n_1, n_2 subject to the condition $\max(n_1, n_2) \geq 3$.

Let us now consider the result of applying the operator $L_\theta^{(k)}$ to the right-hand side of (3.2), i. e., the expression

$$L_\theta^{(k)} \alpha\theta^{-\frac{n_2}{2}} (1+\theta)^{-N+\frac{1}{2}}.$$

Applying the considerations explained in the begining of this section, we find

$$L_\theta^{(k)} \alpha\theta^{-\frac{n_2}{2}} (1+\theta)^{-N+\frac{1}{2}} = H \frac{\Gamma(N+k)}{\left(D_0 - \frac{\Delta}{2} - 1\right)^{N+k}}. \tag{6.12}$$

If Δ is chosen sufficiently small and then, for fixed Δ, k is taken sufficiently large, then (6.11) contradicts (6.12), which proves our theorem.

We remark that for the case that the sample sizes are of different parity and N is an integer, it is possible considerably to simplify and shorten the proof of the theorem, using the argument of note [4]. However, this reasoning does not lead to the desired result in the case of samples with equal parity.

BIBLIOGRAPHY

[1] E. Lehmann, *Testing statistical hypotheses*, Wiley, New York, and Chapman Hall, London, 1959; Russian transl., "Nauka", 1964. MR 21 #6654.

[2] Ju. V. Linnik, *Application of complex variable to the study of the Behrens-Fisher problem*, Dokl. Akad. Nauk SSSR **149** (1963), 252–255 = Soviet Math. Dokl. 4 (1963), 359–362. MR 27 #4306.

[3] ———, *Complex variables in problems with nuisance parameters and finite rank sufficient statistics*, Dokl. Akad. Nauk SSSR **149** (1963), 1026–1028 = Soviet Math. Dokl. **4** (1963), 512–514. MR 27 #3041.

[4] Ju. V. Linnik and O. V. Šalaevskiĭ, *On the analytic theory of tests for the Behrens-Fisher problem*, Dokl. Akad. Nauk SSSR **150** (1963), 26–27 = Soviet Math. Dokl. **4** (1963), 580–581. MR 27 #3042.

Translated by:
M. Machover

ANALYSIS OF LARGE DEVIATIONS IN BOUNDARY-VALUE PROBLEMS WITH ARBITRARY BOUNDARIES. I

A. A. BOROVKOV

In this paper we shall investigate the probabilities of large deviations in a problem whose formulation goes back to Kolmogorov's [1] well-known work. Let $s_0 = 0$, $s_1 = \xi_1$, $s_2 = \xi_1 + \xi_2, \cdots$ be a sequence of sums of independent, equally distributed, random variables $\xi_1, \xi_2, \cdots$, which fulfill the two conditions of the Cramér type to be formulated below, and the relations $\mathbf{M}\xi_k = 0$, $\mathbf{D}\xi_k = 1$. Further, let $g_1(t) > 0 > g_2(t)$ be two functions defined on the closed interval $[0, 1]$; let $g(g_1, g_2)$ be the strip bounded by the curves $g_1(t)$ and $g_2(t)$, and let $x = x(n)$ be an increasing function of n growing not slower than $\sqrt{n}$ and not faster then n. In many applications of the theory of probability, in particular in mathematical statistics (sequential analysis, nonparametric methods, etc.) there arise problems of determining the probabilities of events related to the question of whether a random trajectory made up by a sequence of points with coordinates $(k/n, s_k/x)$, $k = \overline{0, n}$, will fall outside the limits of the strip $g(g_1, g_2)$. Let η_g be the time of the first exit from strip g: $\eta_g = k$ if for the first time $s_k/x \geq g_1(k/n)$ or $s_k/x \leq g_2(k/n)$, and let χ_g be the unnormalized magnitude of the first jump across the boundary of strip g: $\chi_g < y$ if the distance from s_{η_g}/x to the boundary of domain g is less than y/x. We shall investigate the asymptotic behavior of the probabilities of the following events:

$$
\begin{aligned}
\mathcal{B}^k_{gx}\mathcal{X}_y &= \{\eta_g = k,\ \ \chi_g < y\}, \qquad 0 < k \leqslant n;\\
\mathcal{C}_{gx}\mathcal{X}_y &= \{\eta_g \leqslant n,\ \chi_g < y\};\\
\mathcal{C}_{gx}\mathcal{X}_y\overline{\mathcal{A}}^n_{ux} &= \{\eta_g \leqslant n,\ \chi_g < y,\ s_n < ux\}.
\end{aligned}
\tag{1}
$$

Limiting values of the probabilities of the events $\mathcal{C}_{gx} = \{\eta_g \leq n\}$ and $\mathcal{C}_{gx}\overline{\mathcal{A}}^n_{ux} = \{\eta_g \leq n,\ s_n < ux\}$ for the case $x = \sqrt{n}$ (normal deviation) have been found in [1], [2], under very general assumptions on the distribution of ξ_k and for a certain smoothness of the functions g_1 and g_2. A great number of investigations are related to various refinements of those results (e.g., see [3–6]).

The present work is devoted to the problem of large deviations, when $x/\sqrt{n} \to \infty$. We shall limit ourselves to the nonlatticed case under the following conditions.

I. The distribution function $F(t) = \mathbf{P}\{\xi_k < t\}$ has a nonzero absolutely continuous component.

II. $F(t)$ for $t \to -\infty$, and $1 - F(t)$ for $t \to \infty$, decrease exponentially.

All that is set forth below is also applicable to the latticed case. The peculiarity of the latticed case is in the appearance of some arithmetical difficulties in the arguments, and in the possibility of obtaining by the same means local theorems for the magnitude of the jump and the position of the last sum.*

Furthermore, it is sufficient, in problems of large deviations, to consider the case of one boundary, since the desired asymptotics are completely determined by the asymptotics of the probabilities of events (1), where for the strips we take the domains $g^1 = g(g_1, -\infty)$ and $g^2 = g(\infty, g_2)$. Thus in the equality

$$\mathbf{P}(\mathscr{C}_{gx}\mathscr{X}_y) = \mathbf{P}(\mathscr{C}_{g^1x}\mathscr{X}_y) + \mathbf{P}(\mathscr{C}_{g^2x}\mathscr{X}_y) - \mathbf{P}(\mathscr{C}_{g^1x}\mathscr{C}_{g^2x}\mathscr{X}_y)$$

the order of smallness of the last term, as $x/\sqrt{n} \to \infty$, exceeds that of the sum of the two preceding ones. This natural circumstance will become clear after the discussion below. Henceforth, consequently, by strip g we shall mean the strip $g(g, -\infty)$, denoting by the letter g both the strip and the function $g_1(t) = g(t) > 0$.

The basis for the asymptotic analysis to be given below is paper [6] which contains, in particular, the solution to the problem for rectilinear boundaries. The basic results of that analysis are stated in Theorems 1–9. The conditions imposed there on the function g have, as a rule, a "local" character, and thus differ essentially from the conditions imposed on the function g in [1], for example, which describe the behavior of the functions on the whole closed interval $[0, 1]$.

In the investigation of the probabilities of the first two events in (1), for the case $x \sim n$, Condition II can be weakened to require only that the function $1 - F(t)$ decreases exponentially for $t \to \infty$. If one sets

$$\varphi(\lambda) = \mathbf{M}e^{-\lambda\xi_k}, \quad \lambda_- = \inf\{\lambda : \varphi(\lambda) < \infty\}, \quad \lambda_+ = \sup\{\lambda : \varphi(\lambda) < \infty\},$$

Condition II can be written in the form

$$\lambda_+ > 0 > \lambda_-,$$

and its weakened version in the form

$$\lambda_- < 0.$$

In the first section of the present article we shall consider the case $x = n(1 + \kappa)$ where $\kappa = \kappa(n) \to 0$ for $n \to \infty$.

---

* In mathematical statistics one is also interested in the behavior of the conditional probability of the event $\{\eta_g \leq n / s_n = ux\}$ for the case when ξ_k are latticed and $\mathbf{P}(s_n = ux) > 0$. A detailed investigation of this problem (basically the case of rectilinear boundaries) can be found in [7]–[9]. The present paper contains all that is necessary to solve the stated problem for the case of large deviations and arbitrary boundaries.

§1. Preliminary analysis

1. **Properties of the function $\mathfrak{m}(\alpha)$.** The function $\phi(\lambda)$ is defined on the whole interval $[\lambda_-, \lambda_+]$. By the well-known Abelian theorems, $\phi(\lambda) \to \phi(\lambda_-)$ $(\phi(\lambda) \to \phi(\lambda_+))$ for $\lambda \downarrow \lambda_-$ $(\lambda \uparrow \lambda_+)$. Hence for any real α there always exists a unique point $\lambda(\alpha)$ of the interval $[\lambda_-, \lambda_+]$ at which the function $e^{\lambda\alpha}\phi(\lambda)$ reaches its minimum. Let

$$\mathfrak{m}(\alpha) = \min_{\lambda \in [\lambda_-, \lambda_+]} e^{\lambda\alpha}\varphi(\lambda) = e^{\alpha\lambda(\alpha)}\varphi(\lambda(\alpha)). \tag{2}$$

As is shown in [6], the functions $\mathfrak{m}(\alpha)$ and $\lambda(\alpha)$ are analytic inside, and the function $\mathfrak{m}(\alpha)$ is also continuous up to and including the boundaries, of the domain

$$\alpha \in [\alpha_+, \alpha_-] = \left[-\lim_{\lambda\uparrow\lambda_+} \frac{\varphi'(\lambda)}{\varphi(\lambda)},\quad -\lim_{\lambda\downarrow\lambda_-} \frac{\varphi'(\lambda)}{\varphi(\lambda)}\right].$$

Outside the interval (α_+, α_-), the function $e^{\lambda\alpha}\phi(\lambda)$ reaches its minimum on the boundary of the closed interval $[\lambda_-, \lambda_+]$, so that $\lambda(\alpha) = \lambda_-$ for $\alpha_- \le \alpha < \infty$, and $\lambda(\alpha) = \lambda_+$ for $\alpha_+ \ge \alpha > -\infty$. For such values of α we have respectively $\mathfrak{m}(\alpha) = e^{\alpha\lambda_\pm}\phi(\lambda_\pm)$ if $\lambda_\pm$ are finite. If $\lambda_\pm = \pm\infty$, then for $\alpha \lessgtr \alpha_\pm$ we have $\mathfrak{m}(\alpha) = 0$. Let us note that the version $\alpha_- < \infty$, $\lambda_- = -\infty$ $(\alpha_+ > -\infty, \lambda_+ = \infty)$ corresponds to the case when the ξ_k are bounded above (below) by the value of $\alpha_-(\alpha_+)$. Moreover, $\mathfrak{m}(\alpha_\pm) = \mathbf{P}(\xi_k = \alpha_\pm)$, so that if this probability is positive, then $\mathfrak{m}(\alpha)$ has a discontinuity at the point $\alpha_\pm$.

Therefore the function $\mathfrak{m}(\alpha)$ is defined on the whole real axis and is made up, in general, of three pieces of regular functions. Differentiating equality (2) with respect to α $(\alpha_+ < \alpha < \alpha_-)$ and making use of the identity $\alpha\phi(\lambda(\alpha)) + \phi'(\lambda(\alpha)) \equiv 0$, we find

$$\mathfrak{m}'(\alpha) = \lambda(\alpha)\mathfrak{m}(\alpha). \tag{3}$$

Since $\mathfrak{m}(0) = 1$ for $\mathbf{M}\xi_k = 0$, then, solving equation (3), we find that for $\alpha_+ < \alpha < \alpha_-$

$$\mathfrak{m}(\alpha) = \exp\int_0^\alpha \lambda(\alpha)\, d\alpha. \tag{4}$$

But for finite $\lambda_\pm$, the function $\mathfrak{m}(\alpha)$ also satisfies equation (3) outside the closed interval $[\alpha_+, \alpha_-]$. It is also continuous at the points $\alpha_\pm$ along with its first derivatives. Therefore (3) and representation (4) are valid for all real α. The remark made above about the behavior of the function $\mathfrak{m}(\alpha)$ permits us to check that formula (4) continues to hold for all α even if one of the numbers $\lambda_\pm$ is infinite. From (4) and from the monotonic decrease of $\lambda(\alpha)$, it follows that the values of function $\mathfrak{m}(\alpha)$ decrease as we move away from the point $\alpha = 0$, $\mathfrak{m}(\alpha) \to 0$ for

$|\alpha| \to \infty$.

2. **First limit theorem for logarithms.** Let (a_n) and (b_n) $(n = 1, 2, \cdots)$ be two sequences of numbers. We shall write $a_n \sim b_n$ if $\lim_{n\to\infty} a_n/b_n = 1$, and $a_n \overset{c}{\sim} b_n$ if $\lim_{n\to\infty} a_n/b_n = \text{const} > 0$. Denote by $\mathcal{A}^k_{gx}$ the event $\{s_k \geq xg(k/n)\}$ and let $\mathcal{A}^k_{gx} = \mathcal{A}_{\alpha x}$, $\mathcal{C}_{gx} = \mathcal{C}_{\alpha x}$, if $g(t) = \alpha$ for all $t \in [0, 1]$. Further let $\tau = \alpha x/n$ and let $\mathcal{D}^n_{\alpha x}$ denote one of the events $\mathcal{A}^n_{\alpha x}$, $\mathcal{C}_{\alpha x}$. In [6] it is shown that for any $0 < \alpha < \alpha_-$

$$\mathbf{P}(\mathcal{D}^n_{\alpha x}) \overset{c}{\sim} \frac{\mathfrak{m}^n(\tau)}{\sqrt{n}}.$$

It follows from this, in particular, that

$$\ln \mathbf{P}(\mathcal{D}^n_{\alpha x}) \overset{c}{\sim} n \ln \mathfrak{m}(\alpha) = n \int_0^\alpha \lambda(\alpha)\, d\alpha. \tag{5}$$

We begin by proving the following proposition (also see [10]).

Theorem 1. *Relation* (5) *holds for any* $\alpha \geq 0$ *for which* $\mathbf{P}(\mathcal{A}^n_{\alpha x}) > 0$.

Proof. a) If Condition II holds and for any $n \geq 1$, αx and λ, $\lambda \in [\lambda_-, \lambda_+]$, we have the inequality of the Čebyšev type

$$\mathbf{P}(s_n \geqslant \alpha x) \leqslant e^{\alpha x \lambda} \varphi^n(\lambda),$$

from which follows

$$\mathbf{P}(s_n \geqslant \alpha x) \leqslant \left[\mathfrak{m}\left(\frac{\alpha x}{n}\right)\right]^n. \tag{6}$$

By the Kolmogorov inequality (see [11]), we have

$$\mathbf{P}(\mathcal{C}_{\alpha x}) \leqslant 2\mathbf{P}(\mathcal{A}^n_{\alpha x - \sqrt{2n}}).$$

Hence, from (6)

$$\limsup_{n\to\infty} \frac{\ln \mathbf{P}(\mathcal{D}^n_{\alpha x})}{n} \leqslant \mathfrak{m}(\alpha).$$

b) We now find a lower estimate for $\mathbf{P}(\mathcal{A}^n_{\alpha x})$. It is sufficient to consider the case $\alpha_- < \alpha < \infty$. First let $\lambda_- > -\infty$. Then $\mathfrak{m}(\alpha)$ is positive everywhere, and the finiteness of α_- implies the finiteness of $\phi(\lambda_-)$. The proof that for any fixed α

$$\liminf_{n\to\infty} \frac{\ln \mathbf{P}(\mathcal{D}^n_{\alpha x})}{n} \geqslant \mathfrak{m}(\alpha),$$

will be complete if for any $\epsilon > 0$ we can construct random variables ξ_k^* such that

$$1)\ \mathbf{P}(s_n \geqslant \alpha x) \geqslant \mathbf{P}(s_n^* \geqslant \alpha x);\quad 2)\ \alpha_-^* = \infty;\quad 3)\ \mathfrak{m}^*(\alpha) > e^{-\varepsilon}\mathfrak{m}(\alpha).$$

Here all the symbols with an asterisk retain their previous meaning, but apply to the random variables ξ_k^* which are, by definition, independent and equally distributed. But conditions 1)–3) can be fulfilled if, for example, we set

$$F^*(t) = \mathbf{P}(\xi_k^* < t) = \begin{cases} F(t), & t \leqslant N, \\ \max(1 - e^{(\lambda_- - \delta)t}, \ F(t)), & t > N, \end{cases} \tag{7}$$

where N and δ are parameters. Indeed, let us check the first condition.

We shall write $\xi_k^* \underset{\text{prob}}{\leq} \xi_k$, if all t

$$1 - F^*(t) \leq 1 - F(t). \tag{8}$$

If (8) holds, then it is easy to verify that $\xi_1^* + \xi_3 \underset{\text{prob}}{\leq} \xi_1 + \xi_2$. Applying this inequality twice, we find $\xi_1^* + \xi_2^* \underset{\text{prob}}{\leq} \xi_1 + \xi_2$ and, therefore, for any $n \geq 1$ we have $s_n^* \underset{\text{prob}}{\leq} s_n$. Since function (7) satisfies (8), condition 1) has been proved.

Condition 2) follows from $\lambda_-^* \leq \lambda_- - \delta$ and $\phi^*(\lambda_- - \delta) = \infty$. The first relation follows from the equality

$$\int_N^\infty e^{-\lambda t}\, dF^*(t) = (1 - F^*(N))\, e^{-\lambda N} - \lambda \int_N^\infty (1 - F^*(t))\, e^{-\lambda t}\, dt, \quad \lambda > \lambda_- - \delta.$$

To verify the second relation, consider all intervals $\omega_1, \omega_2, \cdots$, situated to the right of point N, on which $1 - F(t) > e^{(\lambda_- - \delta)t}$ Since $\int_{-\infty}^{\infty} (1 - F(t))\, e^{-\lambda t}\, dt$ diverges for any $\lambda < \lambda_-$, then either there exists among the ω_i a rightmost interval of the form $(\cdot, \infty)$, or else there are infinitely many intervals ω_i and among them there will be some of arbitrary length. From this follows

$$\int_N^\infty (1 - F^*(t))\, e^{-(\lambda_- - \delta)t}\, dt \geqslant \sum_i \int_{\omega_i} dt = \infty.$$

Integrating the left-hand side of this inequality by parts (between finite limits), we arrive at the divergence of $\int_N^\infty e^{-(\lambda_- - \delta)t}\, dF^*(t)$.

Let us show, finally, that for any $\epsilon > 0$ we can select parameters $N > 0$ and $\delta > 0$ in such a way that condition 3) is fulfilled. We have

$$\frac{\mathfrak{m}^*(\alpha)}{\mathfrak{m}(\alpha)} = e^{\alpha(\lambda^*(\alpha) - \lambda_-)}\, \frac{\varphi^*(\lambda^*(\lambda))}{\varphi(\lambda_-)}.$$

Let us select and fix a number N such that the following inequality holds:

$$\int_{-\infty}^{N} e^{-\lambda_- t}\, dF(t) > \varphi(\lambda_-) e^{-\frac{\varepsilon}{4}}.$$

Then

$$\frac{\mathfrak{m}^*(\alpha)}{\mathfrak{m}(\alpha)} \geqslant \frac{\varphi^*(\lambda^*(\alpha))}{\varphi(\lambda_-)} e^{-\delta\alpha} > \frac{\int_{-\infty}^{N} e^{-\lambda^*(\alpha)t}\, dF(t)}{\int_{-\infty}^{N} e^{-\lambda_- t}\, dF(t)} e^{-\frac{\varepsilon}{4}-\delta\alpha}, \tag{9}$$

and there remains to pick $\delta < \epsilon/2\alpha$ so small that the ratio of the integrals in (9) is not less then $e^{-\epsilon/4}$.

In the case $\lambda_- = -\infty$, $\alpha_- < \infty$, the values of ξ_k are bounded above by the number α_-, and (5) is clearly satisfied for $\alpha = \alpha_-$, $\alpha x < n\alpha_-$. Q.E.D.

Note that in our case the behavior of $\lim_{n\to\infty} \ln \mathbf{P}(\mathfrak{D}^n_{\alpha x})/n$ in some interval belonging to $(0, \alpha_-)$ uniquely determines the distribution of ξ_k.

3. **Level curves in the family $\mathfrak{A}$ and their properties.** Making use of the equality $\alpha x = \tau n$, we can write

$$\ln \mathbf{P}(\mathcal{A}^n_{\alpha x}) \sim n \ln \mathfrak{m}(\tau).$$

It is natural to ask what must the level a be in order for the probabilities of the events $\mathfrak{A}^k_{an}$ and $\mathfrak{A}^n_{\tau n}$, $k < n$, to be in some sense equivalent. For instance, in order to have

$$\ln \mathbf{P}(\mathcal{A}^k_{an}) \sim \ln \mathbf{P}(\mathcal{A}^n_{\tau n}), \tag{10}$$

or equivalently

$$k \ln \mathfrak{m}\left(\frac{an}{k}\right) \sim n \ln \mathfrak{m}(\tau).$$

Consider the equation

$$t \ln \mathfrak{m}\left(\frac{a}{t}\right) = \ln \mathfrak{m}(\tau), \quad 0 < t \leqslant 1, \tag{11}$$

whose positive solution $a = a_\tau(t)$ (if there exists) for $t = k/n$ satisfies (10). For simplicity of formulation, we shall also call solutions of (11) those values of a for which

$$t \ln \mathfrak{m}\left(\frac{a}{t}\right) \geqslant \ln \mathfrak{m}(\tau), \quad t \ln \mathfrak{m}\left(\frac{a+0}{t}\right) \leqslant \ln \mathfrak{m}(\tau). \tag{12}$$

These solutions which satisfy (12) but are not solutions in the usual sense will be called boundary solutions.

Lemma 1. *Equation* (11) *for any* $0 < t \leq 1$, $0 < \tau < \bar{\alpha} = \sup(t\colon F(t) < 1)$ *has a unique positive solution* $a_\tau(t) \geq \tau t$. *The function* $a_\tau(t)$, *defined for every fixed* τ *in* *the domain* $t \in (0, 1]$, *increases and results, in general, from the fusion of two regular functions. The fusion takes place and originates at the point* $t_\tau = \min(1 - \ln \mathfrak{m}(\tau)/\ln \mathfrak{m}(\alpha_-))$, *if* $0 < t_\tau < 1$. *For* $t \in (0, t_\tau)$ *we have*

$$\mathfrak{a}_\tau(t) = \frac{\ln \mathfrak{m}(\tau) - t \ln \varphi(\lambda_-)}{\lambda_-}, \tag{13}$$

where for $\lambda_- = -\infty$ *the ratio* $\ln \phi(\lambda_-)/\lambda_-$ *should be replaced by the boundary value* α_-. *The straight line* (13) *is the extension of the tangent to the curve* $\mathfrak{a}_\tau(t)$ *at the point* $t_\tau - 0$. *The solution* $\mathfrak{a}_\tau(t)$ *is a boundary solution provided* $t < t_\tau$, ξ_k *are bounded on the right,* $\mathbf{P}(\xi_k = \alpha_-) > 0$.

Proof. The existence and uniqueness of the solution follows from the fact that $t \ln \mathfrak{m}(\mathfrak{a}/e)$ is a nonincreasing function of $\mathfrak{a}$, $\lim_{\alpha\to\infty} \ln \mathfrak{m}(\alpha) = -\infty$, and for $\mathfrak{a} = \tau t$

$$t \ln \mathfrak{m}\left(\frac{\mathfrak{a}}{t}\right) = t \ln \mathfrak{m}(\tau) \geqslant \ln \mathfrak{m}(\tau) > -\infty.$$

Now let $t_\tau < 1$ and $t \in (t_\tau, 1]$. Then

$$t \ln \mathfrak{m}(\alpha_-) > \ln \mathfrak{m}(\tau) \tag{14}$$

and hence $\mathfrak{a}_\tau(t) < \alpha_- t$. The regularity of $\mathfrak{a}_\tau(t)$ in the domain under consideration is a consequence of the implicit function theorem and the last inequality.

Assume further that at least one of the relations $\alpha_- = \infty$, $\lambda_- > -\infty$ holds. Then $\mathfrak{m}(\alpha)$ is continuously differentiable on the whole positive semiaxis, and so is the function $\mathfrak{a}_\tau(t)$ in the interval $(0, 1)$. Differentiating (11) with respect to t and making use of (3), we find

$$\frac{\partial}{\partial t} \mathfrak{a}_\tau(t) = \mathfrak{a}_\tau'(t) = -\frac{\ln \varphi\left(\lambda\left(\frac{\mathfrak{a}_\tau(t)}{t}\right)\right)}{\lambda\left(\frac{\mathfrak{a}_\tau(t)}{t}\right)} > 0. \tag{15}$$

For $t < t_\tau$ (if $t_\tau > 0$, then $\mathfrak{m}(\alpha_-) > 0$ and, hence, $\alpha_- < \infty$), by virtue of the inequality which is the reverse of (14), $\mathfrak{a}_\tau(t) = \alpha_- t$. Hence equation (11) takes the form

$$\ln \mathfrak{m}(\tau) = t\left(\lambda_- \frac{\mathfrak{a}_\tau(t)}{t} + \ln \varphi(\lambda_-)\right),$$

equivalent to (13).

There remains to consider the case $\alpha_- < \infty$, $\lambda_- = -\infty$. For $\mathfrak{m}(\alpha_-) = 0$ we have $t_\tau = 0$. If $\mathfrak{m}(\alpha_-) > 0$, the function $\mathfrak{a}_\tau(t) = \alpha_- t$ is a boundary solution for $t < t_\tau$ and moreover

$$\lim_{t \downarrow t_\tau} \mathfrak{a}_\tau'(t) = -\lim_{\lambda \to -\infty} \frac{\ln \varphi(\lambda)}{\lambda} = \alpha_-.$$

Since the case $\alpha_- < \infty$, $\lambda_- = -\infty$, $\mathfrak{m}(\alpha_-) > 0$ is unique when $\mathfrak{m}(\alpha)$ has a discontinuity for $\alpha > 0$, the proof of the lemma is complete.

Let us describe a series of important properties of the functions $a_\tau(t)$. In order to consider them defined on the whole closed interval $[0, 1]$, we set $a_\tau(0) = \lim_{t\downarrow 0} a_\tau(t)$. As τ varies, the one-parameter family $\mathfrak{A}$ of curves $a_\tau(t)$ fills the domain situated above the closed interval $[0, 1]$ and bounded above by the straight line $a = \bar{a}t$. The value of t_τ is the abscissa of the point of intersection of the curve $a_\tau(t)$ with the straight line $a = \alpha_- t$, which is the line of fusion of the regular pieces. The family is characterized by a remarkable property arising from (15): all lines of the family intersect the straight line $a = ct$, $0 < c < \bar{a}$, under the same angle, whose tangent is equal to $a'_c(1)$. Hence all the straight lines in $\mathfrak{A}$ situated above a particular one can be obtained from it by a similarity transformation. In other words, the family of curves $ua_\tau(t/u)$, where $u > 1$ is fixed, $0 < \tau < \bar{a}$, bounded to the right of the straight line $t = 1$, coincides with $\mathfrak{A}$.

Since $a_\tau(1) = \tau$ and $\lambda'(\tau) < 0$ for $\tau < \alpha_-$, it follows from (15) that

$$a_\tau(1) = a_\tau = -\frac{\ln \varphi(\lambda(\tau))}{\lambda(\tau)} > 0, \tag{16}$$

$$\frac{da_\tau}{d\tau} = \left(\tau - \frac{\ln \mathfrak{m}(\tau)}{\lambda(\tau)}\right)' = \frac{\ln \mathfrak{m}(\tau)}{\lambda^2(\tau)}\lambda'(\tau) > 0.$$

But $a_\tau(t) > \tau t$ for $t < 1$. Hence*

$$a'_\tau(t) = a_{\frac{a_\tau(t)}{t}} > a_\tau \quad (t < 1,\ \tau < \alpha_-).$$

Hence, making use of a similarity transformation, we find for any two points $t_1 < t_2$ in the interval $(t_\tau, 1)$ that

$$a'_\tau(t_1) > a'_\tau(t_2).$$

This means that the curves of $\mathfrak{A}$ are convex in the domain $a < \alpha_- t$.

Let us, furthermore, clarify the behavior of the curves of $\mathfrak{A}$ in the neighborhood of the straight line $a/\alpha_- = t$. Let $\alpha_- < \infty$. For $\tau < \alpha_-$ the curves $a_\tau(t)$ intersect this straight line under an angle whose tangent is

$$\lim_{\tau\uparrow\alpha_-} a_\tau = -\lim_{\lambda\downarrow\lambda_-}\frac{\ln\varphi(\lambda)}{\lambda} = \alpha_- - \lim_{\tau\uparrow\alpha_-}\frac{\ln \mathfrak{m}(\tau)}{\lambda(\tau)}.$$

This angle is less than $\arctan \alpha_-$, if λ_- is finite, and equal to $\arctan \alpha_-$ if

---

* For the family of straight lines $g(t) = a + a(t-1)$ the value $a = a_\alpha$ is "transitional": for $a = a_\alpha$ the form of the asymptotics of $\mathbf{P}(\mathfrak{C}^n_{gn})$ changes (see [6], Theorem 10). This is easily verified by making use of the identity $\phi(\lambda(a)) \exp a_\alpha \lambda(a) = 1$.

$\lambda_- = -\infty$. In the last case the point of contact is at the origin if and only if $\mathfrak{m}(\alpha) = \mathbf{P}(\xi_k = \alpha_-) = 0$.

For $\alpha_- = \infty$ the classification of the behavior of the curves of $\mathfrak{A}$ is determined by the following statement. $\alpha'_\tau(+0)$ is finite and $\alpha_\tau(0)$ is positive if and only if $\phi(\lambda_-)$ and λ_- are correspondingly finite. We begin with the derivative. The sufficiency of the condition $\phi(\lambda_-) < \infty$ is clear from (16). To prove the necessity, we first show that the finiteness of $\lim_{\lambda \downarrow \lambda_-} \ln \phi(\lambda)/\lambda$ leads to the finiteness of λ_-. Indeed it follows from the unboundedness of α_- that for any N there exists an $\epsilon > 0$ such that $\ln \phi(\lambda) > \ln \epsilon e^{-\lambda N}$. Assuming the unboundedness of λ_-, we will find that the limit

$$\lim_{\lambda \downarrow \lambda_-} \frac{\ln \varphi(\lambda)}{\lambda} \geqslant N$$

and is therefore infinite. The finiteness of $\phi(\lambda_-)$ follows from the finiteness of the limit considered and of λ_-. The equivalence of the conditions $\alpha_\tau(0) > 0$ and $\lambda_- > -\infty$ follows from considering the asymptotic behavior of both sides of the equality

$$\int_0^{\frac{\alpha_\tau(t)}{t}} \lambda(\alpha)\, d\alpha = \frac{1}{t} \ln \mathfrak{m}(\tau),$$

which is a consequence of (4) and (11). The same equality leads to

$$\alpha_\tau(0) = \frac{\ln \mathfrak{m}(\tau)}{\lambda_-}.$$

All the qualitative types of curves $\alpha_\tau(t)$ are shown in Diagram 1.

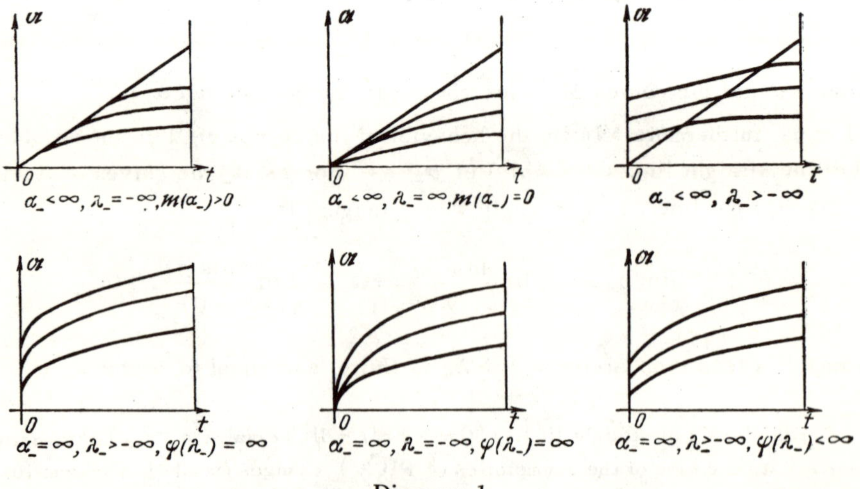

Diagram 1.

We conclude by noting that the function of two variables $a(\tau, t) = a_\tau(t)$ can be described in the same way as the solution of the Cauchy problem for the equation

$$a''_{\tau t} + \frac{a'_\tau}{a/t - a'_t} a''_{tt} = 0 \tag{17}$$

with boundary conditions $a(\tau, 1) = \tau$, $a'_t(\tau, 1) = a_\tau$. In order to derive the equation we make use of the characteristic properties of our curves: they intersect the straight line $a = t\,a(\tau_0, t_0)/t_0$ under a constant angle. Let $t_0 + \delta$ be the abscissa of the intersection of that straight line with the curve $a = a(\tau_0 + \Delta, t)$. Then

$$a'_t(\tau_0, t_0) = a'_t(\tau_0 + \Delta, t_0 + \delta). \tag{18}$$

Since

$$\delta = \Delta \frac{a'_\tau(\tau_0, t_0)}{\dfrac{a(\tau_0, t_0)}{t_0} - a'_t(\tau_0, t_0)} + O(\Delta^2),$$

then, expanding the right-hand side of (18) into a series we arrive at problem (17). With the help of the Cauchy-Kovalevskiĭ theorem [12] we can verify that this problem has a unique solution in the class of analytic functions in the domain $0 < \tau < a_-$, $t_\tau < t < 1$.*

4. **Second limit theorem for logarithms.** We consider the joint distribution of $\max_{0 \le k \le n} s_k$ and s_n. But first we introduce some notation. Let $\lambda_\pm(z)$ ($\lambda_+(z) \ge \lambda_-(z)$) be the real zeros of the function $1 - z\phi(\lambda)$ determined for $z \in [r_\pm, 1]$ respectively, where $r_\pm = \phi^{-1}(\lambda_\pm)$, so that $\lambda_\pm(r_\pm) = \lambda_\pm$. The functions $\lambda_\pm(z)$ are analytic inside the closed intervals $[r_\pm, 1]$ and have a common branch point at the point $z = 1$. Let $r = \max(r_+, r_-)$ and let $L_{\tau,\theta}(z) = \tau\lambda_-(z) - \theta\lambda_+(z) - \ln z$ be a function defined on $[r, 1]$. It is shown in [6] that if $\lim_{n\to\infty}\theta = \theta_0$ exists and if the following condition holds

$$-\alpha \frac{\varphi(\lambda_-(r))}{\varphi'(\lambda_-(r))} + \theta_0 \frac{\varphi(\lambda_+(r))}{\varphi'(\lambda_+(r))} < 1, \tag{19}$$

then at some point $z_{\tau,\theta} \in (r, 1)$

$$\min_{z \in (r,1)} L_{\tau,\theta}(z) = \ln m(\tau, \theta),$$

* In the plane (a, t) one can therefore introduce some metric induced by the negative logarithms of transition probabilities, in which the curves of $\mathfrak{A}$ are aggregates of points equidistant from the origin. The triangle inequality for that metric apparently follows from Lemma 3, §2.

is attained, and we have the relation

$$\mathbf{P}(\mathscr{C}_{\alpha x}\overline{\mathscr{A}}^n_{\alpha x-\theta n}) = \mathbf{P}(\max_{0\leqslant k\leqslant n} s_k \geqslant \alpha x,\ s_n < \alpha x - \theta n) \overset{c}{\sim} \frac{\mathfrak{m}^n(\tau, \theta)}{\sqrt{n}},$$

where $\tau = \alpha x/n$. Let $\lambda_\pm(z) = \lambda_\pm$ for $z \leq r_\pm$. It is clear then that if (19) holds, we have

$$\ln \mathfrak{m}(\tau, \theta) = \min_{0\leqslant z\leqslant 1} L_{\tau,\theta}(z)$$

and

$$\ln \mathbf{P}(\mathscr{C}_{\tau n}\overline{\mathscr{A}}^n_{(\tau-\theta)n}) \sim n \ln \mathfrak{m}(\tau, \theta). \tag{20}$$

Let

$$\mathfrak{M}(\tau, \rho, t) = t \ln \mathfrak{m}\left(\frac{\tau}{t}\right) + (1-t)\ln \mathfrak{m}\left(\frac{\rho-\tau}{1-t}\right),$$

$$\mathfrak{M}(\tau, \rho) = \sup_{0<t<1} \mathfrak{M}(\tau, \rho, t). \tag{21}$$

Theorem 2. *Relation* (20) *holds for any* $\tau > 0$, $\theta > 0$. *Moreover*, $\ln \mathfrak{m}(\tau, \theta) = \mathfrak{M}(\tau, \tau-\theta)$.

Proof. To avoid unessential stipulations, we shall restrict ourselves to those τ and θ for which $\mathfrak{M}(\tau, \tau-\theta) > -\infty$. We have

$$\mathscr{C}_{\tau n}\overline{\mathscr{A}}^n_{(\tau-\theta)n} = \bigcup_{j=1}^{n} \mathscr{A}^j_{\tau n}\overline{\mathscr{A}}^n_{(\tau-\theta)n},$$

$$\max_j \ln \mathbf{P}(\mathscr{A}^j_{\tau n}\overline{\mathscr{A}}^n_{(\tau-\theta)n}) < \ln \mathbf{P}(\mathscr{C}_{\tau n}\overline{\mathscr{A}}^n_{(\tau-\theta)n}) < \ln n + \max_j \ln \mathbf{P}(\mathscr{A}^j_{\tau n}\overline{\mathscr{A}}^n_{(\tau-\theta)n}). \tag{22}$$

Moreover, for any $\epsilon > 0$ and $1 \leq j \leq n-1$, we have

$$\mathbf{P}(\mathscr{A}^j_{\tau n} - \mathscr{A}^j_{(\tau+\varepsilon)n})\mathbf{P}(\overline{\mathscr{A}}^{n-j}_{-(\theta+\varepsilon)n}) < \mathbf{P}(\mathscr{A}^j_{\tau n}\overline{\mathscr{A}}^n_{(\tau-\theta)n}) < \mathbf{P}(\mathscr{A}^j_{\tau n})\mathbf{P}(\overline{\mathscr{A}}^{n-j}_{-\theta n}). \tag{23}$$

Let $n\delta < j < n(1-\delta)$ for some $\delta > 0$. Then for $n \to \infty$ $\mathbf{P}(\mathfrak{A}^j_{\tau n} - \mathfrak{A}^j_{(\tau+\epsilon)n}) \sim \mathbf{P}(\mathfrak{A}^j_{\tau n})$, and the logarithms of the extreme terms of inequality (23) will be respectively equivalent to

$$j \ln \mathfrak{m}\left(\frac{\tau n}{j}\right) + (n-j)\ln \mathfrak{m}\left(-\frac{(\theta+\varepsilon)n}{n-j}\right)$$

and

$$j \ln \mathfrak{m}\left(\frac{\tau n}{j}\right) + (n-j)\ln \mathfrak{m}\left(-\frac{\theta n}{n-j}\right).$$

Therefore if for $n \to \infty$ the sequence of points $-\theta/(1-j/n)$ lies in a closed domain which does not contain the point of discontinuity of the function $\mathfrak{m}(\alpha)$, making use of the arbitrariness of ϵ, we find

$$\ln \mathbf{P}(\mathcal{A}^j_{\tau n}\bar{\mathcal{A}}^n_{(\tau-\theta)n}) \sim n\mathfrak{M}(\tau, \rho, t), \tag{24}$$

where $t = j/n$, $\rho = \tau - \theta$. In the case when α_+ is the discontinuity point of $\mathfrak{m}(\alpha)$ ($\xi_k \geq \alpha_+$, $\mathbf{P}(\xi_k = \alpha_+) > 0$) and the sequence $-\theta/(1 - j/n)$ approaches α_+ unboundedly, the truth of (24) follows immediately. By virtue of (23), and (6), along with (24) for any $1 \leq j \leq n - 1$ we have

$$\ln \mathbf{P}(\mathcal{A}^j_{\tau n}\bar{\mathcal{A}}^n_{\rho n}) < n\mathfrak{M}(\tau, \rho, t). \tag{25}$$

Assume that the function $\mathfrak{M}(\tau, \rho, t)$ reaches its maximum at an interior point of the closed interval $t \in [0, 1]$. Then from (24), (25) follows

$$\max_j \ln \mathbf{P}(\mathcal{A}^j_{\tau n}\bar{\mathcal{A}}^n_{\rho n}) \sim \max_{0 \leqslant t \leqslant 1} \mathfrak{M}(\tau, \rho, t). \tag{26}$$

Let us now consider the "extreme" cases. Let, for instance, $\mathfrak{M}(\tau, \rho, +0) \geq \mathfrak{M}(\tau, \rho, t)$ for any $0 \leq t \leq 1$. In that case $\mathfrak{M}(\tau, \rho, +0)$ and λ_- are finite. Since $t \ln \mathfrak{m}(\tau/t) = \min_z (\tau\lambda_-(z) - t \ln z)$, $\mathfrak{M}(\tau, \rho, t)$ is continuous at the point $t = 0$. Therefore, taking into account (24) and (25), we again arrive at (26). One considers similarly the case where $\mathfrak{M}(\tau, \rho, t)$ reaches its maximum at $t = 1$.

Comparing (26) and (22), we find that as $n \to \infty$

$$\ln \mathbf{P}(\mathscr{C}_{\tau n}\bar{\mathcal{A}}^n_{\rho n}) \sim n \max_{0 \leqslant t \leqslant 1} \mathfrak{M}(\tau, \rho, t).$$

We now prove the equality

$$\max_t \{\min_z (\tau\lambda_-(z) - t \ln z) + \min_z (-\theta\lambda_+(z) - (1 - t) \ln z)\} = \ln \bar{\mathfrak{m}}(\tau, \theta), \tag{27}$$

equivalent to the second proposition of the theorem. Let t_0 be the point (or one of the points) at which $\mathfrak{M}(\tau, \rho, t)$ reaches its maximum. Then

$$\mathfrak{M}(\tau, \tau - \theta) \leqslant \min_z \{\tau\lambda\ (z) - t_0 \ln z - \theta\lambda_+(z) - (1 - t_0) \ln z\} = \ln \mathfrak{m}(\tau, \theta). \tag{28}$$

To obtain the reverse inequality, we shall verify that there exists at least one value $t = t^0$ for which $\min_z (\tau\lambda_-(z) - t \ln z)$ and $\min_z (-\theta\lambda_+(z) - (1 - t) \ln z)$ are reached at a common point z^0. Let $z_1(t)$ and $z_2(t)$ be the corresponding minimum points. The functions $z_1(t)$ and $-z_2(t)$ are continuous, they do not decrease on $(0, 1)$, $z_1(+0) = r_-$, $z_1(1) = \phi^{-1}(\lambda(\tau))$, $z_2(0) = \phi^{-1}(\lambda(-\theta))$, $z_2(1 - 0) = r_+$. Hence if $\phi^{-1}(\lambda(-\theta)) > r_-$ and $\phi^{-1}(\lambda(\tau)) < r_+$, then the existence of t^0, for which $z_1(t^0) = z_2(t^0)$, is obvious. Let $z_2(0) < r_-$ or $z_1(1) > r_+$. The value of $z_1(0)$ ($z_2(1)$) can be any z of the closed interval $[0, r_-]$ ($[0, r_+]$). Consequently, one can take for z^0 in the first case $z_2(0)$, in the second $z_1(1)$, so that the desired values of t^0 are 0 and 1, respectively. One can therefore write:

$$\mathfrak{M}(\tau, \tau-\theta) \geqslant \mathfrak{M}(\tau, \tau-\theta, t^0) = \min_z (\tau\lambda_-(z) - t^0 \ln z)$$
$$+ \min_z (-\theta\lambda_+(z) - (1-t^0)\ln z) = L_{\tau,\theta}(z^0) \geqslant \ln \mathfrak{m}(\tau, \theta).$$

A comparison with (28) proves (27) and the possibility to choose t^0 (if it is not unique) such that $t^0 = t_0$. Q.E.D.

5. **Level curves of the families $\mathfrak{B}$ and $\mathfrak{C}$ and their properties.** For some $1 \le k \le n-1$ let us find a level $\mathfrak{b}$ such that the probability of the events $\mathfrak{A}^k_{\mathfrak{b}n}\overline{\mathfrak{A}}^n_{\rho n}$ and $\mathfrak{C}_{\tau n}\overline{\mathfrak{A}}^n_{\rho n}$, $\rho < \tau$, be of the same logarithmic order of smallness. To that effect note that for $\mathfrak{b} \ge 0$ (24) gives

$$\ln \mathbf{P}(\mathcal{A}^k_{\mathfrak{b}n}\overline{\mathcal{A}}^n_{\rho n}) \sim k \ln \mathfrak{m}\left(\frac{\mathfrak{b}n}{k}\right) + (n-k)\ln \mathfrak{m}_-\left(\frac{(\rho-\mathfrak{b})n}{n-k}\right), \tag{29}$$

where

$$\mathfrak{m}_-(\alpha) = \begin{cases} 1 & \text{for } \alpha > 0, \\ \mathfrak{m}(\alpha) & \text{for } \alpha \leqslant 0. \end{cases}$$

This relation still holds for $\mathfrak{b} < 0$ satisfying the inequality $\mathfrak{b} > k\rho/n$. Indeed, in that case for any sufficiently small $\epsilon > 0$

$$\ln \mathbf{P}(\mathcal{A}^k_{\mathfrak{b}n}\overline{\mathcal{A}}^n_{\rho n}) \geqslant \ln \mathbf{P}(\mathcal{A}^k_{\mathfrak{b}n}\overline{\mathcal{A}}^k_{(\mathfrak{b}+\varepsilon)n}) + \ln \mathbf{P}(\overline{\mathcal{A}}^{n-k}_{(\rho-\mathfrak{b}-\varepsilon)n})$$
$$\sim k \ln \mathfrak{m}\left(\frac{(\mathfrak{b}+\varepsilon)n}{k}\right) + (n-k)\ln \mathfrak{m}\left(\frac{(\rho-\mathfrak{b}-\varepsilon)n}{n-k}\right);$$

$$\mathbf{P}(\mathcal{A}^k_{\mathfrak{b}n}\overline{\mathcal{A}}^n_{\rho n}) < \sum_{j=0}^{\left[-\frac{\mathfrak{b}}{\varepsilon}\right]} \mathbf{P}(\mathcal{A}^k_{(\mathfrak{b}+\varepsilon j)n}\overline{\mathcal{A}}^k_{(\mathfrak{b}+\varepsilon j+\varepsilon)n})\,\mathbf{P}(\overline{\mathcal{A}}^{n-k}_{(\rho-\mathfrak{b}-\varepsilon j)n}) + \mathbf{P}(\overline{\mathcal{A}}^{n-k}_{\rho n})$$
$$< 2\sum_{j=0}^{\left[-\frac{\mathfrak{b}}{\varepsilon}\right]} \mathfrak{m}^k_-\left(\frac{(\mathfrak{b}+\varepsilon j+\varepsilon)n}{k}\right)\mathfrak{m}^{n-k}\left(\frac{(\rho-\mathfrak{b}-\varepsilon j)n}{n-k}\right). \tag{30}$$

But

$$\frac{\partial}{\partial u}\left(t \ln \mathfrak{m}_-\left(\frac{\mathfrak{b}+\varepsilon+u}{t}\right) + (1-t)\ln \mathfrak{m}\left(\frac{\rho-\mathfrak{b}-u}{1-t}\right)\right)$$
$$= \begin{cases} \lambda\left(\frac{\mathfrak{b}+\varepsilon+u}{t}\right) - \lambda\left(\frac{\rho-\mathfrak{b}-u}{t}\right) & \text{for } \mathfrak{b}+\varepsilon+u \leqslant 0, \\ -\lambda\left(\frac{\rho-\mathfrak{b}-u}{1-t}\right) & \text{for } \mathfrak{b}+\varepsilon+u > 0. \end{cases} \tag{31}$$

Since $\mathfrak{b} > \rho t$, then $(\mathfrak{b} + \epsilon + u)/t > (\rho - \mathfrak{b} - u)/(1 - t)$ for any $u \geq 0$. Hence the derivative in (31) is negative and the maximum term in sum (30) corresponds to $j = 0$. Together with the preceding this shows that

$$k \ln \mathfrak{m}\left(\frac{(\mathfrak{b}+\varepsilon)\,n}{k}\right) + (n-k)\ln \mathfrak{m}\left(\frac{(\rho-\mathfrak{b}-\varepsilon)\,n}{n-k}\right) \lesssim \ln \mathbf{P}\,(\mathcal{A}^k_{\mathfrak{b}n}\overline{\mathcal{A}}^n_{\rho n})$$

$$\lesssim k \ln \mathfrak{m}\left(\frac{(\mathfrak{b}+\varepsilon)\,n}{k}\right) + (n-k)\ln \mathfrak{m}\left(\frac{(\rho-\mathfrak{b})\,n}{n-k}\right).$$

Here the relation $a_n \lesssim \mathfrak{b}_n$ means that $a_n < b_n$ for all n starting from some value. Making use of the arbitrariness of ϵ we arrive, as before (compare with §4) at (29).

Therefore the desired level $\mathfrak{b}$ must satisfy the relation (evidently it cannot lie in the domain $\mathfrak{b} < k\rho/n$)

$$k \ln \mathfrak{m}\left(\frac{\mathfrak{b}n}{k}\right) + (n-k)\ln \mathfrak{m}_{-}\left(\frac{(\rho-\mathfrak{b})\,n}{n-k}\right) \sim n\mathfrak{M}(\tau,\ \rho),$$

which leads to the equation

$$t \ln \mathfrak{m}\left(\frac{\mathfrak{b}}{t}\right) + (1-t)\ln \mathfrak{m}_{-}\left(\frac{\rho-\mathfrak{b}}{1-t}\right) = \quad (\tau,\ \rho),\ \ \rho < \tau. \tag{32}$$

However, we have already investigated the solution of that equation for the case when the second term on the left-hand side reduces to 0. We shall consider the more symmetric equation

$$t \ln \mathfrak{m}\left(\frac{\mathfrak{b}}{t}\right) + (1-t)\ln \mathfrak{m}\left(\frac{\rho-\mathfrak{b}}{1-t}\right) = \mathfrak{M}(\tau,\ \rho) \tag{33}$$

or, equivalently,

$$\mathfrak{M}(\mathfrak{b},\ \rho,\ t) = \mathfrak{M}(\tau,\ \rho) = \mathfrak{M}(\tau,\ \rho,\ t_0),$$

which gives the solution to the problem which we shall call "local" (and establish the equivalence, for instance, of the probabilities of the events $\mathfrak{A}^k_{\mathfrak{b}n}\overline{\mathfrak{A}}^n_{\rho n}\mathfrak{A}^n_{\rho n-1}$ and $\mathfrak{C}_{\tau n}\overline{\mathfrak{A}}^n_{\rho n}\mathfrak{A}^n_{\rho n-1}$). As will become clear in the sequel, the solution of the original "integral" problem is obtained from the solution of the local problem by replacing a portion of the latter by a piece of a function of known form.

Just as in §3, we define the concept of a boundary solution of equation (33) as a value of $\mathfrak{b}$ for which the left-hand side of (33) has a jump which covers the value of $\mathfrak{M}(\tau, \rho)$. Clearly, boundary solutions can exist only if there exists at least one value of $\overline{a}$ or $\underline{a}$ such that $\mathbf{P}(\xi_k = \overline{a}) > 0$, $\mathbf{P}(\xi_k > \overline{a}) = 0$, $\mathbf{P}(\xi_k = \underline{a}) > 0$, $\mathbf{P}(\xi_k < \underline{a}) = 0$.

Lemma 2. *For any $t \in (0, 1)$ and $\tau > \rho$ such that $\mathfrak{M}(\tau, \rho) > -\infty$, equation* (33) *has a unique positive solution* $\mathfrak{b}_{\tau,\rho}(t)$ *lying above the broken line* $\mathfrak{b} = \tau t/t_0$ *for* $0 \leq t \leq t_0$ *and* $\mathfrak{b} = (\rho(t-t_0) + \tau(1-t))/(1-t_0)$ *for* $t_0 \leq t \leq 1$. *The curve* $\mathfrak{b}_{\tau,\rho}(t)$, $t \in (0, 1)$, *is convex and is made up from portions of, in general, three regular functions. The derivative* $\partial\mathfrak{b}_{\tau,\rho}(t)/\partial t$ *is continuous on the whole interval* $(0, 1)$, *nonpositive for* $t > t_0$, *and nonnegative for* $t < t_0$; $\mathfrak{b}_{\tau,\rho}(t_0) = \tau$ *(see Diagram 2).*

Proof. Assume that a solution exists. Since $\rho < \tau$, it follows that $\mathfrak{b}_{\tau,\rho}(t) > \rho t$. Moreover, the following inequalities are always true:

$$\mathfrak{b}_{\tau,\rho}(t) > t\alpha_+, \quad \mathfrak{b}_{\tau,\rho}(t) > \rho - \alpha_-(1-t).$$

Indeed, assume for instance that the intersection of the domains $\mathfrak{b} > \rho t$ and $\mathfrak{b} \leq \rho - \alpha_-(1-t)$ is not empty and that there exists a solution there. Then $\rho > \alpha_-$ and λ_- is finite (from the condition $\mathfrak{M}(\tau, \rho) > -\infty$). On the stated intersection $\mathfrak{b}/t > (\rho - \mathfrak{b})/(1-t) \geq \alpha_-$ and hence there exists

$$\frac{\partial \mathfrak{M}(\mathfrak{b}, \rho, t)}{\partial \mathfrak{b}} = \lambda\left(\frac{\mathfrak{b}}{t}\right) - \lambda\left(\frac{\rho - \mathfrak{b}}{1-t}\right) = 0.$$

Therefore, for any point of the intersection, we have

$$\mathfrak{M}(\mathfrak{b}, \rho, t) = \mathfrak{M}(t\rho, \rho, t) = \ln \mathfrak{m}(\rho) > \mathfrak{M}(\tau, \rho).$$

Similarly, we verify that points of intersection of the domains $\mathfrak{b} > \rho t$ and $\mathfrak{b} \leq t\alpha_+$ cannot be solutions.

On the other hand, the solution can lie only in the domain of finiteness of $\mathfrak{M}(\mathfrak{b}, \rho, t)$. Hence it is sufficient to prove the existence and uniqueness of the solution in the domain $A(t)$ defined by the inequality

$$\mathfrak{b} > \max(\rho t, \alpha_+ t, \rho - \alpha_-(1-t)), \quad t \in (0, 1),$$

and bounded above in the case of infinite $\lambda_-(\lambda_+)$ by the inequality $\mathfrak{b} < \alpha_- t$ $((\rho - \mathfrak{b})/(1-t) \geq \alpha_+)$. In that domain

$$\frac{\mathfrak{b}}{t} > \frac{\rho - \mathfrak{b}}{1-t}, \quad \frac{\rho - \mathfrak{b}}{1-t} < \alpha_-, \quad \frac{\mathfrak{b}}{t} > \alpha_+,$$

and consequently there exists

$$\frac{\partial \mathfrak{M}(\mathfrak{b}, \rho, t)}{\partial \mathfrak{b}} = \lambda\left(\frac{\mathfrak{b}}{t}\right) - \lambda\left(\frac{\rho - \mathfrak{b}}{1-t}\right) < 0, \tag{34}$$

so that the function $\mathfrak{M}(\mathfrak{b}, \rho, t)$ decreases monotonically in the domain $A(t)$, $\mathfrak{M}(\mathfrak{b}, \rho, t) \to -\infty$ as $\mathfrak{b} \to \infty$, $t \in (0, 1)$.

Further, let $\gamma(f(t)) = (\rho - f(t))/(1-t)$. Then

$$\frac{\partial \mathfrak{M}(at,\ \rho,\ t)}{\partial t}=\int_0^a \lambda(\alpha)\,d\alpha-\int_0^{\gamma(at)}\lambda(\alpha)\,d\alpha-\frac{a-\rho}{1-t}\lambda(\gamma(at)).$$

If $at \in A(t)$, then $\gamma(at) < a$ and $(a-\rho)\lambda(\gamma(at))/(1-t) > \int_{\gamma(at)}^a \lambda(\alpha)\,d\alpha$. From this we find

$$\frac{\partial \mathfrak{M}(at,\ \rho,\ t)}{\partial t}<0. \tag{35}$$

Since $\tau t/t_0 \in A(t)$ for $t \in (0,\ t_0)$, it follows from the above inequality that $\mathfrak{M}(\tau t/t_0,\ \rho,\ t) > \mathfrak{M}(\tau,\ \rho,\ t_0)$ for $t \in (0,\ t_0)$. Together with (34), this means that there exists a unique solution of the local problem $\mathfrak{b}_{\tau,\rho}(t) > \tau t/t_0,\ t \in (0,\ t_0)$.

In order to limit ourselves here and in the sequel to the case $t \in (0,\ t_0)$, let us note that in the domains $(0,\ t_0)$ and $(t_0,\ 1)$ the function $\mathfrak{b}_{\tau,\rho}(t)$ behaves qualitatively in a completely symmetric fashion. This is easily verified by making the change of variable $u = 1-t$ in equation (33). It turns out then that the function $\tilde{\mathfrak{b}}_{\tau-\rho,-\rho}(u) = \mathfrak{b}_{\tau,\rho}(1-u) - \rho$ is a solution of the local problem which corresponds to the random variables $\tilde{\xi}_k = -\xi_k,\ k = 1, 2, \cdots$, and to the parameters $\tilde{\tau} = \tau - \rho$, $\tilde{\rho} = -\rho$. The above remark completes the proof of the existence and uniqueness of the solution of equation (33) in the domain indicated in the lemma.

From the aforesaid, it follows also that

$$\mathfrak{b}_{\tau,\rho}(t)\leqslant\tau,\quad t\in(0,\ 1),\quad \mathfrak{b}_{\tau,\rho}(t_0)=\tau. \tag{36}$$

From now on, we shall simply write $\mathfrak{b}(t)$ or $\mathfrak{b}$ instead of $\mathfrak{b}_{\tau,\rho}(t)$ in cases where it will not lead to confusion. If $\mathfrak{b}$ is not a boundary solution, then $\mathfrak{M}(\mathfrak{b},\ \rho,\ t)$ is continuously differentiable with respect to t and $\mathfrak{b}$. For such $(t,\ \mathfrak{b})$ there exists

$$\frac{\partial \mathfrak{b}_{\tau,\rho}(t)}{\partial t}=\mathfrak{b}'=-\frac{\ln\varphi\left(\lambda\left(\frac{\mathfrak{b}}{t}\right)\right)-\ln\varphi\left(\lambda\left(\frac{\rho-\mathfrak{b}}{1-t}\right)\right)}{\lambda\left(\frac{\mathfrak{b}}{t}\right)-\lambda\left(\frac{\rho-\mathfrak{b}}{1-t}\right)}. \tag{37}$$

If $\mathfrak{b}$ is a boundary solution for $t \in (0,\ t_0)$ $(t \in (t_0,\ 1))$, then $\mathfrak{b} = \alpha_- t$ $(\mathfrak{b} = \rho - \alpha_+(1-t))$, so that the derivatives for boundary solutions are equal to α_- and α_+, respectively, on the intervals $(0,\ t_0)$ and $(t_0,\ 1)$, and coincide with the limiting values of the right-hand side of (37) for $\lambda(\mathfrak{b}/t) \to \lambda(\alpha_-) = -\infty$ and $\lambda((\rho-\mathfrak{b})/(1-t)) \to \lambda(\alpha_+) = \infty$.

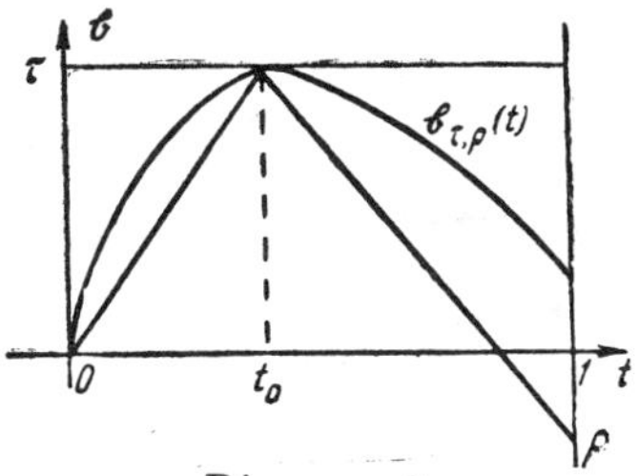

Diagram 2

The straight lines $\mathfrak{b} = \alpha_- t$ and $\mathfrak{b} = \rho - \alpha_+(1-t)$ divide the domain $\mathfrak{b} \geq t\rho,\ 0 \leq t \leq 1$, in general, into four sectors $D_1 - D_4$ shown in

Diagram 3. In the course of the variation of τ $(\rho < \tau < \infty)$, as a function of the quantities $\bar{\alpha}$ and $\underline{\alpha}$ (boundaries of the possible values of ξ_k), the family $\mathfrak{B}$ of curves $\mathfrak{b}_{\tau,\rho}(t)$ fills up either the sector D_1, or D_1 and D_2, or D_1 and D_3, or the whole domain $\mathfrak{b} > t\rho$. As is shown by (37), the curves $\mathfrak{b}(t)$ are analytic in each of the sectors $D_1 - D_4$. The sector boundaries serve as fusion lines of those curves which have equal first derivatives at the points of fusion. In sector D_4 (which is not empty and is filled up by curves of the family $\mathfrak{B}$ if α_-, α_+, $\lambda_\pm$ are finite), the $\mathfrak{b}(t)$ assume the form of straight lines $\mathfrak{b}' = (1/(\lambda_+ - \lambda_-))\ln \phi(\lambda_-)/\phi(\lambda_+)$. In sectors D_2 and D_3 the curves of $\mathfrak{B}$ possess the property characteristic of curves of the family $\mathfrak{A}$: they intersect the straight lines $(\rho - \mathfrak{b})/(1 - t) = \text{const}$ and $\mathfrak{b}/t = \text{const}$, respectively, under constant angles.

Furthermore, let us show that the curve $\mathfrak{b}(t)$ is convex. Since $\mathfrak{b}(t_1)t/t_1 \in A(t)$ for $t \in (0, t_1)$, then, by virtue of (35),

$$\mathfrak{M}\left(\frac{\mathfrak{b}(t_1)}{t_1}t,\ \rho,\ t\right) > \mathfrak{M}(\mathfrak{b}(t_1),\ \rho,\ t_1) = \mathfrak{M}(\tau,\ \rho),$$

and therefore $\mathfrak{b}(t_1)/t_1 < \mathfrak{b}(t)/t$. Hence it follows that

$$\mathfrak{b}' - \frac{\mathfrak{b}}{t} < 0. \tag{38}$$

From symmetry considerations we also see that for $1 < t < t_1 < 1$, $(\rho - \mathfrak{b}(t_1))/(1 - t_1) > (\rho - \mathfrak{b}(t))/(1 - t)$ and therefore

$$\mathfrak{b}' - \frac{\rho - \mathfrak{b}}{1 - t} < 0. \tag{39}$$

Let $\mathfrak{b}(t)/t = b$, $(\rho - \mathfrak{b}(t))/(1 - t) = \gamma$. Then

$$\mathfrak{b}'' = \frac{\partial \mathfrak{b}'}{\partial \lambda(b)} \frac{\partial \lambda(b)}{\partial b} \frac{\partial b}{\partial t} + \frac{\partial \mathfrak{b}'}{\partial \lambda(\gamma)} \frac{\partial \lambda(\gamma)}{\partial \gamma} \frac{\partial \gamma}{\partial t}. \tag{40}$$

We have

$$\frac{\partial \mathfrak{b}'}{\partial \lambda(b)} = \frac{b - \mathfrak{b}'}{\lambda(b) - \lambda(\gamma)}, \tag{41}$$

$$\frac{\partial \mathfrak{b}'}{\partial \lambda(\gamma)} = \frac{\gamma - \mathfrak{b}'}{\lambda(\gamma) - \lambda(b)}. \tag{42}$$

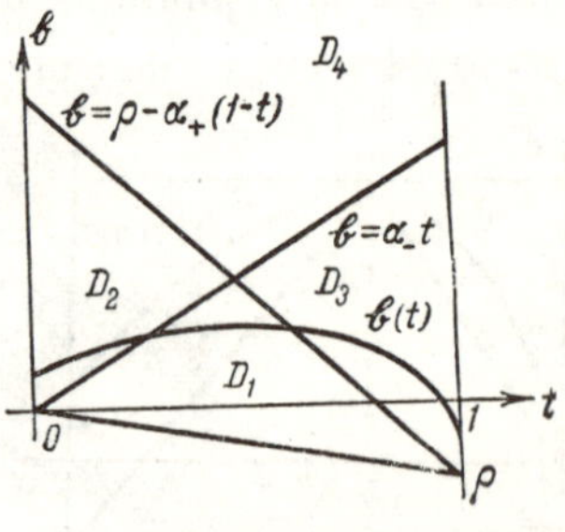

Diagram 3

Since $bt \in A(t)$, $t \in (0,1)$, the difference $\lambda(b) - \lambda(y) < 0$ and is finite. From this and from inequalities (38), (39), it follows that the right-hand side is negative in (41) and positive in (42). Furthermore $\lambda'(\alpha) \leq 0$, where the equality sign is possible only for $\alpha > \alpha_-$ or $\alpha < \alpha_+$. Finally by virtue of (38), (39), $\partial b/\partial t < 0$, $\partial y/\partial t > 0$. Returning to (40), we find that the derivative $\mathfrak{b}'' \leq 0$ for all $t \in (0, 1)$, has a discontinuity at the intersection of the sector boundaries, and reduces to 0 only at points of domain D_4. The proof of the lemma, including also some additional information, is complete.

The investigation of the properties of family $\mathfrak{B}$ on the boundary of the sectors and for $t \to 1, 0$, brings us to a classification quite similar to that indicated in Diagram 1. For $\alpha_- < \infty$ tangency between the straight line $\mathfrak{b} = \alpha_- t$ and curves of the family occurs only if $\lambda_- = -\infty$. Furthermore, just as in §3, we can prove that the finiteness of $\phi(\lambda_-)$ is a necessary and sufficient condition for the finiteness of $\mathfrak{b}'(+0)$. The equivalence of the conditions $\lambda_- > -\infty$ and $\mathfrak{b}(+0) > 0$ follows from the relation

$$\lim_{t \downarrow 0} t \int_0^{\mathfrak{b}/t} \lambda(\alpha)\, d\alpha = \ln \mathfrak{m}(\tau, \tau - \rho) - \ln \mathfrak{m}(\rho - \mathfrak{b}(+0)).$$

Thus we obtain the equation

$$\lambda_- \mathfrak{b}(+0) = \ln \mathfrak{m}(\tau, \tau - \rho) - \ln \mathfrak{m}(\rho - \mathfrak{b}(+0))$$

for $\mathfrak{b}(+0)$. From symmetry considerations we easily obtain similar statements concerning the behavior of $\mathfrak{b}(t)$ in the neighborhood of the straight lines $(\mathfrak{b} - \rho)/(1 - t) = \alpha_+$ and $t = 1$.

Returning to equation (32) corresponding to the integral problem, let us note that its solution is obtained from the solution of the local problem by replacing the portion of the curve $\mathfrak{b}_{\tau,\rho}(t)$ lying in the domain $0 \leq \mathfrak{b} < \rho$ (empty for $\rho \leq 0$) by the portion of the curve $\mathfrak{c}_{\tau,\rho}(t)$ defined by the solution of the equation

$$t \ln \mathfrak{m}\left(\frac{\mathfrak{c}}{t}\right) = \mathfrak{M}(\tau, \rho).$$

We investigated such equations in §3. At the points of fusion of the parts of $\mathfrak{c}_{\tau,\rho}(t)$ and $\mathfrak{b}_{\tau,\rho}(t)$, their first derivatives are equal. We shall denote by $\mathfrak{C}^*$ the newly obtained family of curves.

---

* If we connect the point (t_0, τ) with the points (0, 0) and (1, ρ), we shall obtain in the metric introduced in the footnote to §3, the "shortest path" connecting the points (0, 0) and (1, ρ) and reaching the level τ. A curve in $\mathfrak{B}$ is the geometric locus of the points, the sum of whose distances to the origin and to the point (1, ρ) remains the same.

§2. Limit theorems. I

6. **Estimate of the probability of the distorted path.** Henceforth we shall denote by the letters c and ϵ, with or without indices, constants and sufficiently small (s. s.) parameters, respectively, not always the same, so that we may write expressions of the form $c\Sigma(1/k^2) \le c$. Consider the sequences (a_n), (b_n), $n = 1, 2, \cdots$, of the form $a_n = a + \epsilon_a$, $b_n = b + \epsilon_b$, where $\epsilon_a = \epsilon_a(n) \to 0$, $\epsilon_b = \epsilon_b(n) \to 0$ for $n \to \infty$. Let $a_0 > 0$, $n_0 > 0$. Denote by Q the set of all such sequences (a_n, b_n), for which $a_n > a_0$, $a_0 < a_n + b_n < \alpha_- - a_0$ for $n \ge n_0$, and set for short $\mathfrak{A}_k = \mathfrak{A}^k_{a_n n, b_n k} = \{s_k \ge a_n n + b_n k\}$.

Lemma 3. *If $(a_n, b_n) \in Q$, then there exists $c > 0$ and $h < 1$ which depend only on a_0 and n_0 and such that for all $k < n$*

$$\mathbf{P}(\mathcal{A}_k/\mathcal{A}_n) < ch^l,$$

where $l = n - k$.

Proof. We note first that it is sufficient to consider the values $b_n \ge 0$, since for $b_n < 0$

$$\mathbf{P}(\mathcal{A}_k/\mathcal{A}_n) < \mathbf{P}(s_k > (a_n + b_n)n/\mathcal{A}_n).$$

For some $\delta > 0$ we have, for $n \ge n_0$,

$$\mathbf{P}(\mathcal{A}_k/\mathcal{A}_n) \leqslant \frac{1}{\mathbf{P}(\mathcal{A}_n)}\left\{\sum_{j=0}^{\left[\frac{b_n}{\delta}-1\right]} \mathbf{P}(s_k \geqslant a_n n + b_n k + \delta j l)\,\mathbf{P}(s_l \geqslant b_n l - \delta l(j+1))\right.$$

$$\left. + \mathbf{P}\left(s_k \geqslant a_n n + b_n k + \delta l\left[\frac{b_n}{\delta}\right]\right)\right\} < \frac{c\sqrt{n}}{\mathrm{m}^n(a_n + b_n)} \sum_{j=0}^{\left[\frac{b_n}{\delta}\right]} \mathrm{m}^k\left(a_n \frac{n}{k} + b_n\right.$$

$$\left. + \delta j \frac{l}{k}\right) \mathrm{m}^l(b_{j,n}) = c\sqrt{n} \sum_{j=0}^{\left[\frac{b_n}{\delta}\right]} \exp\left\{k \int_{a_n+b_n}^{a_n+b_n+\frac{l}{k}(\delta j + a_n)} \lambda(\alpha)\,d\alpha\right.$$

$$\left. - l \int_{b_{j,n}}^{a_n+b_n} \lambda(\alpha)\,d\alpha\right\},$$

where $c = c(a_0, n_0)$,
$$b_{j,n} = \begin{cases} 0 & \text{for } j = \left[\frac{b_n}{\delta}\right], \\ b_n - \delta(j+1) & \text{for } j < \left[\frac{b_n}{\delta}\right]. \end{cases}$$

Applying the mean-value theorem of the difference of integrals inside the curly brackets, we find the following expression for it:

$$l\{(\delta j + a_n)\lambda(\alpha_1) - \lambda(\alpha_2)\min(a_n + b_n,\ a_n + \delta(j+1))\}, \tag{44}$$

where α_1 and α_2 belong to the corresponding intervals of integration. From the properties of the class Q and of the function $\lambda(\alpha)$ there follows the existence of an $\epsilon > 0$ such that $\lambda(\alpha_1) < \lambda(\alpha_2) - \epsilon$ for all $(a_n, b_n) \in Q$ and l. Therefore (44) will not exceed the value

$$-l\{(\delta j + a_n)\varepsilon + \delta\lambda(\alpha_2)\} \leqslant -l(a_0\varepsilon + \delta\lambda(\alpha_- - a_0)).$$

Set $\delta = -a_0\epsilon/2\lambda(\alpha_- - a_0)$; then

$$\mathbf{P}(\mathcal{A}_k/\mathcal{A}_n) < c\sqrt{n}\exp\left(-\frac{la_0\varepsilon}{2}\right). \tag{45}$$

If l increases sufficiently fast with n, for example $l \geq a_0 n/2\alpha_-$, then (45) proves the desired inequality. But if $l < a_0 n/2\alpha$, then for $n \geq n_0$

$$(a_n + b_n)\frac{n}{k} < \frac{a_n + b_n}{1 - \dfrac{a_0}{2\alpha_-}} < \frac{\alpha_-(a_n + b_n)}{(\alpha_- - a_0)\left(1 + \dfrac{a_0}{2\alpha_-}\right)} < \frac{\alpha_-}{1 + \dfrac{a_0}{2\alpha_-}}$$

and to calculate the probability $\mathbf{P}(s_k > a_n n + b_n k + \delta jl)$ in (43) we can make use of a well-known limit theorem. We find

$$\mathbf{P}(\mathcal{A}_k/\mathcal{A}_n) < \frac{c}{\mathfrak{m}^n(n_n + b_n)} \sum_{j=0}^{\left[\frac{b_n}{\delta}\right]} \mathfrak{m}^k\left(a_n\frac{n}{k} + b_n + \delta j\frac{l}{k}\right)\mathfrak{m}^l(b_{j,n}).$$

The subsequent arguments remain the same. Q.E.D.

7. Theorem of the time of first passage.

Lemma 4. *Let $\pi(t) = a_n + b_n t$ and $\pi_1(t) = a_n + \epsilon_h + bt$ ($\pi(1) = \pi_1(1)$) be straight lines for which $(a_n, b_n) \in Q$. Then*

$$\mathbf{P}(\mathcal{B}^n_{\pi n}\mathcal{X}_y) = \mathbf{P}(\mathcal{B}^n_{\pi_1 n}\mathcal{X}_y)(1 + O(\varepsilon_b)).$$

Proof. The statement of the lemma follows the fact that for s.s. ϵ, independent of n, we have the following equality for the straight lines $\pi_2(t) = a_n - \epsilon + (b + \epsilon)t$ and $\pi_3(t) = a_n + bt$:

$$\mathbf{P}(\mathcal{B}^n_{\pi_2 n}\mathcal{X}_y) = \mathbf{P}(\mathcal{B}^n_{\pi_3 n}\mathcal{X}_y)\Xi\left(\frac{1}{n}\right),$$

where the coefficients of the asymptotic expansion $\Xi(1/n)$ in powers of $1/n$ are analytic functions of ϵ. We can show that this statement is true by going over to the random variables $\xi_k^2 = \xi_k - b - \epsilon$ and $\xi_k^3 = \xi_k - b$, respectively, and making use of formula (52) of the proof of Theorem 5 in [6].

We shall need the following for the statement of a theorem.

Definition. Let $R(t)$ be a function defined in ω_ϵ, a neighborhood or half-neighborhood of the point $t = 0$, contained in the interval $(-\epsilon, \epsilon)$. We shall say that $R(t)$ has a power majorant if there exist positive numbers r, ϵ, c_1, c_2 and a function $r(t)$ defined in ω_ϵ such that

1) $|R(t)| < c_1 r(t)$ for $t \in \omega_\epsilon$;

2) if $t \in \omega_\epsilon$ and $|t| \downarrow 0$, then $r(t) \downarrow 0$;

3) $r(dt) \geq c_2 d^r r(t)$ for any $t \in \omega_\epsilon$ and $0 < d \leq 1$.

We shall call the number r in the last inequality the degree of the majorant.

Let us note that the class of majorants M_r of degree r includes, for instance, all functions sufficiently smooth at the origin, and decreasing for $t \to 0$ more slowly than $|t|^r$. It is also clear that $M_{r_1} \subset M_{r_2}$ if $r_1 \leq r_2$.

Theorem 3. *Assume the following hypothesis:*

1) $k/n \to v$ *for* $n \to \infty$, v *is independent of* n *and belongs to the half-open interval* $(0, 1]$;

2) *in the neighborhood of point* v, *the function* $g(t)$ *has* $p \geq 1$ *derivatives, and in the representation*

$$g(t) = g(u) + \ldots + (t-u)^p \frac{g^{(p)}(u)}{p!} + (t-u)^p R_u^1(t-u) \tag{46}$$

the remainder term $R_u^1(t-u)$ *has the same power majorant for all* u, $|u - v| < \epsilon_1$;

3) $g'(v) < g(v)/v < \alpha_-$;

4) *for any* $\epsilon_2 > 0$ *one can find* $\epsilon_3 > 0$ *such that* $g(t) > g(v)t/v + \epsilon_3$ *for all* $t < v - \epsilon_2$.

Then for any fixed $y > 0$ *we have*

$$\mathbf{P}(\mathcal{B}_{gx}^k \mathcal{X}_y) = \mathbf{P}(\mathcal{B}_{\pi x}^k \mathcal{X}_y)\left(1 + O\left(r_2\left(\frac{1}{n}\right)\right)\right),$$

where $\pi(t) = \pi(k/n, t) = g(k/n) + (t - k/n)g'(k/n)$, $r_2(u-t)$ *is the majorant of the function* $R_u^2(t-u) = (g(t) - g(u) - (t-u)g'(u))/(t-u)$, *the estimate is uniform in* k *for* $|k/n - v| < \epsilon_1$.

Remark. Although it is not very restrictive, the condition of the existence of the power majorant of the function $R_u^1(t-u)$ is not essential to the matter, and is introduced only to simplify the estimates in the following calculations.

Proof. We shall make use of the following relations:

a) $\mathbf{P}(\mathcal{B}^k_{\pi x}\mathcal{X}_y) \lesssim \mathbf{P}(\mathcal{B}^k_{\pi x}) \lesssim \mathbf{P}(\mathcal{A}^k_{\pi x})$;

b) for any $\delta > 0$, $x_1 > 0$ and s.s. $\epsilon_4 > 0$ there exists a constant c such that for all j, $v - \epsilon_4 < j/n < k/n$ we have the inequality

$$\mathbf{P}((\mathcal{A}^j_{\pi x} - \mathcal{A}^j_{\pi x+\delta})\mathcal{A}^k_{x_1}) < c\,\delta\mathbf{P}(\mathcal{A}^j_{\pi x}\mathcal{A}^k_{x_1}). \tag{47}$$

The first relation follows from Lemma 4 and paper [6]. The second follows from the inequalities

$$\mathbf{P}(\mathcal{A}^k_{x_1}/\mathcal{A}^j_{x_2} - \mathcal{A}^j_{x_2+\delta}) \leqslant \mathbf{P}(\mathcal{A}^k_{x_1}/\mathcal{A}^j_{x_2}),\ \mathbf{P}(\mathcal{A}^j_{x_2} - \mathcal{A}^j_{x_2+\delta}) < c\,\delta\mathbf{P}(\mathcal{A}^j_{x_2}),$$

the last one of which is a corollary of the theorem of large deviations for sums (for example see [6], if $\epsilon_5 < x_2/j < \alpha_- - \epsilon_5$ for some $\epsilon_5 > 0$ and $j \to \infty$.

We have

$$\mathbf{P}(\mathcal{B}^k_{gx}\mathcal{X}_y) = \mathbf{P}(\mathcal{B}^k_{\pi x}\mathcal{X}_y) + \sum_{j=1}^{k-1}\mathbf{P}(\mathcal{B}^j_{\pi x}\mathcal{B}^k_{gx}\mathcal{X}_y) - \sum_{j=1}^{k-1}\mathbf{P}(\mathcal{B}^j_{gx}\mathcal{B}^k_{\pi x}\mathcal{X}_y). \tag{48}$$

Consider the first summation. Let us divide the range of summation into two parts: I_1 the interval of values $1 \leq j \leq [n(v-\epsilon)] - 1$, and I_2 the interval $[n(v-\epsilon)] \leq j \leq k-1$. Here $\epsilon > 0$ is picked such that (47) is valid for $j \in I_2$. Since $\mathbf{P}(\mathcal{B}^j_{\pi x}\mathcal{B}^k_{gx}\mathcal{X}_y) \leq \mathbf{P}(\mathcal{A}^j_{\pi x}\mathcal{A}^k_{gx})$ and since there exists $a_0 > 0$ for which $2a_0 < g(v)/v < \alpha_- - 2a_0$, then it is natural to make use of Lemma 3 to estimate the summation over I_1. By this lemma, we can find $h < 1$ and $c > 0$ such that

$$\sum_{I_1}\mathbf{P}(\mathcal{A}^j_{\pi x}\mathcal{A}^k_{gx}) < ch^n\,\mathbf{P}(\mathcal{A}^k_{gx}).$$

Further, assuming for simplicity that $g(j/n) > \pi(j/n)$ for $j \in I_2$, we find

$$\sum_{I_2}\mathbf{P}(\mathcal{B}^j_{\pi x}\mathcal{B}^k_{gx}\mathcal{X}_y) < \sum_{I_2}\mathbf{P}((\mathcal{A}^j_{\pi x} - \mathcal{A}^j_{gx})\mathcal{A}^k_{gx}) < c\sum_{I_2}x\left(g\left(\frac{j}{n}\right)\right.$$
$$\left. - \pi\left(\frac{j}{n}\right)\right)\mathbf{P}(\mathcal{A}^j_{\pi x}\mathcal{A}^k_{gx}) < c\sum_{I_2}x\left(g\left(\frac{j}{n}\right) - \pi\left(\frac{j}{n}\right)\right)h^{k-j}\,\mathbf{P}(\mathcal{A}^k_{gx}).$$

But the relation

$$\frac{g(t) - \pi(t)}{t - \dfrac{k}{n}} = \frac{g(t) - g\left(\dfrac{k}{n}\right) - \left(t - \dfrac{k}{n}\right)g'\left(\dfrac{k}{n}\right)}{t - \dfrac{k}{n}} = R^2_{\frac{k}{n}}\left(t - \frac{k}{n}\right),$$

obviously has, together with $R^1_{k/n}(t - k/n)$, a power majorant. Denote it by $r_2(k/n - t)$ and let r_2 be its degree. Then

$$\left| x \sum_{I_2} \left(g\left(\frac{j}{n}\right) - \pi\left(\frac{j}{n}\right) \right) h^{k-j} \right| \leqslant \frac{x}{n} \sum_{I_2} (k-j) \left| R_{\frac{k}{n}}^{2}\left(\frac{j-k}{n}\right) \right| h^{k-j}$$

$$\leqslant c \sum_{I_2} (k-j)^{\gamma_2+1} h^{k-j} r_2\left(\frac{1}{n}\right) \leqslant c r_2\left(\frac{1}{n}\right),$$

and therefore

$$\sum_{j \in I_2} \mathbf{P}(\mathcal{B}^j_{\pi x} \mathcal{B}^k_{gx} \mathcal{X}_y) \leqslant c r_2 \left(\frac{1}{n}\right) \mathbf{P}(\mathcal{A}^k_{gx}), \text{ for all } k, \left|\frac{k}{n} - v\right| < \varepsilon_1.$$

If the condition $g(j/n) > \pi(j/n)$ does not hold at all the points of an ϵ-neighborhood of point v, then the sum in question will only decrease. The second summation in (48) is estimated in an identical fashion. Making use of relation a) completes the proof of the theorem.

Remark. The proof of the theorem shows that if we consider only the values $k \leq nv$, then it is sufficient to require that condition 2) be valid only from the left: for $t, u \leq v$. On the other hand, for $k \geq nv$ the behavior of the function g for $t \leq v$ can substantially influence the asymptotics of $\mathbf{P}(\mathcal{B}^k_{gx} \mathcal{X}_y)$. Assume (46) holds only for $t, u > v$, and

$$\liminf_{t \uparrow v} g(t) > \lim_{t \downarrow v} g(t). \tag{49}$$

Then if the remaining hypothesis of Theorem 3 is true, where the symbol $g^{(i)}(v)$ is to be interpreted as $\lim_{t \downarrow v} g^{(i)}(t)$, we have

Theorem 3A. $\mathbf{P}(\mathcal{B}^k_{gx} \mathcal{X}_y) = \mathbf{P}(\mathcal{B}^k_{\pi x} \mathcal{X}_y)(1 + O(r_2(1/n), h^{k-nv}))$,* *where the estimate is uniform with respect to* k *for* $k/n - v < \epsilon_1$, $h < 1$.

Proof. If we let, in the present case, I_3 denote the interval of values $1 \leq j \leq [nv]$ and I_4, the interval $[nv] < j < k$, we shall obtain the following estimates:

$$\sum_{I_4} \mathbf{P}(\mathcal{B}^j_{\pi x} \mathcal{B}^k_{gx} \mathcal{X}_y) < c r_2 \left(\frac{1}{n}\right) P(A^k_{gx}); \quad \sum_{I_3} \mathbf{P}(\mathcal{B}^j_{\pi x} \mathcal{B}^k_{gx} \mathcal{X}_y) < c h^{k-nv} \mathbf{P}(\mathcal{A}^k_{gx}).$$

The statement of the theorem follows from this and from (48).

Similarly $\mathbf{P}(\mathcal{B}^k_{gx} \mathcal{X}_y)$ can be estimated with the help of $\mathbf{P}(\mathcal{B}^k_{\pi x} \mathcal{X}_y)$, and in the case when the expansions (46) exist for values $t, u < v$ and $t, u > v$, but are different.

Let us now find an explicit form for the asymptotic behavior of the probability $\mathbf{P}(\mathcal{B}^k_{gx} \mathcal{X}_y)$. Let

---

* Here and in the sequel $O(a_n, b_n) = O(a_n) + O(b_n)$.

$$\pi_1(t) = \frac{x}{n} g\left(\frac{k}{n}\right) - \frac{k}{n} g'(v) + tg'(v).$$

Then, by virtue of Lemma 4,

$$\mathbf{P}(\mathcal{B}^k_{\pi x} \mathcal{X}_y) = \mathbf{P}(\mathcal{B}^k_{\pi_1 n} \mathcal{X}_y)\left(1 + O\left(\frac{x}{n} g'\left(\frac{k}{n}\right) - g'(v)\right)\right).$$

We have $(x = n(1 + \kappa))$

$$\left|\frac{x}{n} g'\left(\frac{k}{n}\right) \leqslant g'(v)\right| \leqslant |\varkappa g'(v)| + 2\left|R^3_v\left(\frac{k}{n} - v\right)\right|,$$

where $R^3_v(k/n - v) = g'(k/n) - g'(v)$. Therefore, if the hypothesis of Theorem 3 holds,

$$\mathbf{P}(\mathcal{B}^k_{gx} \mathcal{X}_y) = \mathbf{P}(\mathcal{B}^k_{\pi_1 n} \mathcal{X}_y)\left(1 + O\left(r_2\left(\frac{1}{n}\right), \varkappa g'(v), R^3_v\left(\frac{k}{n} - v\right)\right)\right).$$

Here, as before, the estimate is uniform with respect to k. Furthermore, we find with the help of formula (52) in [6]

$$\mathbf{P}(\mathcal{B}^k_{\pi_1 n} \mathcal{X}_y) = \frac{\mathfrak{m}(s)^{k+1/2}\,\psi(z_s, y)}{\sqrt{2\pi k \varphi''_\lambda(\lambda(s), s)|\lambda'_-(z_s)|}}\, \Xi\left(\frac{1}{n}\right) = \frac{\mathfrak{m}(s)^k}{\sqrt{k}} \Psi(s, g'(v), y)\, \Xi\left(\frac{1}{n}\right), \tag{50}$$

where $s = (x/k)g(k/n)$, $\phi(\lambda, s) = e^{\lambda s}\phi(\lambda)$, $\bar{\phi}(\lambda) = \phi(\lambda, g'(v))$, $z_s = \bar{\phi}^{-1}(\lambda(s))$, $\Xi(1/n)$ is an asymptotic expansion in powers of $1/n$. The coefficient of $\psi(z, y)$ is defined in [6] and possesses properties by virtue of which the function $\Psi(s, g'(v), y)$, defined by equality (50), is positive and analytic in s in $(0, \alpha_-)$; $\Psi(s, g'(v), \infty) < \infty$ and for any $s \in (0, \alpha_-)$ the ratio $\Psi(s, g'(v), y)/\Psi(s, g'(v), \infty)$ is a distribution function with respect to y with a bounded density.

We now transform the right-hand side of (50). Set $x/n = 1 + \kappa = \sigma$ and denote by τ_σ the solution of the equation

$$\sigma g(v) = \alpha_\tau(v). \tag{51}$$

Then

$$\mathfrak{m}^k(s) = \mathfrak{m}^{nt}\left(\frac{g(t)\sigma}{t}\right)\Bigg|_{t=\frac{k}{n}},$$

$$\mathfrak{m}^t\left(\frac{g(t)\sigma}{t}\right) = \mathfrak{m}(\tau_\sigma) \exp T(\Delta, \sigma, v),$$

where $\Delta = t - v$,

$$T(\Delta, \sigma, v) = t \int_A^B \lambda(\alpha) d\alpha, \; A = \frac{\alpha_{\tau_\sigma}(t)}{t}, \quad B = \frac{g(t)\sigma}{t}. \tag{52}$$

Since $\partial \alpha_\tau(t)/\partial \tau = \lambda(\tau)/\lambda(\alpha_\tau(t)/t)$, we have

$$\frac{\partial}{\partial \sigma} T(\Delta, \sigma, v) = g(t)\lambda\left(\frac{g(t)\sigma}{t}\right) - g(v)\lambda\left(\frac{g(v)\sigma}{v}\right).$$

Hence, setting

$$T_0(t) = T(\Delta, 1, v), \; T_k(t) = \frac{g(t)}{k!}\lambda^{(k-1)}\left(\frac{g(t)}{t}\right)\left(\frac{g(t)}{t}\right)^{k-1}, \; k = 1, 2, \ldots,$$

we find

$$T(\Delta, \sigma, v) = T_0(t) + \sum_{k=1}^{\infty} \varkappa^k [T_k(t) - T_k(v)].$$

It is clear that in the neighborhood of the point v the functions $T_0(t)$ and $T_k(t) - T_k(v)$, $k = 1, 2, \cdots$, have a representation of the form

$$\sum_{j=1}^{p} \Delta^j T_{k,j}(v) + \Delta^p R^4_{k,v}(\Delta),$$

where $|R^4_{k,v}(\Delta)| < c|R^1_v(\Delta)|$, $T_{k,j}(v) = T_k^{(j)}(v)/j!$ (in order to calculate the derivatives $\lambda^{(k)}(\alpha)$ one should use the identity $\alpha\phi(\lambda(\alpha)) + \phi'(\lambda(\alpha)) \equiv 0$). Hence

$$T(\Delta, \sigma, v) = \sum_{k=0}^{\infty} \varkappa^k \left[\sum_{j=1}^{p} \Delta^j T_{k,j}(v) + \Delta^p R^4_{k,v}(\Delta)\right]. \tag{53}$$

In particular, if at the point v the functions $g(t)$ and $\alpha_{\tau_1}(t)$ have $0 \le q \le p$ equal first derivatives, then

$$T_{0,j}(v) = 0 \text{ for } j \leqslant q, \; T_{0,q+1}(v) = \frac{1}{(q+1)!}\lambda\left(\frac{g(v)}{v}\right)\left[g^{(q+1)}(v) - \alpha^{(q+1)}_{\tau_1}(v)\right]. \tag{54}$$

Consequently we can formulate the statements of Theorems 3 and 3A as follows.

Theorem 4. *Assuming the hypotheses of Theorem 3, we have the equality*

$$\mathbf{P}(\mathcal{B}^k_{gx}\mathcal{X}_y) = \frac{m^n(\tau_\sigma)}{\sqrt{k}} \Psi(G, g'(v), y) e^{nT(\Delta, \sigma, v)} \left\{1 + O\left(\frac{1}{n}\right)\Delta, \varkappa, r_2\left(\frac{1}{n}\right), R^3_v(\Delta)\right\}, \tag{55}$$

where τ_σ *is the solution of equation* (51), $\kappa = \sigma - 1 = x/n - 1$, $\Delta = k/n - v$, $G = g(v)/v$, $R_v^3(\Delta) = g'(v + \Delta) - g'(v)$. *The functions* $\Psi(a, b, y)$ *and* $T(\Delta, \sigma, v)$ *are defined in* (50) *and* (52)–(54) *respectively.*

Theorem 4A. *If under the hypothesis of Theorem* 3, *relation* (46) *holds only for* $t, u > v$ *and if* (49) *is valid, then the value of* $\mathbf{P}(\mathfrak{B}^k_{gx}\mathfrak{X}_y)$ *is given by the right-hand side of* (55) *with, in addition, a remainder term* $O(h^{\Delta n})$ *for some* $h < 1$.

Corollary. *Under the hypothesis of Theorem* 3, *there exists a limiting conditional distribution of the magnitude* χ_g *of the first jump across the boundary* $xg(t)$:

$$E_G(y) = \lim_{h\to\infty} \mathbf{P}(\chi_g < y/\mathfrak{B}^k_{gx}) = \frac{\psi(z_G, y)}{\psi(z_G, \infty)}$$

(*see* (50)), *which has a bounded density and coincides with the limiting distribution* $\lim_{n\to\infty} \mathbf{P}(\chi_f < y/\mathfrak{B}^n_{fx})$, *where* $f(t) = G$ (*see the corollary of Theorem* 5 *in* [6]; *in the notation of that paper:* $\int e^{i\lambda y}\, dE_G(y) = \mathfrak{B}_{z_G+}(\lambda)/\mathfrak{B}_{z_G+}(0)$). *Under the hypothesis of Theorem* 4A, *these relations hold provided* $k - nv \to \infty$ *together with* n.

Thus we have found the asymptotic behavior of the first probability in (1). We shall consider the solution of the two other problems in (1) first for the simplest and clearest case of $x = n$ (§§8, 9, 11). The case $x \neq n$ will be considered separately.

8. Theorem on the probability that a trajectory will fall outside the limits of a domain. We shall determine the probability of event $\mathcal{C}_{gx}\mathfrak{X}_y$ in which the broken line with vertices at the points $(k/n, s_k/x)$ crosses the curve $g(t)$ and the first jump across the boundary is less than y/x.

Let the following quantity be defined and finite:

$$\tau_g = \sup\{\tau: \inf_{0\leq t\leq 1}(g(t) - a_\tau(t)) > 0\}. \tag{56}$$

Let V_ϵ be the set of all points t for which $g(t) \leq a_{\tau_g}(t) + \epsilon$, and let V^g be the closure of the set, on which $a''_{\tau_g}(t) = 0$. The set V^g is clearly of the form $0 \leq t \leq t_{\tau_g}$ and, by definition, it contains the point $t = 0$ if $t_{\tau_g} = 0$. We shall assume that there exists an $\epsilon_g > 0$ for which V_{ϵ_g} has no common point with V^g, and shall consider the sets V_ϵ only for $\epsilon \leq \epsilon_g$.

In the sequel, only the properties of the function $g(t)$ on the sets V_ϵ, and properties of the sets V_ϵ themselves, will be essential for us. Here the interest lies in the fact that the desired probabilities depend only on values of the functions $g(t)$ at rational points.

Let us adopt the following notation: let $\epsilon_k > 0$ ($k = 1, 2, \cdots, \epsilon_1 \leq \epsilon_g$) denote a sequence monotonically converging to zero for $k \to \infty$; denote the closure of set A by $\overline{A}$, and complementation with respect to the closed interval $[0, 1]$ by CA.

In order to avoid exceptions, when speaking of intervals belonging to A we shall have in mind intervals, the ends of which belong to $\overline{CA}$. We shall call a collection of sets depending on a parameter a sequence, regardless of whether the parameter is discrete or continuous.

Definition. A sequence of imbedded sets $L_\epsilon \subset [0, 1]$, $L_{\epsilon_k} \subset L_{\epsilon_{k-1}}$, $k = 2, 3, \cdots$, possesses the D-property if for every $\delta > 0$ we can find ϵ_δ and n_δ such that for any $n \geq n_\delta$ the number of points of the form j/n, $0 \leq j \leq n$, belonging to the set L_{ϵ_δ}, is less than δn.

Let $V = V_0 = \bigcap_{k=1}^{\infty} V_{\epsilon_k}$ and let Ω be the union of all intervals belonging to V.

Theorem 5. *If the sequence* $V_\epsilon - \Omega$ *has the D-property, then (see* (50))

$$\mathbf{P}(\mathscr{C}_{gn}\mathscr{X}_y) \sim \sqrt{n}\,\mathfrak{m}^n(\tau_g) \int_\Omega t^{-1/2}\,\Psi\left(\frac{g(t)}{t}, g'(t), y\right) dt. \tag{57}$$

Corollary. *Under the stated hypothesis there exists a limiting distribution* $\lim_{n\to\infty} \mathbf{P}(\chi_g < y/\mathcal{C}_{gn})$ *with a bounded density.*

We shall preface the proof of the theorem with a lemma giving a sufficient condition for the existence of the D-property.

Lemma 5. *A sequence of imbedded sets* L_ϵ *for which the Lebesgue measure* $\operatorname{mes} \overline{L}_\epsilon \to 0$ *for* $\epsilon \to 0$ *possesses the D-property.*

Proof. Assume the contrary. Then we can find $\delta > 0$ and sequences $n_k \to \infty$, $\epsilon_k \to 0$ for $k \to \infty$ such that the number of points of the form j/n_k, $0 \leq j \leq n_k$, contained in L_{ϵ_k} is larger than δn_k for all k. Choose k_0 such that

$$\operatorname{mes} \overline{L}_{\varepsilon_{k_0}} < \delta/2, \tag{58}$$

and denote by S_k, $k = 1, 2, \cdots$, the sets of points of the closed interval $[0, 1]$ whose distance to points of the set $\overline{L}_{\epsilon_k}$ is less than or equal to $1/2n_k$, $t \in S_k$, if $\inf_{s \in L_{\epsilon_k}} |t - s| \leq 1/2n_k$. S_k is measurable since it is a collection of closed intervals and by assumption has the property $\operatorname{mes} S_k \geq \delta$. Let $S = \limsup S_k$ be the set of points belonging to the infinite number of sets of the sequence $S_1, S_2, \cdots$. The set S is obviously measurable. Let us show that

$$\operatorname{mes} S \geq \delta/2. \tag{59}$$

Consider a collection of some N sets S_{k_j}, $j = \overline{1, N}$, and denote by S_N^l the set of all points t belonging to $s_t \geq l$ sets of the chosen collection, not necessarily to the same ones. Then

$$\operatorname{mes} S_N^l = p_l = \mathbf{P}(\zeta_1 + \ldots + \zeta_N \geqslant l),$$

where the ζ_j are defined as the characteristic functions of the sets S_{k_j} on the space of elementary events $[0, 1]$ with a Lebesgue probability measure. But

$$N\delta \leqslant \mathbf{M}(\zeta_1 + \ldots + \zeta_N) \leqslant l(1 - p_l) + Np_l.$$

Hence

$$p_l = \text{mes}\, S_N^l \geqslant \frac{N\delta - l}{N - l} \geqslant \frac{\delta}{2}$$

for all $N > l(2 - \delta)/\delta$. Therefore the set S_∞^l of points t which belong simultaneously to some $s_t \geq l$ sets of the collection $S_1, S_2, \cdots$, also has the measure

$$\text{mes}\, S_\infty^l \geq \delta/2. \tag{60}$$

Since the sequence $S_\infty^1, S_\infty^2, \cdots$, is a sequence of imbedded sets, inequality (60) holds for any l and $S = \bigcap_{l=1}^{\infty} S_\infty^l$, and (59) has been proved. Finally, by definition of set S, any one of its points is a limit point for some sequence belonging to $L_{\epsilon k_0}$. This leads to the relation $S \subset \overline{L}_{\epsilon k_0}$ in contradiction with (59) and (58). Q.E.D.

Proof of Theorem 5. We have

$$\mathbf{P}(\mathscr{C}_{gx}\mathscr{X}_y) = \sum_{k=1}^{n} \mathbf{P}(\mathscr{B}_{gx}^k \mathscr{X}_y). \tag{61}$$

There exists a sequence of imbedded sets Ω_k, $k = 1, 2, \cdots$, consisting of a finite number k of intervals belonging to Ω such that $\text{mes}\,\overline{(\Omega - \Omega_k)} \to 0$ for $k \to \infty$. By the previous lemma, the sequence $\Omega - \Omega_k$ possesses the D-property. Since the union of two sequences with the D-property also has the D-property, for any $\delta > 0$ we can find n^δ, ϵ_δ and n_δ such that the set $\Omega^\delta = V_{\epsilon_\delta} - \Omega_{n^\delta}$ contains for any $n \geq n_\delta$ not more than δn points of the form j/n, $0 \leq j \leq n$. Choose some $\delta > 0$ and denote by $\omega_1, \omega_2, \cdots, \omega_{n^\delta}$ the intervals contained in Ω_{n^δ}. From (61) and properties of the functions $a_\tau(t)$, we have for some $h = h(\epsilon_\delta) < 1$

$$\mathbf{P}(\mathscr{C}_{gn}\mathscr{X}_y) = \sum_{\frac{k}{n} \in V_{\varepsilon_\delta}} \mathbf{P}(\mathscr{B}_{gn}^k \mathscr{X}_y) + O(h^{-n}\mathfrak{m}^n(\tau_g)).$$

Let us replace the summation over domain V_{ϵ_δ} in this equality by separate summations over domains $\omega_1, \cdots, \omega_{n^\delta}$ and Ω^δ. Since for any sufficiently large n and any j such that $j/n \in V_{\epsilon_\delta}$ we have

$$\mathbf{P}(\mathscr{B}_{gn}^j \mathscr{X}_y) < 2\sup_{t \in V_{\varepsilon_g}} t^{-1/2}\Psi\left(\frac{a_{\tau_g}(t)}{t}, a'_{\tau_g}(t), y\right) \frac{\mathfrak{m}^n(\tau_g)}{\sqrt{n}} < c\,\frac{\mathfrak{m}^n(\tau_g)}{\sqrt{n}},$$

then

$$\limsup_{n\to\infty} \sum_{\frac{k}{n}\in\Omega^\delta} \mathbf{P}(\mathcal{B}^k_{gn}\mathcal{X}_y)/\sqrt{n}\,\mathfrak{m}^n(\tau_g) < c\,\delta.$$

Furthermore, making use of Theorem 4 (with $x = n$, $g = a_{\tau_g}$) it is easy to see that for any $1 \le j \le n^\delta$

$$\lim_{n\to\infty} \sum_{\frac{k}{n}\in\omega_j} \mathbf{P}(\mathcal{B}^k_{gn}\mathcal{X}_y)/\sqrt{n}\,\mathfrak{m}^n(\tau_g) = J(\omega_j) = \int_{\omega_j} t^{-1/2}\Psi\left(\frac{g(t)}{t}, g'(t), y\right) dt.$$

Hence we find

$$\limsup_{n\to\infty} \mathbf{P}(\mathcal{C}_{gn}\mathcal{X}_y)/\sqrt{n}\,\mathfrak{m}^n(\tau_g) < J(\Omega_{n^\delta}) + c\,\delta,$$

$$\liminf_{n\to\infty} \mathbf{P}(\mathcal{C}_{gn}\mathcal{X}_y)/\sqrt{n}\,\mathfrak{m}^n(\tau_g) \geqslant J(\Omega_{n^\delta}).$$

Since δ is arbitrary and $\operatorname{mes}(\Omega - \Omega_{n^\delta}) \to 0$ for $\delta \to 0$. Q.E.D.

It is natural to expect that the hypothesis of Theorem 5 holds if, for instance, g is continuous in $\overline{V}_\epsilon$. We can make a stronger statement.

Lemma 6. *The following three conditions are equivalent*:

I. V *is measurable*, $\operatorname{mes}(\overline{\overline{V}_\epsilon - V}) \to 0$ *for* $\epsilon \to 0$;

II. V *is measurable*, $\operatorname{mes}(\overline{\overline{V}_\epsilon - V}) = 0$, $\operatorname{mes}(\overline{V}_\epsilon - V) \to 0$ *for* $\epsilon \to 0$;

III. $\operatorname{mes}(\overline{V_\epsilon - \Omega}) \to 0$ *for* $\epsilon \to 0$ (*condition of Lemma* 5 *for the sequence* $V_\epsilon - \Omega$).

If g is continuous in $\overline{V}_\epsilon$ for all $\epsilon > 0$ smaller than a certain one, then the sets $\overline{V} - V$ and $\overline{V}_\epsilon - V$ are empty. As is indicated by Condition II, we allow the existence of points of discontinuity, the measure of whose closure tends to zero for $\epsilon \to 0$.

We shall require the following statement in order to prove Lemma 6.

Lemma 7. *If* V *is measurable and* $\operatorname{mes}(\overline{\overline{V} - V}) = 0$, *then* $\operatorname{mes}(V - \Omega) = 0$.

Proof. We know that set V can be represented in the form $V = \Omega_1 + e' - e''$, where $\operatorname{mes}(e') = \operatorname{mes}(e'') = 0$, and Ω_1 is a system of intervals. We can obviously assume that in this representation $e' \cap e''$ is empty, $e'' \subset \Omega_1$. Let $t \in e''$. Then $t \notin V$, but there are points of V in any neighborhood of point t. Therefore $e'' \subset \overline{V} - V$, $\overline{e''} \subset \overline{\overline{V} - V}$, $\operatorname{mes}(\overline{e''}) = 0$. The set $\Omega_2 = \Omega_1 - \overline{e''}$ is also a system of intervals. Since $V = \Omega_2 + e' + (\overline{e''} \cap \Omega_1 - e'')$ and $\operatorname{mes}(\overline{e''} \cap \Omega_1 - e'') = 0$. Q.E.D.

Proof of Lemma 6. 1) If Condition II holds, we have

$$\operatorname{mes}(\overline{V_\varepsilon - \Omega}) \leqslant \operatorname{mes}(\overline{\overline{V}_\varepsilon - \Omega}) = \operatorname{mes}(\overline{V}_\varepsilon - \Omega) = \operatorname{mes}(\overline{V}_\varepsilon - V) \to 0$$

for $\epsilon \to 0$. Conversely, it follows from Condition III that V is measurable,

$\operatorname{mes}(\overline{V - \Omega}) = 0$. Since $\overline{V} = \overline{\Omega} + (\overline{V - \Omega})$, we have

$$\overline{\overline{V} - V} \subset \overline{\overline{V} - \Omega} = \overline{V} - \Omega = (\overline{V} - \overline{\Omega}) + (\overline{\Omega} - \Omega) \subset (\overline{V - \Omega}) + (\overline{\Omega} - \Omega)$$

and therefore, $\operatorname{mes}(\overline{\overline{V} - V}) \leq \operatorname{mes}(\overline{V - \Omega}) + \operatorname{mes}(\overline{\Omega} - \Omega) = 0$. Furthermore, $\overline{V}_\epsilon = \overline{V} + (\overline{V_\epsilon - V})$, $\overline{V}_\epsilon - V \subset \overline{V}_\epsilon - \Omega \subset (\overline{V} - \Omega) + (\overline{V_\epsilon - V})$. Since $\overline{V} - \Omega \subset (\overline{\Omega} - \Omega) + (\overline{V - \Omega})$ and $\overline{V}_\epsilon - V \subset \overline{V}_\epsilon - \Omega$, then we find $\operatorname{mes}(\overline{V}_\epsilon - V) \leq \operatorname{mes}(\overline{V - \Omega}) + \operatorname{mes}(\overline{V_\epsilon - \Omega}) \to 0$ for $\epsilon \to 0$. Therefore Conditions II and III are equivalent.

2) The equivalence of Conditions II and III follows from the following relations. If II is fulfilled,

$$\operatorname{mes}(\overline{\overline{V}_\varepsilon - V}) \leqslant \operatorname{mes}(\overline{\overline{V}_\varepsilon - \Omega}) = \operatorname{mes}(\overline{V}_\varepsilon - \Omega) \to 0 \quad \text{for} \quad \varepsilon \to 0;$$

if III is fulfilled,

$$\operatorname{mes}(\overline{\overline{V} - V}) \leqslant \operatorname{mes}(\overline{\overline{V}_\varepsilon - V}) \to 0 \quad \text{for} \quad \varepsilon \to 0.$$

9. **Theorem on the probability that a trajectory will fall outside the limits of a domain for the case** $\operatorname{mes} V = 0$. If $\operatorname{mes} V = 0$, the integral in the right-hand side of (57) is equal to 0 and the estimate of $\mathbf{P}(\mathcal{C}_{gn}\mathfrak{X}_y)$ requires special consideration. We shall limit ourselves here to only one important case, where for all $\epsilon > 0$, smaller than a certain one, the set V_ϵ consists of a finite number of half-open intervals or closed intervals $\omega_1(\epsilon), \cdots, \omega_N(\epsilon)$ on which g is piecewise continuous (at interval endpoints we have in mind left-hand or right-hand differentiation. Our assumption does not exclude the case of points of discontinuity of the function g lying in V_ϵ). As before, we find

$$\mathbf{P}(\mathscr{C}_{gn}\mathscr{X}_y) = \sum_{\frac{k}{n} \in \omega_1(\varepsilon)} \mathbf{P}(\mathscr{B}^k_{gn}\mathscr{X}_y) + \ldots + \sum_{\frac{k}{n} \in \omega_N(\varepsilon)} \mathbf{P}(\mathscr{B}^k_{gn}\mathscr{X}_y) + O(h^n \mathfrak{m}^n(\tau_g)),$$

where $h = h(\epsilon) < 1$. Hence in order to describe the behavior of $\mathbf{P}(\mathcal{C}_{gn}\mathfrak{X}_y)$ for $n \to \infty$ it is sufficient to study the case when V_ϵ consists of only one half-interval or closed interval $\omega(\epsilon)$. On the closure $\overline{\omega}(\epsilon)$ there always exists a unique point v such that

$$\lim_{\substack{t \to v \\ t \in \omega(\varepsilon)}} g(t) = \alpha_{\tau_g}(v). \tag{62}$$

We shall call such a v the point of tangency of curve g with the level curve α_{τ_g}. An interior point of tangency divides $\omega(\epsilon)$ into two parts each of which has property (62). We can therefore limit ourselves to the case when v is a boundary point of $\omega(\epsilon)$. For instance, let v be a left boundary point. By our assumptions, $g(t)$ has a representation of the form (46) in the half-interval $\omega(\epsilon)$ of point v. Let us assume that in that representation the function $R^1_v(t - v)$ fulfills the hypothesis of Theorem 3. First let us assume that

$$p \geq q+1, \tag{63}$$

where q is the number of right-hand derivatives of $g^{(k)}(v)$ coinciding with $\alpha_{\tau_g}^{(k)}(v)$. Then there exists (from the right) $g^{(q+1)}(v) > \alpha_{\tau_g}^{(q+1)}(v)$ so that the main term of the function $T(\Delta, v) = T(\Delta, 1, v)$, equal to $T_{0,q+1}(v)\Delta^{q+1}$, is negative.

Furthermore, if t exceeds v by a quantity of order $n^{-\delta}$, then $ng(t)$ exceeds $n\alpha_{\tau_g}(t)$ by a quantity of order $n^{1-\delta(q+1)}$. Pick $\delta > 0$ such that $\nu = 1 - \delta(q+1) > 0$. Then we can write

$$\sum_{\frac{k}{n}\in\omega(\varepsilon)} \mathbf{P}(\mathcal{B}_{gn}^{k}\mathcal{X}_y) = \sum_{\frac{k}{n}\in\omega_n} \mathbf{P}(\mathcal{B}_{gn}^{k}\mathcal{X}_y) + O(h^{n^\nu}\mathfrak{m}^n(\tau_q)) + \mathbf{P}(\mathcal{B}_{gn}^{[nv]}\mathcal{X}_y)\,\Delta((nv-[nv])\,w), \tag{64}$$

where $h < 1$, ω_n is the interval $(v, v+n^{-\delta})$, ω is the number of points in set $\bar{\omega}(\epsilon) - \omega(\epsilon)$,

$$\Delta(t) = \begin{cases} 1, & t = 0 \\ 0, & t \neq 0 \end{cases}.$$

By Theorem 4A (its hypothesis is obviously fulfilled) we have

$$\sum_{\frac{k}{n}\in\omega_n} \mathbf{P}(\mathcal{B}_{gn}^{k}\mathcal{X}_y) = \frac{\mathfrak{m}^n(\tau_g)}{\sqrt{nv}}\Psi(G, g'(v), y) \sum_{\frac{k}{n}\in\omega_n} e^{nT\left(\frac{k}{n}-v,\, v\right)}(1+R_{k,n}), \tag{65}$$

where $R_{k,n} < c\max(k/n - v,\ h^{k-nv},\ r_2(1/n),\ R_v^3(k/n - v))$. Let the function g be such that

$$q \geq 1.$$

Then, by virtue of (63), $|R_v^3(\Delta)| < c\Delta$ for all s. s. Δ, $r_2(1/n) < c/n$. Furthermore, a bound for the error made in replacing the sum of decreasing terms by an integral is given by the maximum term. Therefore

$$\sum_{\frac{k}{n}\in\omega_n} e^{nT\left(\frac{k}{n}-v,\, v\right)} = n\int_0^{n^{-\delta}} e^{nT(t,\, v)}dt + O(1), \tag{66}$$

and the absolute value of the sum $\Sigma_{k/n\in\omega_n} R_{k,n}\, e^{nT(k/n-v,v)}$ does not exceed

$$cn\left[\int_0^{n^{-\delta}} te^{nT(t,\, v)}dt + \int_0^{n^{-\delta}} h^{nt}e^{nT(t,\, v)}dt\right].$$

Consider first the integral in (66). Let us write the function $T(t, v)$ in the form

$$T(t, v) = T_{0, q+1}(v)t^{q+1} + t^{q+1}R_T(t, v), \tag{67}$$

where the remainder factor $R_T(t, v)$ has, together with $R_v^1(t)$, a power majorant. Denote it by $r_T(t)$. (Clearly $r_T(t) < ct$ for $p > q + 1$.) Let us form the sequence $t_n = \min(n^{-\delta}, \inf\{t : nt^{q+1} r_T(t) \geq 1\})$ and assume first that $t_n < n^{-\delta}$. Then, making use of the inequality $|e^z - 1| \leq e|z|$ valid for $|z| \leq 1$, and writing for short $T_{0,q+1}(v) = T$, we shall have for sufficiently large n

$$\left| n \int_0^{n^{-\delta}} e^{nTt^{q+1}} (e^{nt^{q+1}R_T(t, v)} - 1)\, dt \right| \leqslant en \int_0^{t_n} e^{nTt^{q+1}} nt^{q+1} r_T(t)\, dt$$

$$+ n \int_{t_n}^{n^{-\delta}} e^{\frac{1}{2}nTt^{q+1}} dt \leqslant en \int_0^{n^{-\delta}} e^{\frac{1}{2}nTt^{q+1}} nt^{q+1} r_T(t)\, dt$$

$$\leqslant 2en^{1-\frac{1}{q+1}} \int_1^{n^{\frac{\nu}{q+1}}} e^{\frac{1}{2}Tu^{q+1}} u^{q+1} r_T(un^{-\frac{1}{q+1}})\, du$$

$$\leqslant cn^{1-\frac{1}{q+1}} r_T(n^{-\frac{1}{q+1}}) \int_1^{\infty} e^{\frac{1}{2}Tu^{q+1}} u^{q+r_T+1} du = cn^{1-\frac{1}{q+1}} r_T(n^{-\frac{1}{q+1}}).$$

In the above formulas r_T is the degree of majorant $r_T(t)$. If the relation $t_n < n^{-\delta}$ should not hold for all n, then the obtained inequality will only be strengthened. Furthermore we have

$$n \int_0^{n^{-\delta}} e^{nTt^{q+1}} dt = n^{1-\frac{1}{q+1}} \frac{1}{q+1} \Gamma\left(\frac{1}{q+1}\right) (-T)^{-\frac{1}{q+1}} \left(1 + O\left(\frac{1}{n}\right)\right),$$

$$n \int_0^{n^{-\delta}} te^{nT(t, v)} dt < cn^{1-\frac{2}{q+1}}, \quad n \int_0^{n^{-\delta}} h^{nt} e^{nT(t, v)} dt < c.$$

Summing up the indicated estimates, we find

$$\sum_{\frac{k}{n} \in \omega_n} e^{nT\left(\frac{k}{n} - v,\, v\right)} (1 + R_{k, n}) = n^{1-\frac{1}{q+1}} \frac{1}{q+1} \Gamma\left(\frac{1}{q+1}\right) (-T)^{-\frac{1}{q+1}} \Big(1$$

$$+ O\left(n^{-\frac{1}{q+1}}, r_T\left(n^{-\frac{1}{q+1}}\right)\right)\Big).$$

Substituting the resulting expression into (65) and taking into account (64), we obtain the value of $\mathbf{P}(\mathcal{C}_{gn} \mathcal{X}_y)$.

The case when the point of tangency is a right-hand boundary of $\omega(\epsilon)$ is handled in exactly the same way.

Consequently, if $g(t)$ is tangent to $\mathfrak{a}_{\tau_g}(t)$ at a finite number of points, the asymptotic behavior of the probability $\mathbf{P}(\mathcal{C}^g_{gn}\mathcal{X}_y)$ is clarified by the following theorem.

Theorem 6. *Let the set V_ϵ, for some $\epsilon > 0$, consist of one half-interval or closed interval, and let the set $\bigcap_{k=1}^{\infty}\overline{V}_{\epsilon_k}$ consist of one point v which is the boundary of $\overline{V}_\epsilon$. Moreover, let condition 2) of Theorem 3 be fulfilled in the half-neighborhood $\overline{V}_\epsilon$ of point v. Then, if the number q of general first derivatives of the functions g and $\mathfrak{a}_{\tau_g}$ at the point v is positive and $p > q+1$, we have*

$$\mathbf{P}(\mathscr{C}_{gn}\mathscr{X}_y) = n^{\frac{1}{2}-\frac{1}{q+1}}\,\mathfrak{m}^n(\tau_g)\,\frac{1}{q+1}\,\Gamma\left(\frac{1}{q+1}\right)\left[\frac{\lambda(G)\left(\mathfrak{a}_{\tau_g}^{(q+1)}(v)-g^{(q+1)}(v)\right)}{(q+1)!}\right]^{-\frac{1}{q+1}}$$
$$\times\,\Psi(G, g'(v), y)\left(1+O\left(n^{-\frac{1}{q+1}}, r_T\left(n^{-\frac{1}{q+1}}\right)\right)\right).$$

Here by derivatives of the function g we mean right-hand or left-hand derivatives, respectively; $r_T(t)$ is the power majorant of function $R_T(t)$ defined in (67); $G = g(v)/v$. For the meaning of the remaining notation, see (50), (56), (53).

Corollary. *Under the same hypothesis as in Theorem 6, the following limiting distribution exists:*

$$\lim_{n\to\infty}\mathbf{P}(\chi_g < y/\mathscr{C}_{gn}) = \frac{\psi(z_G, y)}{\psi(z_G, \infty)}$$

(see (50) and the corollary to Theorem 4).

Let us note that if v is an interior point of the closed interval $\overline{V}_\epsilon$ and if $g^{(q+1)}(v)$ exists, then it is necessary that $q+1$ be even. In that case the asymptotics of $\mathbf{P}(\mathcal{C}_{gn}\mathcal{X}_y)$ are obviously obtained by simply adding the respective asymptotic formulas.

In proving Theorem 6 we made use of the conditions $q \geq 1$ and $p \geq q+1$. Let us consider several cases where those conditions are violated.

1) Let $p = q \geq 1$, and in the representation

$$g(t) = \sum_{k=1}^{q}\frac{\Delta^k}{k!}\,\mathfrak{a}_{\tau_g}^{(k)}(v) + \Delta^q R_v^1(\Delta), \quad \Delta = t - v,$$

let the function $R_v^1(\Delta)$ have the form

$$R_v^1(\Delta) = |\Delta|^\nu (g_v + R_v^5(\Delta)),$$

where $g_v > 0$, $0 < \nu < 1$, and $R_v^5(\Delta)$ has a power majorant. Then one easily sees that (see (53))

$$T(\Delta, v) = g_v\lambda(G)|\Delta|^{q+\nu} + |\Delta|^{q+\nu}R_v^6(\Delta), \quad |R_v^6(\Delta)| < c|R_v^5(\Delta)|.$$

Without departing essentially from the proof of Theorem 6, and assuming for simplicity that the difference between the derivatives is also majorized by the function $r_2(\Delta)$ (see Theorem 3): $|R_v^3(\Delta)| < cr_2(\Delta) < c|\Delta|^v$, we find

$$\mathbf{P}(\mathscr{C}_{gn}\mathscr{X}_y) = n^{\frac{1}{2}-\frac{1}{q+\nu}}\mathfrak{m}^n(\tau_g)\frac{1}{q+\nu}\Gamma\left(\frac{1}{q+\nu}\right)[-\lambda(G)g_v]^{-\frac{1}{q+\nu}}\Psi(G, g'(v), y)(1$$
$$+ O\left(r_2\left(n^{-\frac{1}{q+\nu}}\right), r_5\left(n^{-\frac{1}{q+\nu}}\right)\right)),$$

where $r_5(\Delta)$ is the majorant of function $R_v^5(\Delta)$.

One can similarly find an explicit form for the dominant part of $\mathbf{P}(\mathcal{C}_{gn}\mathcal{X}_y)$ for the other cases when a sufficiently regular dominant part can be extracted from the function $R_v^1(\Delta)$.

2) If $q = 0$, $p \geq 1$, the terms in (64) decrease very rapidly. If $v < 1$, the difference $n[g(t_n) - a_{\tau_g}(t_n)]$, where $t_n = [nv]/n$ or $t_n = [nv] + 1/n$, oscillates between limits congruent to 1. Hence in this case

$$\mathbf{P}(\mathscr{C}_{gn}\mathscr{X}_y) = \frac{\mathfrak{m}^n(\tau_g)}{\sqrt{n}}c_{n,\,g},$$

where $c_{n,g}$ is a positive bounded function of n such that

$$0 < c_1 < \liminf_{n\to\infty} c_{n,\,g} < \limsup_{n\to\infty} c_{n,\,g} < c_2 < \infty.$$

The constants c_1 and c_2 are easily evaluated with the help of formulas (64), (65), and Theorems 4, 4A. If $v = 1$ and $g(1-0) = g(1)$, then $\tau_g = g(1)$, $T_{0,1}(1) = \lambda(g(1))(g'(1) - a'_{\tau_g}(1))$, so that

$$\mathbf{P}(\mathscr{C}_{gn}\mathscr{X}_y) = \frac{\mathfrak{m}^n(g(1))}{\sqrt{n}}\Psi(g(1), g'(1)y)[1 - e^{-T_{0,1}(1)}]^{-1}$$
$$\times\left(1 + O\left(r_2\left(\frac{1}{n}\right), r_5\left(\frac{1}{n}\right)\right)\right).$$

If $g(1) > g(1-0)$, a factor $e^{-T_{0,1}(1)}$ appears in the right-hand side of the last formula above; the quantities $g(1)$, $g'(1)$ should be construed as the corresponding limits.

§3. Limit theorems. II (with regard to the position of the last sum)

10. **Asymptotic behavior of the probability $\mathbf{P}(\mathcal{B}^k_{gx}\mathcal{X}_y\overline{\mathcal{A}}^n_{pn})$.** Here, in order to estimate the remainder terms in the same form as in §7, we must have recourse to

additional information (form of the expansion of probability $\mathbf{P}(\mathcal{B}^k_{gx}\mathcal{X}_y\overline{\mathcal{A}}^n_{\rho n})$ for rectilinear boundaries). Consequently we shall limit ourselves to finding the main asymptotic terms.

Assume that the hypothesis of Theorem 3 holds and, moreover, that $0 < v < 1$ and $\gamma(g, v) = (\rho - g(v))/(1 - v) > \alpha_+ + \epsilon$ for some $\epsilon > 0$. If event $\mathcal{B}^k_{gx}$ has occurred and if y is the magnitude of the jump across the boundary, the conditional probability of event $\overline{\mathcal{A}}^n_{\rho n}$ is given by

$$\mathbf{P}\left(s_{n-} < \rho n - g\left(\frac{k}{n}\right)x - y\right) = \frac{\mathfrak{m}^{n-k}(s)}{\sqrt{n-k}}\Lambda(s)\left(1 + O\left(\frac{1}{n}\right)\right),$$

where

$$s = \frac{\rho n - xg\left(\frac{k}{n}\right) - y}{n-k}, \quad \Lambda(s) = \frac{1}{\lambda(s)\sqrt{2\pi\varphi''_\lambda(\lambda(s), s)/\varphi(\lambda(s), s)}}. \tag{68}$$

Again setting $x/n = \sigma$, we find

$$\mathfrak{m}^{n-k}(s) = \mathfrak{m}^{n(1-t)}\left(\gamma(\sigma g, t) - \frac{y}{n(1-t)}\right)\Big|_{t=\frac{k}{n}} \sim \mathfrak{m}^{n(1-t)}(\gamma(\sigma g, t))\, e^{-y\lambda(\gamma(\sigma g, t))}\Big|_{t=\frac{k}{n}}.$$

This, together with formula (50), leads to

$$\mathbf{P}(\mathcal{B}^k_{gx}\mathcal{X}_y\overline{\mathcal{A}}^n_{\rho n}) = \frac{\Lambda(\gamma)}{n\sqrt{v(1-v)}}\int_0^y e^{-y\lambda(\gamma)}d_y\Psi(G, g'(v), y)$$

$$\times\left[\mathfrak{m}^t\left(\frac{\sigma g(t)}{t}\right)\mathfrak{m}^{1-t}\left(\frac{\rho - \sigma g(t)}{1-t}\right)\right]^n_{t=\frac{k}{n}}(1 + o(1)), \tag{69}$$

where $\gamma = \gamma(g, v)$. This equality clearly holds if, assuming the hypothesis of Theorem 3A, we also have $k - nv \to \infty$ for $n \to \infty$.

Let us transform the expression $\exp\mathfrak{M}(\sigma g(t), \rho, t)$ between square brackets in (69). Denote by τ_σ the solution of the equation $\mathfrak{b}_{\tau,\rho}(v) = \sigma g(v)$, $A^\sigma_1 = \sigma g(t)/t$, $A^\sigma_2 = \gamma(\sigma g, t)$. Then

$$\exp\mathfrak{M}(\sigma g(t), \rho, t) = \exp\left\{t\int_0^{A^\sigma_1}\lambda(\alpha)\,d\alpha + (1-t)\int_0^{A^\sigma_2}\lambda(\alpha)\,d\alpha\right\}$$

$$= \exp\{\mathfrak{M}(\tau_\sigma, \rho) + U(\Delta, \sigma, v)\},$$

where $\Delta = t - v$,

$$U(\Delta, \sigma, v) = t \int_{B_1^\sigma}^{A_1^\sigma} \lambda(\alpha)\, d\alpha + (1-t) \int_{B_2^\sigma}^{A_2^\sigma} \lambda(\alpha)\, d\alpha, \tag{70}$$

$B_1^\sigma = \mathfrak{b}_{\tau_\sigma, \rho}(t)/t$, $B_2^\sigma = \gamma(\mathfrak{b}_{\tau_\sigma,\rho}, t)$. Using the relations

$$\frac{\partial \mathfrak{b}_{\tau,\rho}(t)}{\partial \tau} = \frac{\lambda\left(\frac{\tau}{t_0}\right) - \lambda\left(\frac{\rho-\tau}{1-t_0}\right)}{\lambda\left(\frac{\mathfrak{b}}{t}\right) - \lambda(\gamma(\mathfrak{b}, t))}, \quad \frac{\partial \tau_\sigma}{\partial \sigma} = g(v) \Bigg| \frac{\partial \mathfrak{b}_{\tau,\rho}(v)}{\partial \tau}\Bigg|_{\tau=\tau_\sigma},$$

we find

$$\frac{\partial}{\partial \sigma} U(\Delta, \sigma, v) = g(t)\left[\lambda\left(\frac{\sigma g(t)}{t}\right) - \lambda(\gamma(\sigma g, t))\right]$$
$$- g(v)\left[\lambda\left(\frac{\sigma g(v)}{v}\right) - \lambda(\gamma(\sigma g, v))\right].$$

Hence

$$U(\Delta, \sigma, v) = U_0(t) + \sum_{k=1}^{\infty} \varkappa^k [U_k(t) - U_k(v)],$$

$$U_0(t) = t \int_{B_1^1}^{A_1^1} \lambda(\alpha)\, d\alpha + (1-t) \int_{B_2^1}^{A_2^1} \lambda(\alpha)\, d\alpha, \tag{71}$$

$$U_k(t) = g(t)\left[\lambda^{(k-1)}\left(\frac{g(t)}{t}\right)\left(\frac{g(t)}{t}\right)^{k-1} - \lambda^{(k-1)}(\gamma(g, t))\left(-\frac{g(t)}{1-t}\right)^{k-1}\right],$$
$$k = 1, 2, \ldots .$$

Here, as in §7, in the neighborhood of point v the functions $U_0(t)$ and $U_k(t) - U_k(v)$, $k = 1, 2, \cdots$, have a representation of the form

$$\sum_{j=1}^{p} \Delta^j U_{k,j}(v) + \Delta^p R^7_{k,v}(\Delta),$$

where $|R^7_{k,v}(\Delta)| < c\,|R^1_v(\Delta)|$,

$$U_{0,j}(v) = 0 \text{ for } j \leqslant q, \; U_{0,q+1}(v) = \frac{[g^{(q+1)}(v) - \mathfrak{b}^{(q+1)}_{\tau_1,p}(v)]\,[\lambda(G) - \lambda(\gamma)]}{(q+1)!}, \tag{72}$$

$q \le p - 1$ is the number of general derivatives of the functions g and $\mathfrak{b}_{\tau_1,\rho}$ at the point v.

Therefore we have the following theorem in the notation defined above.

Theorem 7. *Under the hypothesis of Theorem 3 where, in addition,* $0 < v < 1$ *and* $\gamma = (\rho - g(v))/(1 - v) > \alpha_+ + \epsilon$, *we have the equality*

$$\mathbf{P}(\mathcal{B}^k_{gx}\mathcal{X}_y\bar{\mathcal{A}}^n_{\rho n})$$

$$= \frac{\Lambda(\gamma)}{n\sqrt{v(1-v)}}\int_0^y e^{-y\lambda(\gamma)}d_y\Psi(G, g'(v), y)\, e^{n(\mathfrak{M}(\tau_\sigma, \rho)+U(\Delta, \sigma, v))}(1+o(1)),$$

where $\Delta = k/n - v$, $G = g(v)/v$, *the function* $U(\Delta, \sigma, v)$ *satisfies relations* (70)–(72), *the functions* $\Psi(a, b, y)$ *and* $\mathfrak{M}(\tau, \rho)$ *are defined in* (50), (21). *Under the hypothesis of Theorem* 3A (*see* (49)), *the theorem remains valid if* $k - nv \to \infty$ *for* $n \to \infty$.

Corollary. *Under the stated hypothesis, the following limiting distribution exists*

$$E_{\gamma, G}(y) = \lim_{n\to\infty}\mathbf{P}(\chi_g < y/\mathcal{B}^k_{gx}\bar{\mathcal{A}}^n_{\rho n}) = \frac{\int_0^y e^{-t\lambda(\gamma)}d\psi(z_G, t)}{\int_0^\infty e^{-t\lambda(\gamma)}d\psi(z_G, t)}$$

(*see* (50)). *In the notation of* [6]

$$\int_0^\infty e^{i\lambda y}dE_{\gamma, G}(y) = \frac{\mathfrak{W}_{z_{G+}}(\lambda + i\lambda(\gamma))}{\mathfrak{W}_{z_{G+}}(i\lambda(\gamma))}.$$

11. Asymptotic behavior of the probability $\mathbf{P}(\mathcal{C}_{gx}\mathcal{X}_y\bar{\mathfrak{A}}^n_{\rho n})$. The evaluation of the probability of event $\mathcal{C}_{gx}\mathcal{X}_y\bar{\mathfrak{A}}^n_{\rho n}$ (consisting of the following: the random broken line with vertices at points $(k/n, s_k/x)$, $k = 0, 1, \cdots, n$, crosses the curve $g(t)$, its first jump across the latter is not smaller than y/x, and the last point of the broken line lies below the level $\rho n/x$) proceeds along lines completely analogous to those in §8. We again assume, for the sake of simplicity, that $x = n$. There arises apparently no ambiguity if, instead of introducing a new notation, we set

$$\tau_g = \sup\{\tau : \inf_{0<t<1}(g(t) - \mathfrak{b}_{\tau, \rho}(t)) > 0\}$$

and denote by V_ϵ the set of all points of the closed interval $[0, 1]$ for which $g(t) \le \mathfrak{b}_{\tau_g, \rho}(t) + \epsilon$ ($\mathfrak{b} \in \mathfrak{S}$), V^g is the closure of the set of point t for which $(t, \mathfrak{b}_{\tau_g, \rho}(t))$ belongs to the sectors $D_2 - D_4$, so that CV^g is always an interval. As before, we assume $\bar{V}_{\epsilon_g} \subset CV^g$ for s. s. ϵ_g, and in the subsequent formulations we consider only sets V_ϵ with $\epsilon \le \epsilon_g$. We have the following theorems.

Theorem 8. *Let* $V = V_0 = \bigcap_{k=1}^\infty V_{\epsilon_k}$, *and let* Ω *be the union of all intervals of set* V. *If the sequence* $V_\epsilon - \Omega$ *possesses the D-property in addition to satisgying the assumed conditions, then we have*

$$\mathbf{P}(\mathscr{C}_{gn}\mathscr{X}_y\overline{\mathscr{A}}^n_{\rho n}) = e^{n\mathfrak{M}(\tau_g,\rho)}\int_\Omega \Phi(t,y)\,dt\,(1+o(1)),$$

where

$$\Phi(t,y) = \frac{\Lambda(\gamma(g,t))}{\sqrt{t(1-t)}}\int_0^y e^{-y\lambda(\gamma(g,t))}d_y\Psi\left(\frac{g(t)}{t}, g'(t), y\right), \gamma(g,t) = \frac{\rho-g(t)}{1-t}.$$

If g is tangent to $b_{\tau_{g,\rho}}$ at a finite number of points, we can evaluate $\mathbf{P}(\mathcal{C}_{gn}\mathcal{X}_y\overline{\mathcal{A}}^n_{\rho n})$ using the following analog of Theorem 6 (see also the remarks which precede Theorem 6).

Theorem 9. *If the hypothesis of Theorem 6 is fulfilled, we have*

$$\mathbf{P}(\mathscr{C}_{gn}\mathscr{X}_y\overline{\mathscr{A}}^n_{\rho n})$$

$$= n^{-\frac{1}{q+1}} e^{n\mathfrak{M}(\tau_g,\rho)}\Phi(v,y)\frac{1}{q+1}\Gamma\left(\frac{1}{q+1}\right)[-U_{0,q+1}(v)]^{-\frac{1}{q+1}}(1+o(1)).$$

The expression $U_{0,q+1}(v)$ is defined in (72).

From Theorems 8, 9 there follows the existence of corresponding limiting conditional distributions for the value of the first jump across the boundary $ng(t)$. Here we have a complete analogy with the results and remarks of §8 relating to the case when the conditions $p \geq q+1 \geq 1$ are not fulfilled.

The proofs of the stated theorems do not differ substantially from those of Theorems 6 and 7. We shall therefore omit them.

BIBLIOGRAPHY

[1] A. N. Kolmogorov, *Eine Verallgemeinerung der Laplace-Ljapounoffschen Satzes*, Izv. Akad. Nauk SSSR, Moscow, 1931, 959–962.

[2] A. Ja. Hinčin, *Asymptotic laws of probability theory*, ONTI, Moscow, 1936.

[3] Ju. V. Prohorov, *Convergence of random processes and limit theorems of probability theory*, Teor. Verojatnost. i Primenen. 1 (1956), 177–238. MR 18, 943.

[4] A. V. Skorohod, *Investigations in the theory of random processes*, Kiev, 1961.

[5] V. S. Koroljuk, *On the asymptotic distribution of maximum deviations*, Dokl. Akad. Nauk SSSR 142 (1962), 522–525 = Soviet Math. Dokl. 3 (1962), 90–93. MR 25 #3555.

[6] A. A. Borovkov, *New limit theorems in boundary-value problems for sums of independent terms*, Sibirsk. Mat. Ž. 3 (1962), 645–694; English transl., Selected Transl. in Math. Stat. and Prob. 5 (1965), 315–372. MR 26 #3099.

[7] I. I. Gihman, B. V. Gnedenko and N. V. Smirnov, *Non-parametric problems of statistics*, Trans. of the Third All-Union Math. Conference 3 (1956), 320–333.

[8] V. S. Koroljuk, *Asymptotic analysis of maximum deviation distributions in the Bernoulli scheme,* Teor. Verojatnost. i Primenen. 4 (1959), 369–397. MR 22 #10051.

[9] A. A. Borovkov, *On the two-sample problem,* Izv. Akad. Nauk SSSR Ser. Mat. 26 (1962), 605–624. MR 27 #897.

[10] H. Chernoff, *A measure of asymptotic efficiency for tests of a hypothesis based on sums of observations,* Ann. Math. Statist. 23 (1952), 493–507. MR 15, 241.

[11] A. N. Kolmogorov, *Das Gesetz des iterierten Logarithmus,* Math. Ann. 101 (1929), 126–135.

[12] I. G. Petrovskiĭ, *Lectures on partial differential equations,* GITTL, Moscow, 1950; English transl., Interscience, New York, 1964. MR 13, 241.

Translated by:
A. N. Rossolimo

ANALYSIS OF LARGE DEVIATIONS IN BOUNDARY-VALUE PROBLEMS WITH ARBITRARY BOUNDARIES. II

A. A. BOROVKOV

This article is the continuation of paper [1]. We shall retain all the notation of the first part.

§4. General case $x \sim n$

Items 8, 9, 11 of this paper contained limit theorems on the probabilities that a random trajectory with vertices at points $(k/n, s_k/n)$, $k = \overline{0, n}$, will fall outside the limits of some domain, with and without regard to the situation of the last sum s_n and under the assumption that $x = n$. This assumption allowed us to simplify the computations greatly and to clarify the essence of the matter. Now we shall assume that $x \sim n$ and that there exists a finite or infinite

$$\lim_{n\to\infty} \frac{\ln n}{\ln\left|\frac{x}{n} - 1\right|}.$$

12. **Basic lemma.** Let $T(t, \beta)$ be a function which is differentiable $p \geq 2$ times with respect to t, and for sufficiently small β and $|t| < \delta$ can be represented as the sum of converging series:

$$T(t, \beta) = \sum_{k=0}^{\infty} \beta^k \sum_{j=1}^{p} t^j T_{k,j} + t^p \sum_{k=0}^{\infty} \beta^k R_{k,p}(t) = L(t, \beta) + t^p \sum_{k=0}^{\infty} \beta^k R_{k,p}(t),$$

where $R_{k,p}(t) \to 0$ for $t \to 0$ and has a power majorant, and where the first nonzero element in the sequence

$$T_{0,1},\ T_{0,2}, \ldots,\ T_{0,p},\ 1$$

is negative. We shall assume that $\delta > 0$ is small enough that for $t < \delta$ we have $(\Sigma_{j=1}^p t^j T_{0,j} + t^p R_{0,p}(t))' \leq 0$. It is required to find the asymptotic behavior of the integral

$$F_N[T(t, \beta)] = \int_0^\delta \exp\{NT(t, \beta)\}(1 + r(t))\,dt$$

where $N \to \infty$ and $0 < \beta \to 0$ such that $\lim_{N\to\infty} - \ln N/\ln \beta \leq \infty$ exists, and the function $r(t)$ has the properties of a power majorant.

A certain amount of notation must be introduced in order to state the lemma. Let q^k be the subscript of the first nonzero element in the sequence $T_{k,1}, \ldots$ $\ldots, T_{k,p}$; $k = 0, 1, \cdots$ (by assumption $q^0 \leq p$, $T_{0,q^0} < 0$) and let q^l be the first minimal element in the sequence $q^0, q^1, \cdots$; $q^k \geq q^l$ for $k > l$ and $q^k > q^l$ for

$k < l$. Clearly l is finite and

$$T(t, \beta) \sim \sum_{k=0}^{l} t^{q^k} \beta^k T_{k, q^k} \quad \text{for} \quad t \to 0.$$

Let us make the substitution $t = \beta^u$ and introduce the function

$$h(u) = \min_{0 \leq j \leq l} (j + uq^j).$$

The semiaxis $u \geq 0$ can be broken up into the intervals $(u_0 = 0, u_1), (u_1, u_2) \cdots$ $\cdots (u_s, u_{s+1} = \infty)$, $s \leq l$, on which $h(u)$ is rectilinear. Let $h(u) = a_k + uq_k$ for $u \in (u_{k-1}, u_k)$. For any such u we have

$$T(t, \beta) \sim T_k \beta^{h(u)} = T_k \beta^{a_k + uq_k}, \tag{73}$$

where T_k is the coefficient of $\beta^{a_k} t^{q_k}$ in the expansion of the function $T(t, \beta)$ ($T_k = T_{a, q^a}$ for $a = a_k$).

Since $h(u)$ is an increasing function, the value of $u_N^* > 0$, which is the solution of the equation

$$N\beta^{h(u)} = 1,$$

for which there exists by assumption a finite or infinite

$$\lim_{N \to \infty} u_N^* - u^* > 0$$

is uniquely determined.

Let u_k be the first point in the sequence $u_0, u_1, \cdots, u_{s+1}$, for which $u^* \leq u_k$. Denote them by

$$a_k^j + uq_k^j \quad (q_k^{j+1} < q_k^j), \quad j = \overline{1, m}, \tag{74}$$

all the straight lines in the family

$$j + uq^j, \quad j = \overline{0, l}, \tag{75}$$

passing through the point $(u^*, h(u^*))$ ($m = 1$ if $u^* < u_k$, and $m \leq q_k - q_{k+1}$ if $u^* = u_k$). Along with $h(u)$, let us introduce the lower envelope $\hat{h}(u)$ of the family of straight lines (75) in which all the straight lines (74) are replaced by the straight lines $a_k^j + uq_k^{j*}$, where $q_k^{j*} > q_k^j$ $(j = \overline{1, m})$ is the number of the second nonzero element $(q_k^j < p + 1)$ in the sequence

$$T_{a_k^j, 1}, \ldots, T_{a_k^j, p}, 1.$$

We shall call the function $\hat{h}(u) \geq h(u)$ the envelope of family (75) raised over point u^*. If $u^* = \infty$, only the straight line $l + uq^l$ is replaced in (75). Further let

$$\psi(N) = N\beta^{h(u^*)}, \quad L_k(t) = \sum_{j=1}^{m} T_{a_k^j, q_k^j} t^{q_k^j}$$

and

$$I_N[L_k(t)] = \beta^{-u^*} \int_0^\infty \exp\{\psi(N) L_k(t)\} \, dt,$$

where $T_{a_k^1, q_k^1} < 0$. The determination of the asymptotic behavior of the last integral for various conditions does not present any difficulty. We shall only consider the case $u^* < u_k$ $(m = 1)$, when

$$I_N[L_k(t)] = (N\beta^{a_k})^{-\frac{1}{q_k}} \frac{1}{q_k} \Gamma\left(\frac{1}{q_k}\right) |T_k|^{-\frac{1}{q_k}}.$$

We shall take this equality as the definition of the functional $I_N[L_k(t)]$ if $u^* = \infty$ $(u_k = u_{s+1} = \infty)$.

Lemma 8. *Let* $[0, u_k]$ $(0 \leq k \leq s+1)$ *be the smaller of the closed intervals* $[0, u_j]$ *which contain point* u^*. *Then either* (I) $T_j < 0$ *for all* $j \leq k$, *or* (II) *a number* $0 < i < k$ *exists such that* $T_{i+1} > 0$, $T_j < 0$ *for all* $j \leq i$.

For case (I)

$$F_N[T(t, \beta)] = I_N[L_k(t)][1 + R_1(N)].$$

For case (II)

$$F_N[T(t, \beta)] = \left(\frac{2\pi}{MT''}\right)^{1/2} \beta^{u_0} \exp\{NT(\beta^{u^0}, \beta)\}[1 + R_2(N)],$$

where $M = N\beta^{h(u_i)}$, u_0 *is the point of the maximum of function* $T(\beta^\mu, \beta)$. *If only two straight lines* $a_i + uq_i$ *and* $a_{i+1} + uq_{i+1}$ *pass through point* u_i, *and if the envelope* $\hat{h}(u)$ *of family* (75) *raised over point* u_i *is rectilinear in the neighborhood of* u_i, *i.e., can be represented in the form* $\hat{a} + u\hat{q}$, *then*

$$u^0 = u_i + \frac{1}{\ln \beta}\left(C_1 + C_2\beta^{\hat{h}(u_i) - h(u_i)} + \ldots + \beta^{pu_i - h(u_i)} R_{u^0}(\beta^{u_i})\right),$$

$$C_1 = \frac{1}{q_{i+1} - q_i} \ln \frac{T_i q_i}{T_{i+1} q_{i+1}}, \quad C_2 = \frac{\hat{T}\hat{q}e^{C_1(\hat{q} - q_i)}}{T_i q_i (q_{i+1} - q_i)}, \tag{76}$$

$$|R_{u^0}(t)| < c\left|\frac{T_t'(t, \beta) - L_t'(t, \beta)}{t^{p-1}}\right|, \quad T'' = -T_i q_i e^{C_1 q_i}(q_i - q_{i+1}).$$

In these formulas $\hat{T}$ *is the coefficient of* $\beta^{\hat{a}} t^{\hat{q}}$ *in the expansion of function* $T(t, \beta)$.

Estimates of the remainder terms R_1 *and* R_2, *converging to* 0 *for* $N \to \infty$, *are contained in the proof* (*see* [82], [85], [86]).

Proof. Consider case (I). Denote by $\check{u}$ an arbitrary number in the interval (u_{k-1}, u^*) if $u^* > 0$, and let $\check{u} = \ln \delta / \ln \beta$ if $u^* = 0$.

Write

$$F_N[T(t, \beta)] = -\ln \beta \int_{\frac{\ln \delta}{\ln \beta}}^{\infty} \exp\{NT(\beta^u, \beta) + u \ln \beta\}(1 + r(\beta^u))\, du \tag{77}$$

as the sum of the integrals

$$I_1 = -\ln\beta \int_{\frac{\ln\delta}{\ln\beta}}^{\check{u}} \quad \text{and} \quad I_2 = -\ln\beta \int_{\check{u}}^{\infty}.$$

By (73), for sufficiently large N and for some $\epsilon_1 > 0$, we have

$$|I_1| < -\ln\beta \int_0^{\check{u}} \exp\left\{-\frac{1}{2} NT_k\beta^{a_k+uq_k} + u\ln\beta\right\} du = (N\beta^{a_k})^{-\frac{1}{q_k}} o(\exp - \beta^{-\varepsilon_1}). \quad (78)$$

Let $M_k(u) = \sum_{j=1}^{m} T_{a_k^j, q_k^j} \beta^{a_k^j + uq_k^j}$. Let us single out the dominant part of integral I_2 in the form

$$-\ln\beta \int_{\check{u}}^{\infty} \exp\{NM_k(u) + u\ln\beta\}\, du. \quad (79)$$

If $u^* < \infty$, after substituting $\beta^u = t\beta^{u^*}$ we find for some $\epsilon_2 > 0$

$$\beta^{u^*}\left[\int_0^{\infty} \exp\{\psi(N) L_k(t)\}\, dt + o(\exp - N^{2\varepsilon_2})\right] = I_N[L_k(t)](1 + o(\exp - N^{\varepsilon_2})). \quad (80)$$

If $u^* = \infty$, we find at once for (79)

$$\int_0^{\beta^{\check{u}}} \exp\{NT_{s+1}\beta^{s+1}t^{q_{s+1}}\}\, dt = I_N[L_{s+1}(t)](1 + o(\exp - \beta^{-\varepsilon_3})) \quad (81)$$

for any $\epsilon_3 > 0$. It remains to estimate the difference between I_2 and (79), bounded to within constant coefficients by the sum of the integrals

$$I_3 = -\ln\beta \int_{\check{u}}^{\infty} \exp\{NM_k(u) + u\ln\beta\}[1 - \exp\{N(T(\beta^u, \beta) - M_k(u))\}]\, du$$

and

$$I_4 = -\ln\beta \int_{\check{u}}^{\infty} r(\beta^u)\exp\{NT(\beta^u, \beta) + u\ln\beta\}\, du.$$

We have

$$|T(\beta^u, \beta) - M_k(u)| \leqslant |T(\beta^u, \beta) - L(\beta^u, \beta)| + |L(\beta^u, \beta) - M_k(u)| \leqslant c\max(\beta^{up}R_{0,p}(\beta^u), \beta^{\hat{h}(u)}),$$

where the maximum of the first number can be attained only for $k = 1$ and $q_1 = q^0 = p$. For convenience, in the following calculations we shall use the symbol $\langle a \rangle$, where $\langle a \rangle = a$ if we have the equalities $k = 1$, $q_1 = p$, and $\langle a \rangle = 0$ otherwise. Let $r_1(t)$ be the power majorant of the function $R_{0,p}(t)$,

$$N(u) = N\max(\langle\beta^{up}r_1(\beta^u)\rangle, \beta^{h(u)})$$

and

$$u_N^{**} = \inf\{u : N(u) < 1\},$$

so that $u_N^{**} < u_N^*$. Then for sufficiently large N we have

$$|I_3| < -\ln\beta\left[\int_{\check{u}}^{u_N^{**}} \exp\left\{\frac{1}{2} NM_k(u) + u\ln\beta\right\} du - c\int_{u_N^{**}}^{\infty} N(u)\exp\{NM_k(u) + u\ln\beta\}\, du\right]$$

$$\leqslant -c\ln\beta \int_{\check{u}}^{\infty} N(u)\exp\left\{\frac{1}{2} NM_k(u) + u\ln\beta\right\} du.$$

If $0 < u^* < \infty$, we shall assume for simplicity that in the neighborhood of point u^* we have $\hat{h}(u) = \hat{a} + u\hat{q}$. Then, after the substitution $\beta^u = t\beta^{u^*}$ for $u^* < \infty$, and the substitution $\beta^u = t$ for $u^* = \infty$ we find respectively

$$|I_3| < c\beta^{u^*}\int_0^{\infty} \max\left(\langle\psi(N)t^p r_1(t\beta^{u^*})\rangle,\ N\beta^{\hat{h}(u^*)}t^{\hat{q}}\right)\exp\left\{\frac{1}{2}\psi(N)L_k(t)\right\} dt$$

and

$$|I_3| < c\int_0^{\infty} \max\left(\langle Nt^p r_1(t)\rangle,\ N\beta^{\hat{a}}t^{\hat{q}}\right)\exp\left\{\frac{1}{2} NL_{s+1}(t)\right\} dt.$$

The asymptotic behavior of the first integral will depend on the behavior of function $\psi(N)$ and on the location of the maximum of function $L_k(t)$. If we set

$$\psi^*(N) = \begin{cases} \psi(N)^{-\frac{1}{q_k}}, & \text{when } u_N^* < u_k, \\ \psi(N)^{-\frac{1}{q_{k+1}}}, & \text{when } u_N^* > u_k \text{ and } L_k(t) < 0 \text{ for all } t > 0, \\ 1, & \text{when } u_N^* > u_k \text{ and } L_k(t) \geqslant 0 \quad \text{for some } t > 0, \end{cases}$$

a direct calculation shows that

$$|I_3| < cI_N[L_k(t)]\max\left(\langle\psi(N)\psi^*(N)^p r_1(\beta^{u^*}\psi^*(N))\rangle,\ \beta^{\hat{h}(u^*)-h(u^*)}\psi(N)\psi^*(N)^{\hat{q}}\right).$$

For $u^* = \infty$ we clearly have

$$|I_3| < cI_N[L_{s+1}(t)]\max\left(\langle r_1(N^{-\frac{1}{p}})\rangle,\ (N\beta^{a_{s+1}})^{1-\frac{\hat{q}}{q_{s+1}}}\right).$$

Similarly we find the estimates

$$|I_4| < cI_N[L_k(t)]\, r(\beta^{u^*}\psi^*(N)) \quad \text{for } u^* < \infty,$$

$$|I_4| < cI_N[L_{s+1}(t)]\, r\left((N\beta^{a_{s+1}})^{-\frac{1}{q_{s+1}}}\right) \quad \text{for } u^* = \infty.$$

Comparing these inequalities with (78), (80), and (81), we arrive at the first proposition of the lemma. Moreover,

$$R_1(N)=\begin{cases} O(e^{-N^{\varepsilon_2}}, \langle\psi(N)\psi^*(N)^p r_1(\beta^{u^*}\psi^*(N))\rangle, \beta^{\hat{h}(u^*)-h(u^*)}, r(\beta^{u^*}\psi^*(N))), \\ \text{if } u^*<\infty, \\ O(\langle r_1(N^{-\frac{1}{p}})\rangle, (N\beta^{a_{s+1}})^{1-\frac{\hat{q}}{q_{s+1}}}, r((N\beta^{a_{s+1}})^{-\frac{1}{q_{s+1}}})), \\ \text{if } u^*=\infty, \end{cases} \tag{82}$$

for some $\epsilon_2 > 0$. The form of the inequality obtained is maintained if $h(u)$ is discontinuous at point u^*.

In case (II) we shall make use of the Laplace method to estimate (77). It follows from the properties of the coefficients T_j that in the neighborhood of point u_i there exists a point u^0 in which $T(\beta^u, \beta)$ reaches its maximum. Formulas (76) relating to the expansion of u^0 are obtained in a direct fashion. The relations

$$T(t, \beta)=\sum_{j=0}^{p} T^{(j)}(\tau, \beta)\frac{(t-\tau)^j}{j!}+o((t-\tau)^p),$$

$$\beta^u-\beta^{u^0}=\beta^{u^0}\sum_{j=1}^{\infty}\frac{(u-u^0)^j}{j!}(\ln\beta)^j$$

show that for $|u-u^0|=o(1/\ln\beta)$ we have the representation

$$T(\beta^u, \beta)=T(\beta^{u^0}, \beta)+\sum_{j=2}^{p}\frac{(u-u^0)^j}{j!}T^{(j)}+(u-u^0)^p\beta^{h(u_i)}(\ln\beta)^p R^{u^0}((u-u^0)\ln\beta), \tag{83}$$

where $R^{u^0}(t)\to 0$ for $t\to 0$ and has a power majorant,

$$T^{(j)}=\frac{\partial^j}{\partial u^j}T(\beta^u, \beta)\big|_{u=u^0}\sim(\ln\beta)^j\beta^{h(u_i)}T_i q_i e^{C_1 q_i}(q_i^{j-1}-q_{i+1}^{j-1}).$$

In particular,

$$T^{(2)}=\ln^2\beta\,(T_i q_i^2\beta^{a_i+q_i u^0}+T_{i+1}q_{i+1}^2\beta^{a_{i+1}+q_{i+1}u^0}+\hat{T}\hat{q}^2\beta^{\hat{h}(u^0)}+\ldots+\beta^{pu^0}R_T^{(2)}(\beta^{u^0})),$$

where

$$R_T^{(2)}(t)=\frac{T_t^{(2)}(t, \beta)-L_t^{(2)}(t, \beta)}{t^{p-2}}=o(1)\quad\text{for}\quad t\to 0.$$

The uniqueness of point u^0 follows from the negative value of $T^{(2)}$ (recall that $T_i<0$, $q_i>q_{i+1}$).

Denote by I the interval $|u-u^0|<(-\ln\beta)^{-3/2}$, and by I_1 its complement up to the domain $(\ln\delta/\ln\beta, \infty)$. Making use of representation (83) and of the substitution $(u-u^0)\sqrt{-NT^{(2)}}=t$, we find (see (77)):

$$-\ln\beta\int_I \exp\{NT(\beta^u,\beta)+u\ln\beta\}\,du$$

$$=-\frac{\ln\beta}{\sqrt{-NT^{(2)}}}\exp\{NT(\beta^{u^0},\beta)+u^0\ln\beta\}$$

$$\times\int_{-(NT^{(2)}\ln^{-3}\beta)^{\frac{1}{2}}}^{(NT^{(2)}\ln^{-3}\beta)^{\frac{1}{2}}}\exp\left\{-\frac{t^2}{2}+\sum_{j=3}^{p}\frac{t^j}{j!}\frac{NT^{(j)}}{(-NT^{(2)})^{j/2}}\right.\tag{84}$$

$$\left.+\frac{t^pN\beta^{h(u_i)}(\ln\beta)^p}{(-NT^{(2)})^{p/2}}R^{u^0}\left(\frac{t\ln\beta}{\sqrt{-NT^{(2)}}}\right)+\frac{t\ln\beta}{\sqrt{-NT^{(2)}}}\right\}dt.$$

Since $\frac{NT^{(j)}}{(-NT^{(2)})^{j/2}}\sim c_1M^{1-j/2}$, $-\frac{\ln\beta}{\sqrt{-NT^{(2)}}}\sim\frac{1}{\sqrt{MT''}}$, $\frac{N\beta^{h(u_i)}}{-\ln\beta}>\beta^{-c_2}$, where $M=N\beta^{h(u_i)}$, $c_2>0$, we can write (84) in the form

$$\sqrt{\frac{2\pi}{MT''}}\exp\{NT(\beta^{u^0},\beta)+u^0\ln\beta\}(1+R_2'(N)),$$

where

$$R_2'(N)=O\left(M^{-1},\ M^{1-\frac{p}{2}}r^{u^0}\left(M^{-\frac{1}{2}}\right),\ \beta^{\hat{h}(u_i)-h(u_i)},\ \beta^{pu_i-h(u_i)}R_T^{(2)}(\beta^{u_i})\right),\tag{85}$$

$r^{u^0}(t)$ is the power majorant of $R^{u^0}(t)$. The addition here of the integral over domain I_1, evidently does not change the expression obtained. In order to find the representation of $F_N[T(t,\beta)]$, it remains to consider the integral in (77) with the coefficient $r(t)$. This leads to the second statement of the lemma with the function

$$R_2(N)=R_2'(N)+O(r(\beta^{u_{i_1}})).\tag{86}$$

13. Case mes $V=0$. Let us note, first of all, that in the case $\sigma=1+\kappa=x/n\neq 1$ the level lines $\alpha_\tau^\sigma(t)$ along which the principal parts of the probabilities $\mathbf{P}(\mathfrak{A}_{gx}^k)$ or $\mathbf{P}(\mathfrak{B}_{gx}^k)$ retain their value are, in general, different from $\alpha_\tau(t)$. Indeed, the functions $\alpha_\tau^\sigma(t)$ must satisfy the relation (see (50))

$$\mathfrak{m}^t\left(\sigma\frac{\alpha_\tau^\sigma(t)}{t}\right)=\mathfrak{m}(\sigma\tau),$$

from which follows $\alpha_\tau^\sigma(t)=\alpha_{\sigma\tau}(t)/\sigma$.

The family $(0<\tau<\infty)$ of these curves coincides with $\mathfrak{A}$ (the families coincide if the curves passing through one point with abscissa $t>0$ coincide) if and only if the ξ_k have a normal distribution. In that case $\alpha_\tau^\sigma(t)=\alpha_\tau(t)=\tau\sqrt{t}$.

Indeed since $\alpha_\tau^\sigma(1)=\tau$, then if the families coincide we must have $\alpha_\tau^\sigma(t)$ –

$\alpha_\tau(t) = \tau\sqrt{t}$.

Indeed since $\alpha_\tau^\sigma(1) = \tau$, then if the families coincide we must have $\alpha_\tau^\sigma(t) = \alpha_\tau(t)$ or $\alpha_\tau(t) = \tau\alpha_{t0}(t)/\tau_0$ for some τ_0. For solutions of equation (17) of the form $a(\tau, t) = a_1(\tau)a_2(t)$ and satisfying appropriate initial conditions, we find $a_2(t) = c_1 t^{c_2}$, $a_1(\tau)c_1 = \tau$, $\tau c_2 = -\ln\phi(\lambda(\tau))/\lambda(\tau)$. Since $\tau = \phi'(\lambda(\tau))/\phi(\lambda(\tau))$, it follows that $\phi(\lambda) = \exp(c_3\lambda c_4)$. The only Laplace transformation of such form corresponding to a random variable with $D\xi_k = 1$ is the transformation $\phi(\lambda) = e^{\lambda^2/2}$ of the normal law. For it $\lambda(\tau) = -\tau$, $\mathfrak{m}(\tau) = e^{-\tau^2/2}$, $c_2 = 1/2$, and therefore $\alpha_\tau(t) = \tau\sqrt{t}$.

In the general case we only have the relation

$$\alpha_\tau^\sigma(t) \to \alpha_\tau(t) \quad \text{for} \quad \sigma \to 1. \tag{87}$$

We shall assume here that the conditions of §9 are fulfilled and that $p \geq q + 1 \geq 2$. In view of (87) we can again limit ourselves to the case of only one right or left half-neighborhood $\omega(\epsilon)$ of some point of tangency v between curves $g(t)$ and $\alpha_{\tau_g}(t)$.

Making use of (55) we find

$$\sum_{\frac{k}{n}\in\omega(\varepsilon)} \mathbf{P}(\mathscr{B}_{gx}^k\mathscr{X}_y)$$

$$= \frac{\mathfrak{m}^n(\tau_\sigma)}{\sqrt{nv}}\Psi(G, g'(v), y)\left[n\int_0^\delta e^{nT(\pm t,\sigma,v)}(1 + f_1 t + f_2\varkappa)\,dt + O(e^{n\max T})\right], \tag{88}$$

where $\delta = \delta(\epsilon) \to 0$ for $\epsilon \to 0$; f_1, f_2 are uniformly bounded functions of the variables n, t, σ, y; $\max_{k/n\in\omega(\epsilon)} T(\Delta, \sigma, v)$. The sign of the argument t of function $T(\pm t, \sigma, v)$ is determined in an obvious way from the location of $\omega(\epsilon)$ relative to point v. Let

$$\beta = |\varkappa|,\ T(t, \beta) = T(\pm t,\ 1 + \beta\operatorname{sign}\varkappa,\ v).$$

The function $T(t, \beta)$ satisfies the hypothesis of the basic lemma. Since in (88) ϵ can be arbitrarily small, we can apply that lemma, in which we should set $r(t) = t$, $q_0 = q + 1$.

Theorem 10. *Assume the hypothesis of Theorem 6 is fulfilled and* $p \geq q + 1 \geq 2$. *Then in the notation of Theorem 4 and Lemma 8 we have*

$$\mathbf{P}(\mathscr{C}_{gx}\mathscr{X}_y) = \sqrt{\frac{n}{v}}\mathfrak{m}^n(\tau_\sigma)\Psi(G, g'(v), y)F_n[T(t, \beta)][1 + O(\beta, \beta^{c_n})],$$

where $c_n = h(u_n^*) - u_n^*$ *in case* (I) *and* $c_n = (1/2)(h(u_n^*) + h(u_i)) - u_i$ *in case* (II).

The problem of $\mathbf{P}(\mathscr{C}_{gx}\mathscr{X}_y\overline{\mathscr{A}_{\rho x}^n})$ is solved in a similar way.

14. Case mes $V>0$. Here the restrictions on the sets V_ϵ are in general different from those of Theorem 5. Denote by $T_\alpha(\Delta, \sigma, v)$ the function $T(\Delta, \sigma, v)$ for $g(t) = \alpha_{\tau_g}(t)$, and note the following.

a) From the remark made in §3 we easily see that $T_\alpha = 0$ in some neighborhood of point v if and only if the ξ_k have a normal distribution. In that case $\alpha_{\tau_\sigma}(t) = \sigma\alpha_{\tau_g}(t)$, and Theorem 5 will hold for $\mathbf{P}(\mathscr{C}_{gx}, \mathscr{X}_y)$ after the replacement of τ_g by τ_σ.

b) If $\kappa n \longrightarrow c$, then

$$nT_\alpha(\Delta, \sigma, v) = c(T_1(t) - T_1(v)) + o(1).$$

Thus here too Theorem 5 will remain valid if we replace τ_g by τ_σ and also replace the integral which appears in the theorem by

$$\int_\Omega t^{-1/2}\Psi\left(\frac{g(t)}{t}, g'(t), y\right) e^{c(T_1(t)-T_1(v))}dt,$$

where $v \in V$ is the point at which the value of τ_σ is determined in (51).

It is therefore sufficient to consider the case

$$|n\varkappa| \to \infty, \quad T_\alpha \not\equiv 0. \tag{89}$$

We have

$$T_\alpha(\Delta, \sigma, v) = \sum_{k=1}^{\infty} \varkappa^k (T_k^\alpha(t) - T_k^\alpha(v)) = \sum_{k=1}^{\infty} \varkappa^k \sum_{i=1}^{\infty} \Delta^i T_{k,i}^\alpha(v).$$

The coefficient of κ in the above expansion, is equal to $T_1^\alpha(t) - T_1^\alpha(v)$ and does not identically reduce to 0 anywhere. Otherwise we would have the following inequalities in some interval (here $\alpha = \alpha_{\tau_g}(t)$)

$$\frac{\partial}{\partial t}\left(\alpha\lambda\left(\frac{\alpha}{t}\right)\right) = \lambda\left(\frac{\alpha}{t}\right)\alpha' + t\left(\frac{\alpha}{t}\right)\left(\frac{\alpha}{t}\right)'\lambda'\left(\frac{\alpha}{t}\right)$$

$$= \lambda\left(\frac{\alpha}{t}\right)\frac{\alpha'\left(\frac{\alpha}{t}\right)' - \alpha''\left(\frac{\alpha}{t}\right)}{\left(\frac{\alpha}{t}\right)'} = \left[-\lambda\left(\frac{\alpha}{t}\right)\frac{\left(\frac{\alpha}{t}\right)^2}{\left(\frac{\alpha}{t}\right)'}\right]\left(\frac{\alpha' t}{\alpha}\right)' = 0,$$

where the coefficient in the square brackets is always positive. Equation $(\alpha'(t)/\alpha)' = 0$ together with equation (17) lead to the solution $\alpha = \tau_g\sqrt{t}$ which corresponds to the normal law and is excluded from consideration.

Making use of the relation

$$\frac{\alpha'(t)}{\alpha} = 1 - \frac{\ln \mathfrak{m}\left(\frac{\alpha}{t}\right)}{\frac{\alpha}{t}\lambda\left(\frac{\alpha}{\lambda}\right)},$$

we find that the sign of $(\alpha' t/\alpha)'$ coincides with that of the derivative

$$\frac{\partial}{\partial \tau}\left(\frac{\int_0^\tau \lambda(\alpha)\,d\alpha}{\tau\lambda(\tau)}\right)\Bigg|_{\tau=\frac{\alpha}{t}},$$

and therefore, as is easily verified, can be either positive or negative. But the function $T_1^{\alpha}(t) = \alpha\lambda(\alpha/t)$ is regular in domain $\overline{V}$ and has only a finite number of extrema there. Hence without loss in generality we can consider just those sets V on whose closure there exists a unique point v where $T_1^{\alpha}(t)$ reaches its maximum if $\kappa > 0$, or minimum if $\kappa < 0$.

We shall assume that in case (89) there exists an interval $\omega \in V$ such that $v \in \overline{\omega}$. Then the asymptotics of $\mathrm{P}(\mathcal{C}_{gx}\mathcal{X}_y)$ are determined up to known factors by the expression

$$\int_{v-\delta}^{v} e^{nT(t,\sigma,v)}dt + \int_{v}^{v+\delta} e^{nT(t,\sigma,v)}dt. \tag{90}$$

1) If for sufficiently small δ the half-neighborhood ω_δ of point v (one of the intervals $(v, v \pm \delta)$) belongs to V, then

$$nT(t, \sigma, v) = nT_{\alpha}(t, \sigma, v) = n\varkappa \sum_{k=0}^{\infty} \varkappa^k \sum_{j=1}^{\infty} t^j T_{k+1,j}^{\alpha}(v),$$

where the first nonzero product in the summation $\sum_{j=1}^{\infty} t^j T_{1,j}^{\alpha}(v)\,\mathrm{sign}\,\kappa$ is negative. We are therefore again in the situation described in the hypothesis of the basic lemma. In addition, we must set $\beta = |\kappa|$, $N = n\beta$, $p = \infty$, $r(t) = 0$,

$$T(t, \beta) = \frac{1}{\beta} T_{\alpha}(\pm t,\ 1 + \beta\,\mathrm{sign}\,\varkappa,\ v),$$

where t is taken with the + sign if $\omega_\delta = (v, v+\delta)$, and with the − sign if $\omega_\delta = (v-\delta, v)$.

2) Let $\omega_\delta \cap V \neq \omega_\delta$ for any δ. If we have

$$g(t) - \alpha_{\tau_g}(t) > c\,|t - v| \quad \text{for } t \in \omega_\delta \tag{91}$$

and some $\delta > 0$, then we can neglect the integral over ω_δ in (90). If (91) is not fulfilled, we must use Theorem 10, imposing on the function $g(t)$ (or its majorant $g_1(t) \le g(t)$) in addition the restrictions of that theorem.

The problem of determining $\mathrm{P}(\mathcal{C}_{gx}\mathcal{X}_y\overline{\mathcal{C}}^n_{px})$ also reduces to the combined application of Lemma 8 and of the analog of Theorem 10. The analysis of the various particular violations of the condition $p \ge q + 1 \ge 2$ can be carried out with the help of unessential changes in Lemma 8.

§5. Case $x = o(n)$

Here, as in §4, it is essentially sufficient to consider only distributions connected with the case that a trajectory falls outside the limits of a domain without regard to the situation of the last sum.

15. **Level curves.** For $x/n = \sigma \to 0$, $x^2/n \to \infty$, we have

$$\mathbf{P}(\mathcal{A}^n_{\alpha x}) = \mathbf{P}(s_n > \alpha x)\, c\, \frac{\sqrt{n}}{x}\, \mathfrak{m}^n\left(\frac{\alpha x}{n}\right).$$

This leads to the following equation for the level curves:

$$t \ln \mathfrak{m}\left(\frac{\alpha\sigma}{t}\right) = \ln \mathfrak{m}(\tau\sigma),$$

or

$$t \int_0^{\frac{\alpha\sigma}{t}} \lambda(\alpha)\, d\alpha = \int_0^{\tau\sigma} \lambda(\alpha)\, d\alpha.$$

Hence, setting $\theta = \sigma\tau$, $l_k = \lambda^{(k-1)}(0)/k!$, we obtain the following equation for the coefficients of the expansion of the solution $\alpha_\tau(t) = \tau\sum_{k=0}^{\infty} a_k \theta^k$:

$$t \sum_{k=2}^{\infty} \frac{\theta^k}{t^k}\, l_k (a_0 + a_1\theta + \ldots)^k = \sum_{k=2}^{\infty} \theta^k l_k.$$

Since $l_2 = \lambda'(0)/2 = -1/2 \neq 0$, we find

$$a_0 = \sqrt{t},\ a_1 = \frac{l_3}{2l_2}(\sqrt{t} - 1), \ldots,$$

so that

$$\alpha_\tau(t) = \tau\sqrt{t} + \frac{\tau^2 l_3}{2l_2}(\sqrt{t} - 1)\sigma + \ldots.$$

Here as before $\alpha_\tau(t) = \tau\sqrt{t}$ only for the normal law.

16. **Time of first passage.** We shall first consider rectilinear boundaries.

Lemma 9. *Let* $x > 0$, $\bar{x} = o(n)$, $\pi(t) = 1 + \bar{x}/xt$. *Then*

$$\mathbf{P}(\mathcal{B}^n_{\pi x}\mathcal{X}_y) = \frac{x}{n^{3/2}}\, \mathfrak{m}^n\left(\frac{x+\bar{x}}{n}\right) \frac{E_0(y)}{\sqrt{2\pi}}\left[1 + \Xi\left(\frac{\bar{x}}{n}\right)\right.$$
$$\left. + \frac{n}{x}\Xi\left(\frac{\bar{x}}{n}, n^{-1}, \frac{x}{n}\right) + O\left(\frac{e^{-\gamma x}}{x}\right)\right],$$

where $E_0(y)$ *is the limiting distribution function of the magnitude of the first jump defined, for instance, by Corollary* 1 *of Theorem* 4 *in* [2]: $\Xi(a)$ *and* $\Xi(a, b, c)$ *are asymptotic expansions in powers of their arguments and do not contain constant terms: in addition,* $\Xi(a, b, c)$ *does not contain the terms* a^k *and* ca^k, $k =$ $0, 1, \cdots$.

The proof of the lemma is found in a remark to some of the arguments of paper [2]. Therefore, and in order to avoid tedious explanations, we shall carry out the proof using the notation of [2]. Let us note, first of all, that we can improve the estimate of the remainder in formula (48) of Theorem 4 in [2]. Making use of the footnote on page 663 of [2], we easily verify that the singularities of the functions $q_x^\gamma(z)$ and $\tilde{q}_x^\gamma(z)$ at $z = z_0$ are second-order branch points. On the other hand estimate (35) of [2] permits us to represent the function $\tilde{q}_x^\gamma(z)$ in the neighborhood of point z_0 in the form $q(z, x, y) \exp[(\lambda_0 - \gamma)x]$, where $q(z, x, y)$ is a function uniformly bounded in x and y. From this we obtain the estimate (compare with p. 152 [3]) $c \exp[(\lambda_0 - \gamma)x]/z_0^n n^{3/2}$ for the coefficients of the expansion of function $\tilde{q}_x^\gamma(z)$, allowing us to replace the last remainder term in formula (48) of [2] by $O(e^{-\gamma x}/x)$ (compare with Lemma 8, [2]).

Furthermore, we have

$$\begin{aligned} \mathbf{P}(s_j < x + \varepsilon j,\ j = 1, 2, \ldots, n-1;\ x + \varepsilon n \leqslant s_n < x + \varepsilon n + y) \\ = \mathbf{P}(s_j^* < x,\ j = 1, 2, \ldots, n-1;\ x \leqslant s_n^* < x + y), \end{aligned} \tag{92}$$

where the s_j^* are sums of the random variables $\xi_k^* = \xi_k - \epsilon$. Since $\min_\lambda e^{\lambda\alpha} \mathrm{M} e^{-\lambda \xi_k^*} = \mathfrak{m}(\alpha + \epsilon)$, and since all the functionals of the distribution of ξ_k^* involved in formula (48) of [2] are, for probability (92), analytic functions of ϵ for sufficiently small ϵ (see also formula (12) in [2] for an integral representation of the factorization components), we obtain the statement of the lemma after setting $\epsilon = \bar{x}/n$.

Corollary. *Let* $x = u\sqrt{n}$, $u > 0$, $\bar{x} = v\sqrt{n}$ *such that* $\pi(t) = 1 + v/ut$. *Then*

$$\mathbf{P}(\mathscr{C}_{\pi x}\mathscr{X}_y) = \frac{uE_0(y)}{\sqrt{2\pi}} \int_0^1 t^{-3/2} e^{-\frac{1}{2}\frac{(u+vt)^2}{t}}\, dt \left(1 + O\left(\frac{1}{\sqrt{n}}\right)\right). \tag{93}$$

Proof. For sufficiently small ϵ and $n \geq k > \sqrt{n}/\epsilon$ we have

$$\mathbf{P}(\mathscr{B}_{\pi x}^k \mathscr{X}_y) = \frac{u\sqrt{n}}{k^{3/2}} \frac{E_0(y)}{\sqrt{2\pi}} \exp\left\{k \int_0^{\frac{\left(u + v\frac{k}{n}\right)\sqrt{n}}{k}} \lambda(\alpha)\, d\alpha\right\} \left(1 + R_1 \frac{\sqrt{n}}{k}\right),$$

where $R_1 < c$ and for arbitrary k

$$\mathbf{P}(\mathscr{B}_{\pi x}^k \mathscr{X}_y) \leqslant \exp\left\{k \int_0^{\frac{\left(u + v\frac{k}{n}\right)\sqrt{n}}{k}} \lambda(\alpha)\, d\alpha\right\}.$$

Hence we find

$$\sum_{k=1}^{\left[\frac{\sqrt{n}}{\varepsilon}\right]} \mathbf{P}(\mathscr{B}_{\pi x}^k \mathscr{X}_y) < c_1 \exp -c_2\sqrt{n},\quad c_2 > 0, \tag{94}$$

$$\sum_{k>\frac{\sqrt{n}}{\varepsilon}}^{n} \mathbf{P}(\mathcal{B}_{\pi x}^{k}\mathcal{X}_{y})$$

$$=\frac{uE_0(y)}{\sqrt{2\pi}}\sum_{k>\frac{\sqrt{n}}{\varepsilon}}^{n}\frac{n^{1/2}}{k^{3/2}}\exp\left\{-\frac{1}{2}\frac{\left(u+v\frac{k}{n}\right)^2 n}{k}\right\}\left(1+R_1\frac{\sqrt{n}}{k}+R_2\frac{n^{3/2}}{k^2}\right),$$

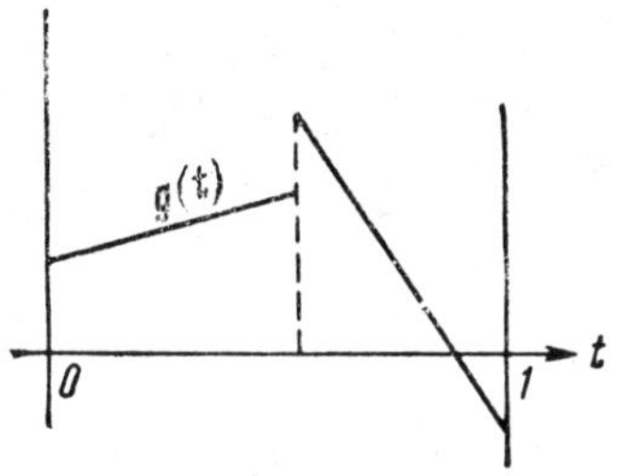

Figure 1

where R_2 is also bounded. Since $t^{-3/2}\exp\{-(u+vt)^2/2t\}$ is an increasing function in the vicinity of the origin, we can write the last sum in the form

$$\int_{\frac{1}{\varepsilon\sqrt{n}}}^{1} t^{-3/2}\exp\left\{-\frac{1}{2}\frac{(u+vt)^2}{t}\right\}\left[1+R_1\frac{1}{t\sqrt{n}}+R_2\frac{1}{t^2\sqrt{n}}\right]dt\left(1+O\left(\frac{1}{n}\right)\right).$$

Together with (94) this yields (93).

Making use of theorems of [2] on the joint distribution of $\max(0, s_1, \cdots, s_n)$ and s_n, we can also find the form of $\mathbf{P}(\mathcal{C}_{g\sqrt{n}}\mathcal{X}_y)$ for any piecewise linear function $g(t)$ (see Figure 1), and verify that in this case too the remainder term is of the form $O(1/\sqrt{n})$.

Lemma 10. *Assume that for some $a_0>0$, $b_0>0$ the elements of the sequences (a_n) and (b_n) fulfill the conditions $a_n>a_0$, $b_0>a_n+b_n>a_0$ for all $n\geq 1$, and let $\mathcal{A}_k=\{s_k/x>a_n+b_n k/n\}$. Then for $x^2/n\to\infty$ uniformly with respect to $k<n$*

$$\mathbf{P}(\mathcal{A}_k/A_n)<c_1\exp\left\{-c_2\frac{x(n-k)}{n}\min\left(1,\frac{x}{k}\right)\right\},\ c_2>0.$$

Proof. For conciseness we shall leave out the index n in the symbols a_n and b_n. If $xa/k\geq\epsilon$ for some $\epsilon>0$, then by virtue of inequality (6),

$$\mathbf{P}(\mathcal{A}_k/\mathcal{A}_n)<\frac{c_1\sqrt{n}\,\mathfrak{m}^k\left(\frac{xa_0}{2k}\right)}{\mathfrak{m}^n\left(\frac{xb_0}{n}\right)}<c_1e^{-c_2x}.$$

If $xa/k\to 0$, then for any $\delta>0$

$$\mathbf{P}(\mathcal{A}_k/\mathcal{A}_n) < \frac{1}{\mathbf{P}(\mathcal{A}_n)} \left\{ \sum_{j=0}^{\left[\frac{b}{\delta}-1\right]} \mathbf{P}\left(s_k \geqslant ax + bx\frac{k}{n} + \frac{j\delta x l}{n} \right) \right.$$

$$\left. \times \mathbf{P}\left(s_l > bx\frac{l}{n} - \frac{(j+1)\delta x l}{n} \right) + \mathbf{P}\left(s_k \geqslant ax + bx\frac{k}{n} + \left[\frac{b}{\delta}\right]\frac{\delta x l}{n} \right) \right\}$$

$$< c \sum_{j=0}^{\left[\frac{b}{\delta}\right]} \mathrm{m}^{-n}\left((a+b)\frac{x}{n} \right) \mathrm{m}^{k}\left(\left(a + b\frac{k}{n} + j\frac{\delta l}{n} \right)\frac{x}{k} \right)$$

$$\times \mathrm{m}^{l}\left(\max\left[0,\ (b-(j+1)\delta)\frac{x}{n} \right] \right)$$

$$= c \sum_{j=0}^{\left[\frac{b}{\delta}\right]} \exp\left\{ \sum_{i=2}^{\infty} l_i \left[\frac{N_1^i}{k^{i-1}} + \frac{\left(N_2 - \frac{\delta x l}{n} \right)^i}{l^{i-1}} - \frac{N^i}{n^{i-1}} \right] \right\}, \tag{95}$$

where

$$l = n - k,\ l_i = \frac{\lambda^{(i-1)}(0)}{i!},\ N_1 = x\left(a + b - \frac{l}{n}(b - j\delta) \right),$$

$$N_2 = x\frac{l}{n}(b - j\delta),\ N = N_1 + N_2.$$

By virtue of the inequality

$$\left| \frac{N_1^i}{k^{i-1}} + \frac{\left(N_2 - \frac{\delta x l}{n} \right)^i}{l^{i-1}} - \frac{N^i}{n^{i-1}} \right| < ci\,\frac{x^i l}{n^i},\ i \geqslant 3, \tag{96}$$

the sum of the terms of the series after the symbol exp in (95) is a quantity $O(x^3 l/n^3)$ for all $i \geq 3$. Further we have

$$\frac{N_1^2}{k} + \frac{N_2^2}{l} - \frac{N^2}{n} = \frac{(kN_2 - lN_1)^2}{kln} = \frac{x^2 l}{kn}(a + j\delta)^2 > \frac{x^2 l a_0^2}{kn}. \tag{97}$$

Since $l_2 = -1/2$, we can choose δ to make the indicated sum in (95) smaller than $-x^2 l a_0^2/4kn$. Q. E. D.

We can now prove the following proposition.

Theorem 11. *For $x = o(n)$ the statement of Theorem 3 remains valid if we replace condition 3 by the condition $g'(v) < g(v)/v$ and the estimate of the remainder $O(r_2(1/n))$ by the estimate $O(r_2(n/x^2))$.*

Proof. Consider a term, for instance, of the second summation in (48) for some value $j > n(v - \epsilon)$, $\epsilon > 0$. It differs from 0 provided $g(t) < \pi(t)$. By Lemma 10 we have

$$\mathbf{P}(\mathcal{B}_{gx}^{j}\mathcal{B}_{\pi x}^{k}\mathcal{X}_y) < \mathbf{P}\left(\mathcal{B}_{\pi_1 x}^{j}\mathcal{X}_{\left(\pi\left(\frac{j}{n}\right) - g\left(\frac{j}{n}\right)\right)x} \right)\mathbf{P}(\mathcal{B}_{\pi_2 x}^{k-j}),$$

where $\pi_1(t)$ is a straight line intersecting the curve $g(t)$ at the point $(j/n, g(j/n))$:

$$\pi_1(t) = g\left(\frac{j}{n}\right) + \left(t - \frac{j}{n}\right)\left(g'\left(\frac{j}{n}\right) - \varepsilon_1\right), \ \varepsilon_1 > 0,$$

$$\pi_2(t) = g\left(\frac{k}{n}\right) - g\left(\frac{j}{n}\right) + \left(t - \frac{k-j}{n}\right) g'\left(\frac{k}{n}\right).$$

Therefore from Lemma 9

$$\mathbf{P}(\mathcal{B}^j_{gx}\mathcal{B}^k_{\pi x}\mathcal{X}_y) = \frac{cx}{j^{3/2}(k-j)^{3/2}} \min\left[1, x\left(\pi\left(\frac{j}{n}\right) - g\left(\frac{j}{n}\right)\right)\right]$$

$$\times \max\left[1, x\left(\pi\left(\frac{j}{n}\right) - g\left(\frac{j}{n}\right)\right)\right] \mathfrak{m}^j\left(\frac{xg\left(\frac{j}{n}\right)}{j}\right) \mathfrak{m}^{k-j}\left(\frac{x\left(g\left(\frac{k}{n}\right) - g\left(\frac{j}{n}\right)\right)}{k-j}\right)$$

$$< c\mathbf{P}(\mathcal{B}^k_{\pi x}\mathcal{X}_y) \frac{x\left(\pi\left(\frac{j}{n}\right) - g\left(\frac{j}{n}\right)\right)}{(k-j)^{3/2}} \mathfrak{m}^{-k}\left(\frac{N}{k}\right) \mathfrak{m}^j\left(\frac{N_1}{j}\right) \mathfrak{m}^{k-j}\left(\frac{N_2}{k-j}\right),$$

where $N_1 = xg\left(\frac{j}{n}\right)$, $N_2 = x\left(g\left(\frac{k}{n}\right) - g\left(\frac{j}{n}\right)\right)$, $N = N_1 + N_2$. Making use of an integral representation of function $\mathfrak{m}$ and of relations analogous to (96), (97), we obtain the following expression for the product of the last three factors for sufficiently small ϵ

$$\exp\left\{-\frac{1}{2}x^2 \frac{\left(kg\left(\frac{j}{n}\right) - jg\left(\frac{k}{n}\right)\right)^2}{j(k-j)k} + O\left(\frac{x^3(k-j)}{k^3}\right)\right\}$$

$$\leqslant \exp\left\{-\frac{1}{3}\frac{x^2(k-j)}{k^2}(g(v) - vg'(v))^2\right\}.$$

From the inequality $\left|\pi\left(\frac{j}{n}\right) - g\left(\frac{j}{n}\right)\right| < \frac{|k-j|}{n} r_2\left(\frac{k-j}{n}\right)$, we obtain the following result:

$$\sum_{j>n(v-\varepsilon)}^{k-1} \mathbf{P}(\mathcal{B}^j_{gx}\mathcal{B}^k_{\pi x}\mathcal{X}_y) < c\mathbf{P}(\mathcal{B}^k_{\pi x}\mathcal{X}_y) \frac{x}{\sqrt{n}} \int_0^{\varepsilon} t^{-1/2} r_2(t)$$

$$\times \exp\left\{-\frac{x^2 t}{3n}\left(\frac{g(v)}{v} - g'(v)\right)^2\right\} dt < c\mathbf{P}(\mathcal{B}^k_{\pi x}\mathcal{X}_y) r_2\left(\frac{n}{x^2}\right).$$

If we estimate the remaining part of the second sum in (48) by means of Lemma 10, we obtain for that sum the expression $\mathbf{P}(\mathcal{B}^k_{\pi x}\mathcal{X}_y) O(r_2(n/x^2))$. We similarly evaluate the first sum in (48). Q. E. D.

Let us consider the exponential part of the probability $\mathbf{P}(\mathcal{B}^k_{gx}\mathcal{X}_y)$ which is of the form $\mathfrak{m}^k(g(k/n)x/k)$. Again let τ_σ be the solution of the equation

$$a_{\tau_\sigma}(v) = g(v).$$

Then

$$\mathfrak{m}^k\left(\frac{g\left(\frac{k}{n}\right)x}{k}\right) = \mathfrak{m}^n(\sigma\tau_\sigma)\exp nT(\Delta, \sigma, v),$$

where $\Delta = t - v$, $t = k/n$, and the function $T(\Delta, \sigma, v)$ is defined by equations (52). Making use of the relations

$$t\lambda\left(\frac{\sigma\mathfrak{a}_{\tau_\sigma}(t)}{t}\right)\frac{\partial}{\partial\sigma}\left(\frac{\sigma\mathfrak{a}_{\tau_\sigma}(t)}{t}\right) = \lambda(\sigma\tau_\sigma)\frac{\partial}{\partial\sigma}(\sigma\tau_\delta),$$

$$\lambda\left(\frac{\sigma g(v)}{v}\right)g(v) = \lambda(\sigma\tau_\sigma)\frac{\partial}{\partial\sigma}(\sigma\tau_\delta),$$

obtained by differentiating the identity $t \ln \mathfrak{m}(\sigma\, \mathfrak{a}_{\tau_\sigma}(t)/t) = \ln \mathfrak{m}(\sigma\tau_\sigma)$, we find

$$\frac{\partial}{\partial\sigma}\left(\frac{\sigma\mathfrak{a}_{\tau_\sigma}(t)}{t}\right) = \frac{g(v)\lambda\left(\frac{\sigma g(v)}{v}\right)}{t\lambda\left(\frac{\sigma\mathfrak{a}_{\tau_\sigma}(t)}{t}\right)},$$

$$\frac{\partial}{\partial\sigma}T(\Delta, \sigma, v) = g(t)\lambda\left(\sigma\frac{g(t)}{t}\right) - g(v)\lambda\left(\sigma\frac{g(v)}{v}\right).$$

Thus we obtain the representations

$$T(\Delta, \sigma, v) = \sum_{k=2}^{\infty}\sigma^k(T_k(t) - T_k(v)),$$

where $T_k(t) = l_k\dfrac{g^k(t)}{t^{k-1}}$, $k = 2, 3, \ldots$; $l_k = \dfrac{\lambda^{(k-1)}(0)}{k!}$, and

$$T(\Delta, \sigma, v) = \sum_{k=2}^{\infty}\sigma^k\left[\sum_{j=1}^{p}\Delta^j T_{k,j}(v) + \Delta^p R^4_{k,v}(\Delta)\right] \tag{98}$$

(compare with (53)), where $T_{k,j}(v) = (l_k/j!)\,(g^k(v)/v^{k-1})^{(j)}$ and the function $R^4_{k,v}(t)$ has the same power majorant as the function $R'_v(t)$ in condition 2 of Theorem 3. Theorem 11 can therefore be stated in the following form:

Theorem 12. *Under the same hypothesis as in Theorem* 11, *we have*

$$\mathbf{P}(\mathcal{B}^k_{gx}\mathcal{X}_y) = \frac{E_0(y)}{\sqrt{2\pi}}\,\frac{x(g(v) - vg'(v))}{(nv)^{3/2}}\,\mathfrak{m}^n(\sigma\tau_\sigma)\exp\{nT(\Delta, \sigma, v)\}$$
$$\times\left[1 + O\left(\sigma, \Delta, r_2\left(\frac{n}{x^2}\right), R^3_v(\Delta)\right)\right],$$

where $\Delta = k/n - v$, $R^3_v(\Delta) = g'(v + \Delta) - g'(v)$, *and the function* $T(\Delta, \sigma, v)$ *has the representation* (98).

It follows from the theorem that there exists, under the hypothesis of Theorem

11, a conditional limiting distribution of the magnitude χ_g of the first jump across the boundary $xg(t)$ corresponding to the unconditional distribution of the magnitude of the jump across an infinitely distant barrier.

17. Exit outside the domain. Let

$$\tau_g = \sup\{\tau : \inf_{0\leqslant t\leqslant 1} (g(t) - \tau\sqrt{t}) > 0\}$$

and let V_ϵ be a set on which $g(t) \leq \tau_g\sqrt{t} + \epsilon$. We assume that for sufficiently small ϵ the set V_ϵ does not contain the point $t = 0$. Then the case of a finite number of points of tangency between the curves $g(t)$ and $\tau_g\sqrt{t}$ is elucidated by the following theorem.

Theorem 13. *Let $g(t)$ be tangent to the parabola $\tau_g\sqrt{t}$ at a unique point $0 < v \leq 1$ at which condition 2 of Theorem 3 holds for some $p \geq 2$, and let $q \leq p$ be the order of the first nonzero derivative in the sequence $(g^2(v))''$, $(g^2(v))'''$, ...*

Then q is even, and in the notation of the basic lemma we have

$$\mathbf{P}(\mathscr{C}_{gx}\mathscr{X}_y) = \frac{E_0(y)}{\sqrt{2\pi}} \frac{x(g(v) - vg'(v))}{n^{1/2}v^{3/2}} \mathfrak{m}^n(\sigma\tau_\sigma)\{\Delta(v) F_{n\sigma^2}[T(t, \sigma)]$$

$$+ F_{n\sigma^2}[T(-t, \sigma)]\}\left[1 + O\left(\sigma, \frac{1}{n\sigma^2}\right)\right],$$

where $\sigma = \dfrac{x}{n}$, $T(t, \sigma) = \dfrac{1}{\sigma^2} T(t, \sigma, v)$, $\Delta(v) = \begin{cases} 1, & \text{if } v < 1, \\ 0, & \text{if } v = 1. \end{cases}$

Proof. Since for $p \geq 2$ we can set $r_2(t) = t$ and $R_v^3(t) < ct$, then

$$\sum_{\frac{k}{n} \in \omega_\delta} \mathbf{P}(\mathscr{B}_{gx}^k\mathscr{X}_y) = \frac{xE_0(y)(g(v) - vg'(v))}{\sqrt{2\pi}(nv)^{3/2}} \mathfrak{m}^n(\sigma\tau_\sigma)\left[n\int_{-\delta}^{\delta\Delta(v)} \exp\{nT(t, \sigma, v)\}\right. \tag{99}$$

$$\left.\times\left(1 + f_1\sigma + f_2 t + f_3\frac{n}{x^2}\right) dt + O(\exp\{n \max T\}),\right.$$

where f_i are bounded, $\max T = \max_{k/n \in \omega_\delta} T(\Delta, \sigma, v)$. In order to make use of the basic lemma to evaluate the integrals in (99), let $\beta = \sigma$, $N = n\sigma^2$, $T(t, \sigma) - T(t, \sigma, v)/\sigma^2$. Then $T_{0,j} = (l_2/j!)\,(g^2(v)/v)^{(j)}$. Since $g(v) = \tau_g\sqrt{v}$, $g'(v) = \tau_g 1/2\sqrt{v}$, then $q^0 \geq 2$ and $q^0 = q$. From the relations

$$0 > -\frac{1}{2}\left(\frac{g^2(t)}{t} - \frac{g^2(v)}{v}\right) \sim (t - v)^q T_{0,q}$$

it follows that q must be even and that $T_{0,q} = -(g^2(v))^{(q)}/2q! < 0$. The hypothesis of Lemma 8 is therefore fulfilled, and there only remains to make use of that lemma. Q. E. D.

Now let $\operatorname{mes} V > 0$. On V we have $g(t) = \tau_g\sqrt{t}$, and therefore $T_2(t) = \text{const}$,

$$T(\Delta, \sigma, v) = \sum_{k=j}^{\infty} \sigma^k \tau_g^k l_k\,(t^{1-k/2} - v^{1-k/2}),$$

where $j = \sup\{k \geq 3: l_3 = l_4 = \cdots = l_{k-1} = 0\}$, so that if $j < \infty$, then l_j is the first nonzero element in the sequence $l_3, l_4, \cdots$.

Theorem 14. *If $n\sigma^j < c$ and if the sequence $V_\epsilon - \Omega$ has the D-property (see Theorem 5), then*

$$\mathbf{P}(\mathscr{C}_{gx}\mathscr{X}_y) \sim \frac{x\tau_g E_0(y)}{2\sqrt{2\pi n}} \mathfrak{m}^n(\sigma\tau_\sigma) \int_\Omega t^{-1} \exp\{n\sigma^j\tau_g^j l_j (t^{1-j/2} - v^{1-j/2})\}\, dt,$$

where $v \in V$ is the value of the parameter in the equation $a_{\tau_\sigma}(v) = \tau_g\sqrt{v}$ for τ_σ.

If $n\sigma^j \to \infty$, $l_j \gtrless 0$, and if for some $\epsilon > 0$ and $\epsilon_1 > 0$ there exists an interval $(v, v \pm \epsilon) \in V$ such that

$$g(t) > \tau_g \min\left(\sqrt{t} + \varepsilon_1, \sqrt{v} + \frac{1 \mp \varepsilon_1}{2\sqrt{v}}(t - v)\right) \tag{100}$$

for $t \lessgtr v$, then

$$\mathbf{P}(\mathscr{C}_{gx}\mathscr{X}_y) \sim \frac{E_0(y)\, x\mathfrak{m}^n(\sigma\tau_\sigma)}{\sqrt{2\pi}\, v\tau_g^{j-1} |l_j| (j-2)\, n^{3/2}\sigma^j}.$$

The proof of the theorem consists in the use of Lemma 8 and of the arguments of Theorem 5.

If condition (100) is violated by the fact that tangency between the curves $g(t)$ and $\tau_g\sqrt{t}$ occurs for $t \lessgtr v$, then we can find $\mathbf{P}(\mathcal{C}_{gx}\mathcal{X}_y)$ making use of the proof of Theorem 13 to evaluate $\sum_{k/n \gtreqless v} \mathbf{P}(\mathcal{B}_{gx}^k \mathcal{X}_y)$.

It follows from the above arguments that $\lim_{n\to\infty} \mathbf{P}(\chi_g < y/\mathcal{C}_{gx}) = E_0(y)$ exists under the hypothesis of Theorems 13 and 14.

BIBLIOGRAPHY

[1] A. A. Borovkov, *Analysis of large deviations in boundary-value problems with arbitrary boundaries*, Sibirsk, Mat. Ž. 5 (1964), 253–289; English transl., Probability and Statistics, Vol. 6, Amer. Math. Soc., Providence, R. I., 1966, pp. MR 29 #645.

[2] ———, *New limit theorems in boundary-value problems for sums of independent terms*, Sibirsk. Mat. Ž. 3 (1962), 645–694. MR 26 #3099.

[3] ———, *Limit theorems on the distribution of the maximum of sums of bounded lattice random variables* (1), Teor. i Primenen. 5 (1960), 137–172. MR 24 #A2414.

Translated by:
A. N. Rossolimo